MATHEMATICAL MODELLING OF ENERGY SYSTEMS

NATO ADVANCED STUDY INSTITUTES SERIES

Proceedings of the Advanced Study Insitute Programme, which aims at the dissemination of advanced knowledge and the formation of contacts among scientists from different countries.

The series published by an international board of publishers in conjunction with NATO Scientific Affairs Division.

A	Life Sciences	Plenum Publishing Corporation
B	Physics	London and New York
C	Mathematical and Physical Sciences	D. Reidel Publishing Company Dordrecht and Boston
D	Behavioural and Social Sciences	Sijthoff & Noordhoff International Publishers B.V.
E	Applied Science	Alphen aan den Rijn, The Netherlands and Rockville, Maryland, USA

Series E: Applied Sciences - No. 37

MATHEMATICAL MODELLING OF ENERGY SYSTEMS

edited by

İBRAHİM KAVRAKOĞLU, Ph.D.

Professor of Engineering
Boğaziçi University
Istanbul, Turkey

SIJTHOFF & NOORDHOFF   1981
Alphen aan den Rijn, The Netherlands
Rockville, Maryland, USA

Proceedings of the NATO Advanced Study Institute on
Mathematical Modelling of Energy Systems
Istanbul, Turkey
June 9-23, 1979

ISBN-13: 978-94-009-8587-2   e-ISBN-13: 978-94-009-8585-8
DOI: 10.1007/978-94-009-8585-8

PREFACE

Prior to the so-called "energy crisis" of 1973, energy played a
relatively minor role in our daily lives and received limited
attention from economists, planners and politicians. As a means
of production its share in the total cost of the average product
was considerably less than 10%. After the decisive events of
1973/74 however, all of this is changed. Energy now affects our
daily lives more than anything else; it is the most current issue
in business circles, the academia, the civil services and politics;
and it is likely to become the most important factor in a potential
international instability.

The jump in oil prices in 1973 did not just lead to global inflation,
but it also made the world a much more complicated environment to
live in. Most decisions now require the analysis of yet another
dimension; the alternatives have increased in number; the penalty
for errors has gone up, and the like.

In contrast to the interwoven, complicated, and mostly incompre-
hensible reality, progress is being made in the realm of mathematical modelling that is comprehensible and has the advantage
that it can be designed to the degree of complication desired.
Viewed in this way, it can be said that the aim of the Advanced
Study Institute held in İstanbul in June 1979 was to try to bridge
the gap between the real system and its model.

As energy systems are so broad-based and complex, it is possible
to approach the subject from many different angles. In this particular ASI, the primary emphasis was the understanding of the
real system through the use of formal models, rather than on mathematical or computational sophistication. In selecting the topics
for the lectures, every effort was made to include important subjects as well as to provide an exposition of the methodology used
in modelling.

The four lectures in the first part address themselves to issues
related to the interaction of the energy system with the rest of
the economy. The lecture on global modelling is followed by those

on regional and national studies. Different techniques are used in these studies.

The second group of lectures essentially deal with national energy systems. The first presentation describes a simulation modelling approach; the second and third lectures describe optimization models, while the fourth and fifth lectures present combinations of techniques.

The basic theme of the third group of lectures is the supply-demand-price relationships of crude oil in the international market. The first presentation essentially deals with the supply side of the picture, the second one involves supply as well as demand and the price resulting from the interaction between the two. The third presentation reports on a more comprehensive model that relates the supply, demand and price of crude oil to the alternatives for oil, and is regionalized.

The fourth part contains models in the field of electricity demand and/or supply. The first presentation involves a disaggregate approach to long term electricity demand, while the second one reports on the integration of wind power within the interconnected system. The last paper in this section is on decision analysis that involves the uncertainties in the demand as well as in the supply of electricity.

The section on methodology contains three presentations. One of these is an exposé of network modelling at the regional level. The second and third presentations deal with linear optimization, the former involving multi-objective linear programming.

The last two lectures are on the subject of cooperation between analyst (or modeller) and decision maker. Experience in the use of models is the theme of the first presentation, while the second one describes the brief history of an effort that brings together modellers and the potential users of models.

One of the most hotly debated topics throughout the meeting was the determinants of the international price of crude oil. A special session was held with the purpose of providing a forum for an unrestricted debate of the theses concerning the pricing mechanism of crude oil. A verbatim account of the discussions are included in the final section of this volume.

During the meeting, considerable time was allowed for the discussions following the lectures, usually lasting for an hour or more. The insights gained from the views expressed sometimes contributed just as much as the lectures to the understanding of the subject matter. Despite their significance, it was not possible to include all the details of the discussions, but a fair representation is given in summary form at the end of each lecture.

A thorough survey of modelling techniques as well as an adequate coverage of the issues involved in energy systems analysis was not the aim of the meeting. It is probably a goal that could not be reached in the span of available time. However, taken as a whole the lectures adequately represent the current state of the art in energy systems modelling and it is hoped that the present volume contributes to further research in the subject.

İ. Kavrakoğlu

İstanbul
November 1979

ACKNOWLEDGEMENTS

Holding an Advanced Study Institute is a rewarding but at the same time quite a demanding task. Its planning, organization, inplementation, and documentation all require cooperation and efforts of many people.

The initial work for the ASI was started in 1977 when I was serving as Advisor to the Marmara Research Institute of the Turkish Scientific and Technical Research Council. Muhittin Oral, then head of the Operations Research Unit at the Institute, fully supported the idea and assisted in its initiation. I benefited considerably from discussions with Wolf Haefele, Alfred Voss, Paul Basile, Leif Ervik, Dale Jorgenson and Jim Sweeney while planning the lectures of the ASI.

The impeccable coordination of the meeting was entirely due to the organization and efforts of Zerrin Abut, Füsun Tezcan and Ateş Abut, who were assisted by Nursuna Erdinç, Ersin Pamuksüzer and Kemal Altınkemer. The help provided by Sibel Bali during various phases of the ASI is gratefully acknowledged. I am also indebted to the Engineering Faculty and the University Press of Boğaziçi University for needed assistance.

The discussions that followed the lectures were summarized by David Nunn and Erhan Topaç, who did an admirable job of capturing the essence of what was said.

Finally, I would like to express my sincere gratitude to Mine Atalay for organizing and typing the final text, and the Ayşe Kutkam for drawing most of the diagrams.

TABLE OF CONTENTS

WHY THIS ADVANCED STUDY INSTITUTE

İ. Kavrakoğlu

Boğaziçi University
Bebek, İstanbul, Turkey

By way of an informal introduction, I would like to say a few
words on the reasons for holding this Advanced Study Institute;
on the dimensions and issues involved in energy systems; and on the
problems that modellers face in trying to comprehend and analyze
energy systems. I will not claim to make an extensive, deep analytical study of the energy problem, but rather, I shall try to
expose certain questions that come to mind. Quite probably, we
will go away not having resolved the real issues, but only those
that we think we have solutions for.

About two years ago, the World Energy Conference was held here,
in İstanbul. I remember leaving the conference with the impression
that actually a considerable amount of data existed on energy
sources, energy technology, and energy production. As a matter
of fact, "energy" is one of the two topics that have attracted
the greatest interest in the last few years; the other is, of
course, "the environment". The accumulation of data has also
brought considerable confusion and controversy among the general
public, and even among the experts. This puts the policy makers,
or the decision makers in a rather awkward position for a sound
decision requires a certain consensus among the experts. If this
cannot be achieved, the controversy should at least be reduced to
a political choice. Presently, there are so many issues that need
clarification that we are not even near the point where the political choices can be made. All this indicates - at least to me -
that greater emphasis is needed for a <u>systems</u> approach to understanding energy.

In order to bring out the essential issues and priorities, one must
somehow try to quantify the problem. Discussions on unquantified

or non-quantifiable subjects may linger on without convergence.
A quantified approach necessitates some sort of a model, a mathematical model. System science is a rather young discipline; it
has not yet matured as some of the classical subjects have, such
as some branches of science, or some branches of social sciences,
for that matter. In the established disciplines the researchers
are sometimes ahead of the problems. In such cases, when the need
arises, there is at hand a methodology, or models that work. But
in the case of energy systems analysis, or energy modelling, the
techniques are only just being developed. First the problem emerges,
then we try to analyze it, and at the same time create certain
tools. We are really trailing behind events. I am under the impression that energy modelling is still in its infancy and that we
still have a long way to go. The strongest evidence for that is
the recent proliferation of models.

While reading the literature one notices that the term "energy
problem" means different things to different groups of people.
For wealthier nations, it is usually a matter of finding the right
choice among a set of alternatives. For the majority of poorer
nations, on the other hand, the problem is mostly to create just
one viable alternative. While the latter is a less desirable
situation to be in, the former is probably a more complex mathematical problem. The fact that a larger number of alternatives
exists makes the problem large. Furthermore, the wealthier nations
are also in a position to shape world affairs, thereby altering
the basic set of assumptions.

I would like to say a few words on the dimensions of the energy
problem. The physical and the fiscal dimensions of energy policy
analysis are well known, but the social and the psychological
aspects are proving to be predominant. Until the events of the
last few years, these have received practically no attention.
More recent occurrences in France, Germany, Sweden, and Austria
- not to mention many others - are reminders that these dimensions
will not simply disappear. It is against this background that we
have to plan our future energy systems.

We use energy for some end use. Frequently this point is missed
and the impression is given that we are trying to secure energy
for its own sake. Although much is known - and is still being
investigated - on how to supply the energy to the consumer, surprisingly little is known about how it is utilized by the user.
One explanation is that energy prices had been falling in real
terms, and that the end-use of energy enters into every conceivable
activity. Recent studies on the thermodynamics of end-use conversions reveal that the actual efficiencies are in order of magnitude smaller than physically possible efficiencies. The fact that
there is so much room for improvement is an indication of where to
channel modelling efforts.

With the end-use of energy, the question of life-styles comes up.
How much are we going to change the technology to suit an existing
or proposed life-style, and how much are we going to change our
life-styles to suit the existing or proposed technology? This
question has technical as well as social dimensions. For example,
the nuclear technology faces socio-psychological problems more
than anything else. But we cannot dismiss the problem because we
do not understand it. I have given the example of the nuclear
case only because it is of current interest. Other technologies
would probably face similar problems.

The energy system may also be a source of potential strain on the
political system. Several years ago, right after the so-called
"oil crisis", one of my colleagues expressed his concern for the
fate of democratic institutions. His main argument was that events
were taking shape rapidly, while democratic decision making was
taking too much time. At the time, I tended to agree with him.
Although the threat has not disappeared completely, the general
public's rate of absorption of the realities of the energy system
is continuously increasing.

During the past year, while I was trying to organize this ASI,
I had the opportunity to contact a number of experts working on
energy systems. We talked about such things as possible oil
price increases, the intangible aspects of energy systems analysis,
and almost every conceivable factor that had any bearing on energy
modelling. In these very fruitful discussions, I came across a
diversity of opinion, sometimes diametrically opposed. For example,
on the point of possible oil price rises, there are some experts
who maintain that it is too high as it is, and that it can only
go down. On the other hand, some experts believe that it can go
up as much as four-fold or even higher without any substantial
change in the present mechanism of world trade. As I spoke with
a rather large number of persons, one may easily guess that all
shades of grey were covered between these two extremes. The fact
that such a diversity of opinion exists on a subject as important
as oil prices is an indication that the tools of energy systems
analysis are not yet fully developed.

I want to draw your attention to possible psychological consequences of rising oil prices. We have all witnessed the high resilience
of the industrialized nations in absorbing the several fold increases in oil prices. A certain fraction of the increase was absorbed by simply raising the prices of the goods they were exporting
to the oil producing nations, but a certain fraction resulted in a
transfer of wealth, in the form of company shares, land, or buildings changing hands. Even though the transferred wealth represents
a negligible portion of the existing wealth, it was sufficient to
set off a wave of reaction by the public at large. It is not
difficult to surmise the consequences of, say, a ten-fold increase

in the value of the wealth changing hands. The problem is: how are we going to predict the point beyond which the psychological factor will dominate?

We also discussed other topics such as the coupling between the energy system and the rest of the economy, and substitution among different forms of energy. For example, how rapid can the energy system shift away from one source of energy to another? Again, the responses I was getting had considerable range. To some experts, it was almost inescapable to be tied to certain forms of energy for several decades because so much had been invested in terms of conversion equipment, transportation facilities or in terms of persons skilled in dealing with that form of energy. Yet some experts believed that it was only a matter of prices: if you increase the price, people would shift away from that source almost overnight, or something close to that. Now which answer is nearer to the truth?

Another point is that persons are looking at different aspects of the energy problem and you can conclude from discussions with them that they attribute importance to different elements of the system. As we all know, we can only see what we look at. The energy problem is a good example of this. Some of us look at certain aspects and see certain truths, while others look at different aspects and see other truths. And sometimes these truths are contradictory to each other. This, I believe, is one of the major problems that we face. What are the really essential elements of the system that we should be studying? Of course, the problems are too numerous so I do not think there can be any single tool that will provide solutions for each of these cases. Of course, there is also the problem of asking the right questions. In the majority of cases, we may be giving the right answer to the wrong question.

There seems to be one other factor that may prove to be quite important in energy system analysis. I remarked earlier that the speed of response of the social system is increasing. This may also influence the theory that attempts to explain the behaviour of the social system. The problem is that if the theory works, and if a large number of persons are aware of that fact, they may change their behaviour, thus necessitating a change in the theory. A recent example of this phenomenon is the emergence of new theories of macroeconomic behaviour. Keynesian economic theory had worked well and was instrumental in explaining the behaviour of market economies. The general public now believe in the correctness of the theory, and take measures according to predictions made by the use of the theory. For example, if the government comes up with an inflationary budget, they go out and spend more than was stipulated, thus causing a greater inflation than predicted. This error in the results is directly attributable to the altered behaviour of the public since the so-called "rational expectations" theories include a modification which is a function of the difference between the proposed inflation rate and the inflation rate expected by the public. The greater the speed by which information flows within the system, the greater will be the influence of the modifying factor.

Long range modelling of energy systems may also pose problems of feedback between the model and the system. Just as we learn from past experience, we may also learn from future experience. Such experience as gained from "scenarios" developed by the model may influence the future behaviour of the very system being modelled. An example of future experience has been provided by Meadows et.al in their Club of Rome study The Limits to Growth. The fact that the message has reached such a vast audience will probably also change the outcome predicted by the model.

Dynamic optimization models are not free of feedback between the model and the system, either. Here, the problem is that the objective function of the model is specified a priori while in the real system, the objectives are continuously changing. We are beginning to witness events which, I believe, may serve as examples of this phenomenon. The recent plebicite in Austria that resulted in the shut-down of the nuclear reactor may be explained by the changing objectives of the system.

I would like to conclude by saying a few words on the choice of the subjects of the lectures. An attempt has been made to cover as great a range of energy systems as possible, while at the same time exposing the different methods that exist. Some of the lectures deal with energy subsystems such as the electricity sector, or regional electricity demand, while others deal with all forms of energy at the national, or at the global level. The methods employed are also quite diverse, and include simulation, optimization, or econometric techniques, or a combination of these. No claim is made that the choice is exhaustive, but I believe that the present state-of-the-art is adequately covered by these lectures. It is also my sincere hope that all those attending – the lecturers as well as the participants – will benefit from the lectures and the ensuing discussions.

PART I

MODELLING ENERGY-ECONOMY RELATIONS

Global Energy Modelling and Implementation
for Planning

Macro: An Aggregate Macroeconomic Model for
the IIASA Set of Energy Models

Energy and Economic Growth: A Conceptual
Framework

The Norwegian Electro-Industry: A Study of
the Economic Impacts of Energy Use

GLOBAL ENERGY MODELLING AND IMPLEMENTATION FOR PLANNING

P. Basile

International Institute for Applied Systems Analysis
(IIASA), Laxenburg
Austria

ABSTRACT. A set of models have been developed at the International
Institute for Applied Systems Analysis for the study of global,
long term energy issues. These models, and a sample of their operation and results, are discussed. In particular, their structure
as an integrated set is elaborated. Finally, a word is offered
on the implementation of such global analytic tools for policy
making.

1. ENERGY MODELLING AT IIASA

1.1 Why We Do what We Do

Future-study, regrettably, is not easy. None of us is a prophet
or forecaster. We see the future as complex, and largely unknown
- at least unknown to most of us to any degree of relevant detail.
The future intrigues, inspires and occasionally frightens us. But,
intelligently perceived, the future is found to be a motivator for
mature actions. When understood, the future awakens, not frightens
us. And then we can take action. In the words of the sage comic-
strip character Pogo, "Why do we stand here confronted by insur-
mountable opportunities?"

Responsible action today requires an informed perception of tomorrow - an intelligent assessment of the real uncertainties and possible outcomes in the future. To an enterprise, or an office, or
a project, or to anyone raising a family, this is not a novelty.
When society operates educational institutions, or when a family
raises children, it acts (or should act) responsibly, with an
informed perception of the future.

The energy studies at the International Institute for Applied
Systems Analysis (IIASA) were motivated by the belief that informed perceptions of the long term future of our energy systems
are essential for responsible energy planning today. For this
purpose, a set of tools has been developed, and will be described
- in conception, purpose, scope,and results. First, a few words
about IIASA's overall energy work.

1.2 About IIASA's Energy Systems Program

Energy modelling at the International Institute for Applied Systems
Analysis is part of the Energy Systems Program, a research program
which focuses its attention on what has been called the energy
transition - the slow but profound shift from the present energy
system to a future sustainable one. The Program's primary considerations are long term ones, spanning a horizon of 15 to 50 years
from now. Within this period, the Program's findings indicate,
many characteristics of the coming energy transition will be seen
and felt.

We have chosen a long time horizon because we believe that the
evolution of the energy infrastructure (oil fields, pipelines,
power plants, energy-using equipment) is a central and governing
factor of the shape of the transition - the lifetimes of energy
systems are long. Not many groups focus on this 1995-2030 period,
it is clear. Their emphasis is on the more near term future, 5 to
15 years from now. This makes sense, given the dynamics of such
relevant factors as energy prices, evolving laws and regulations,
start-ups of new oil fields, etc. But we want to consider how
long term strategic investments in energy systems could change the
transitional picture. We want to look at the direction that we
might be headed, and ask what mix of market forces and government
plans might get us there. Because of it, we must take a long look,
a fifty-year look.

The study's considerations are also necessarily global ones; the
present and future large scale supply and use of energy mandates
a degree of global interdependence that is unprecedented.

The modelling is, in a very real sense, the synthesis of the several tasks within the Energy Systems Program. The intention is to
bring the elements together in order to identify overall energy
strategies for the long term, and to evaluate the possibilities
for the integration of such strategies into the economy, the environment, and the society. The complexity of the energy transition demands careful analysis of all of the interrelationships.
Such analysis could be seen as the central purpose and strength
of energy modelling.

1.3. Purpose and Goals of Energy Modelling

The general purposes of computer modelling may be three-fold. First, and perhaps foremost, the real value of models is in the insight, not the numbers, that they provide. Excessive attention to the plethora of data or to the multi-digit accuracy that a computer can provide tends to both miss the message and mislead the user. Models should be designed for gaining insight and understanding, not (necessarily) for mathematical sophistication; informal "mental models", indeed, are essential prerequisites to formal mathematical models.

The second purpose of computer modelling is not unrelated to the first. Computer models provide results which should be reproducible from basic logic; model results, once seen, should be obvious. Modelling does not, after all, replace careful thinking - it seeks to enhance it.

Finally, computer models are useful in that they provide calculational consistency. For highly complex and quantitative subjects, modelling provides an essential accounting framework - a necessary classification scheme - to aid in the otherwise laborious if not impossible task of simultaneous calculations with hundreds or thousands of variables.

The objectives or goals of IIASA's particular set of energy models are perhaps four-fold:

1. to study the long term, dynamic (transitional), and strategic dimensions of regional and global energy systems; indeed, one cannot solve a problem until one understands it.
2. to explore the embedding of future energy systems and strategies into the economy, the environment, and society; (is there sufficient time? enough capital? to achieve a given energy strategy).
3. to develop a global framework to enable the assessment of the global implications of long term regional or national energy policies.
4. to evaluate alternative strategies- to compare options- of a physical and technological kind, including their economic impacts.

With these several goals in mind, the energy models at IIASA have been developed. The aim has been to organize and extend the debate on the impacts of future energy alternatives -to evaluate plausible energy strategies in a full systems context, and to do so with a truly global perspective.

1.4 The Set of Energy Models

General Structure

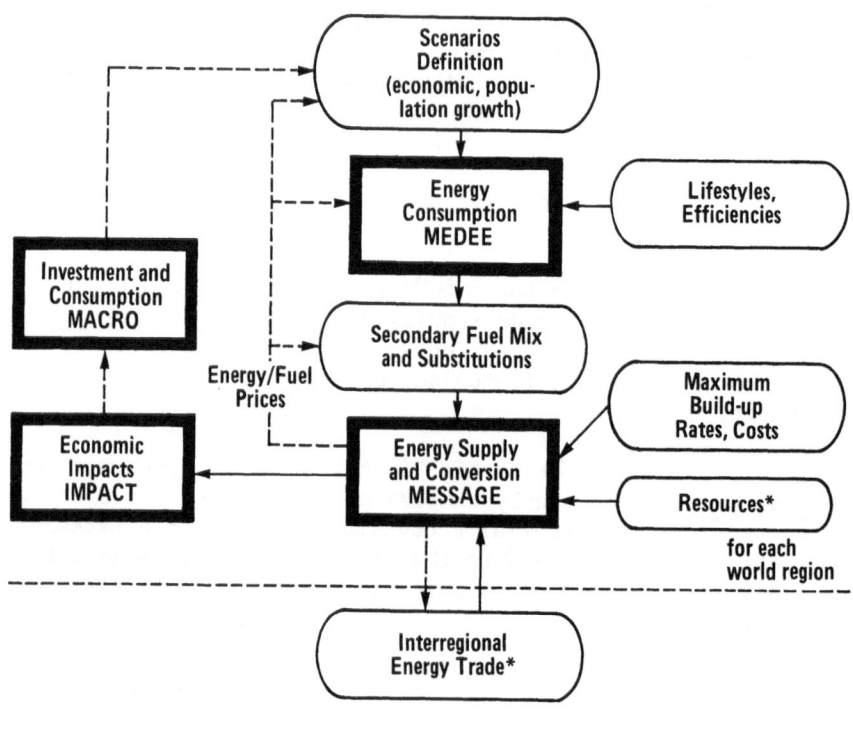

FIGURE 1 - IIASA's Set of Energy Models: A Simplified Representation

The energy modelling approach at IIASA is a highly iterative one
(Fig. 1). Initiating assumptions and judgments lead to
calculations and results which feed back and modify those assumptions and judgments. Most of the feedbacks are manual.
Iteration here is meant to be real, human learning: an original

assumption about the relative rate of penetration of gaseous fuels into residential markets (for example) is increased by the analyst as relative fuel prices (stemming from supplies) show an advantage for gas. While the flow of information is mechanized, the impacts on one set of inputs from changes in another are not.

The energy modelling activity begins with scenario definitions (top of Fig. 1).

In the IIASA energy research, two scenarios are selected - two plausible futures believed to span a reasonable-to-expect range. They are defined by "high" and "low" economic growth within regions, and consequent high and low energy demand growth. Population growth is also a scenario-defining parameter, although at present just one projection of population is used in IIASA energy studies. Other factors vary from scenario to scenario according to judgments about internal consistency.

The scenario projections of economic and population growth for each world region provide the basic inputs for detailed calculations of future final energy consumption consistent with the scenarios. Disaggregation of these overall economic and demographic projections enables detailed consideration of economic and energy-consuming activities in three macro sectors: transportation, household/ service, and industry (agriculture, construction, mining, manu- facturing). An array of judgments about lifestyle developments, improvements in efficiencies of energy-using devices, and the rate of penetration of new and/or improved energy-using equipment augment the disaggregated economic and demographic assumptions for each region. All of these details are meant to be consistent with the general scenario parameters and are recorded in a model called MEDEE-2[*] where calculations lead to estimates of useful energy in the macro sectors. This useful energy is for example high/low temperature steam, space heat, water heat, air condition- ing. Following considerations of the penetrations of certain en- ergy sources such as solar and district heating and electricity, the final energy consumption patterns for each scenario are cal- culated. Final energy here is of two types: non-substitutable (motor fuel, coal for steam trains, coking coal, petrochemical feedstocks, and electricity for special purposes), and substitutable (electricity, solar and fossil fuels used for heating; and district heat). The substitutable categories can be - and are, in practice - adjusted for relative price differentials as iterations around the modelling loop are made.

---

[*] MEDEE stands for Modele d'Evolution de la Demande d'Energie. MEDEE was developed at the University of Grenoble and brought to IIASA by B. Lapillonne. It has been expanded and is now operated at IIASA by A. Khan and A. Hoelzl.

A further step divides the fossil fuels used for heating among liquids, gaseous and solid fuels. This step then completes the secondary energy demand calculations required as input to the energy supply and conversion model MESSAGE[*] (Fig. 1).

MESSAGE calculates the required supplies of primary fuels to meet the secondary energy demand, at lowest cost and within sometimes quite tight constraints on resource availabilities, technological development, and the rates of build-up of new energy facilities. Resource constraints are specified as maximum pools of oil, natural gas, coal and uranium available at specified costs. As prices rise, several high cost alternatives can compete. Limits on the maximum rate of build-up of energy facilities reflect the inherent lead times, as well as limitations of manpower, materials, etc. in a region.

Interregional energy trade considerations provide time profiles of imports and exports of fuels for each regional MESSAGE run. Relatively simple allocation rules distribute available exports of fuels (e.g. oil) from exporting regions (e.g. the Middle East) to competing importing regions (e.g. Western Europe and Africa). Allocations are done iteratively with MESSAGE runs (Fig. 1) so that a globally consistent balance is achieved.[**]

MESSAGE gives fuel production over time and the patch of different primary fuels through conversion processes to a fixed set of secondary demands. In addition, MESSAGE provides the marginal production costs of primary fuels, leading to estimates of time trajectories of fuel and electricity prices. These prices are fed back (Fig. 1) to several points in the loop, in order to iteratively modify initiating assumptions and judgments.

To be specific, prices in this procedure affect three calculations (Fig. 1). Firstly, price changes alter macro-economic growth patterns: increased prices can constrain overall growth and/or can shift activities from more- to less-energy-intensive sectors. These changes are made judgmentally at present, based on estimates of experts inside and outside of IIASA.

---

[*] MESSAGE stands for Model for Energy Supply Systems and their General Environment Impact. MESSAGE is an extended version of the Haefele-Manne model (Haefele and Manne, 1979) and has been developed at IIASA which has been revised and operated by M. Agnew, L. Schrattenholzer and A. Voss.

[**]This procedure will be formalized in the near future through use of a gaming model developed originally at the Siberian Power Institute.

Secondly, prices alter lifestyles and technological efficiencies of energy-using devices. Such alterations can at best be informed guesses; as prices increase, efficiencies tend more toward the technical potentials, and lifestyles adapt more toward lowered energy use.

Finally, relative price changes among different fuels and electricity can cause the mix of secondary energy types demanded to change; relative increases in liquid prices induce shifts toward gaseous fuels, for example. No formal or precise elasticities of substitution are used here; again, "best informed judgments" describe the approach.

The addition of energy facilities required to meet the energy supply scenarios of MESSAGE have direct costs - capital, manpower, and materials costs. An IMPACT* model (Fig. 1) calculates the required direct and indirect costs of new energy facilities, and thus provides the basic information for assessing whether or not an economy can afford a given energy scenario. Exogenous assumptions about facility-specific size, material, and manpower requirements are made for IMPACT in order to calculate the direct and indirect (energy-related) requirements of a given energy strategy.

With IMPACT-calculated costs, we can begin to ask if energy will absorb unacceptably high shares of economic product. What forms of capital and financial aid will be required by developing countries? What level of non-energy exports are necessary to pay for large energy imports?

Finally, a MACRO** economic model (Fig. 1) accepts exogenous assumptions about demographics, and institutional parameters such as productivity, taxes, trade, etc. and calculates investment and consumption rates consistent with the costs from IMPACT. This allows assessment of the magnitude of change in, for example, the capital/output ratio if and when energy becomes increasingly capital intensive. This in turn enables a re-check of the original GNP estimates for each region and a re-entering of the iterative process.

This last feedback is one toward which much of the energy modelling design and implementation work at IIASA has been leading. The critical question is: Can economies afford the capital or the time to achieve energy strategies if, during the transition 15 to 50 years from now, energy becomes increasingly capital intensive?

---

* IMPACT was developed at the Siberian Power Institute and extensively modified for use at IIASA by Y. Kononov.
**MACRO was developed at IIASA by M. Norman. It is currently being operated and extended by H.H. Rogner.

A Highlighting of Methods

Each of the four main models of the set (Fig.1) will be described
here, with necessary if brutal brevity.

The MEDEE-2 model is a simulation model for calculating the useful
and/or final energy demand of major end use categories such as
space heating/cooling, water heating, cooking, etc. in the residential and service sectors, or space/water heating, steam generation, furnace operation, etc. in industry, based on exogenous
assumptions about the long term evolution of the main determinants
of each of these energy-using activities.

Required for the energy demand calculations is the GDP formation
by kind of economic activity. The GDP share of each sector can be
specified exogenously or can be calculated from econometric equations in the model, based on assumptions about GDP expenditure.

Energy demands are calculated in some detail, as the following
examples illustrate.

Total demand for intercity (or rural) passenger transportation
(measured in passenger-kilometers) is calculated from population
and average distance traveled per person per year. Car travel is
calculated from population, car ownership, average distance
traveled per car per year, and an average load factor (passenger-
kilometers per vehicle-kilometer). The remainder is allocated to
public transportation modes (rail, bus, plane) according to exogenously specified shares. The corresponding vehicle-kilometers
are calculated from average load factors for each mode. The energy
intensities (per vehicle-kilometer) also have to be specified.
Only liquid fuels are assumed to be used, except for railways.

Three types of industrial energy end use categories are considered:
specific uses of electricity, thermal uses, and motor fuel use.

Required for the energy demand calculations are first of all the
activity level (value added) and the energy intensities (per unit
value added) for each sector. The energy intensities have to be
specified in terms of final energy for motor fuel and electricity,
and in terms of "electricity equivalent" for thermal uses. The
breakdown of thermal uses (space/water heating, low/high temperature steam generation, furnace operation) is assumed to be constant.

The change in the housing stock of the residential sector is determined from average family size and population, demolition of
existing dwellings by type, and construction of new dwellings by
type. Allowance is made for reduction of heat loss in old dwellings
through retrofitting; the heat loss of post-1975 dwellings is calculated from the average size and the specific heat loss (per m^2)
for each type of dwelling.

Energy demand for water heating, cooking, air conditioning, and the electricity consumption of secondary appliances in the residential and commercial sectors is calculated from exogenously specified ownership fractions and average annual consumption rates.

The central tool used for analyses of the energy supply and conversion system is a dynamic linear programming model called MESSAGE.

In MESSAGE, a number of primary energy sources and their associated conversion technologies are considered (Fig. 2). Primary energy sources in several categories are directly (e.g. by crude oil refining) or indirectly (e.g. by electric power plants fired by liquid fuels) converted into secondary energy. Secondary energy is exogenous to MESSAGE and is provided as time series by type: electricity, "soft" solar, solid, liquid, gaseous fuels and district heat. The planning horizon in MESSAGE (50 years in these analyses) is divided into n time periods of equal length. The variables of the model are expressed in period-averages of annual quantities.

The objective function of the MESSAGE model is the sum of discounted costs of capital, operating/maintenance, and fuels (primary energy):

$$\sum_{t=1}^{n} \beta(t) \; 5\{b^T r(t) + c^T x(t) + d^T y(t)\} \; ,$$

where t   = current index of time period
      n   = number of time periods
      $\beta(t)$ = discount factor
      5   = number of years per period
      b   = vector of energy resources costs
      r   = resource activities (LP variables)
      c   = vector of operating/maintenance costs
      x   = vector of energy conversion activities (LP variables)
      d   = vector of capital (investment) costs
      y   = vector of capacity increments (LP variables)

The discount factor is calculated from an annual discount rate of six percent. As MESSAGE is intended to minimize societal costs this discount rate is to be understood as a pre-tax one.

The cost of increments to capacity still operating at the end of the planning horizon is corrected by a "terminal valuation factor":

$$tv^t = (1 - \beta^{5(n+1-t)}) \; ,$$

e.g., the terminal valuation factor for the last period is

$$tv^n = 1 - \beta^5 .$$

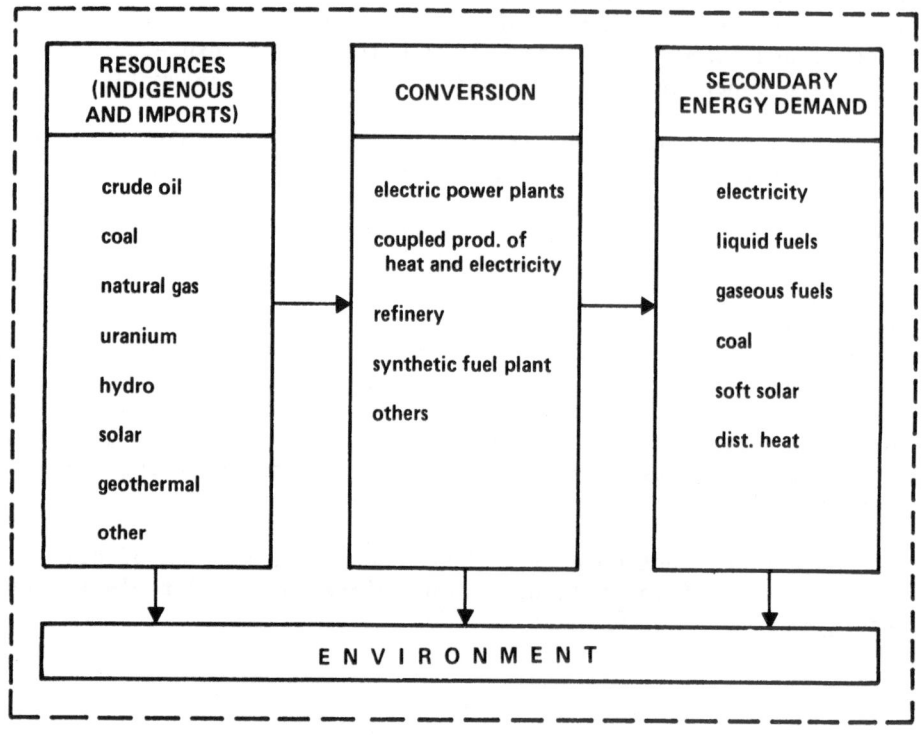

FIGURE 2 - Simplified Schematic of MESSAGE

The IMPACT model belongs to the set of energy oriented dynamic input-output models, explicitly accounting for lags between the start of investment and the putting into operation of production capacities.

For each given energy strategy, the model determines:
- investment in energy system development;
- the required putting into operation of capacities in related branches of industry and corresponding (indirect) capital investment;
- the required output of different types of materials, equipment and services to provide operational and construction requirements of the energy system and related branches;
- direct and indirect WELMM*requirements.

All these indicators are evaluated for each year of the period considered.

---

* WELMM = Water, Energy, Land, Materials, Manpower

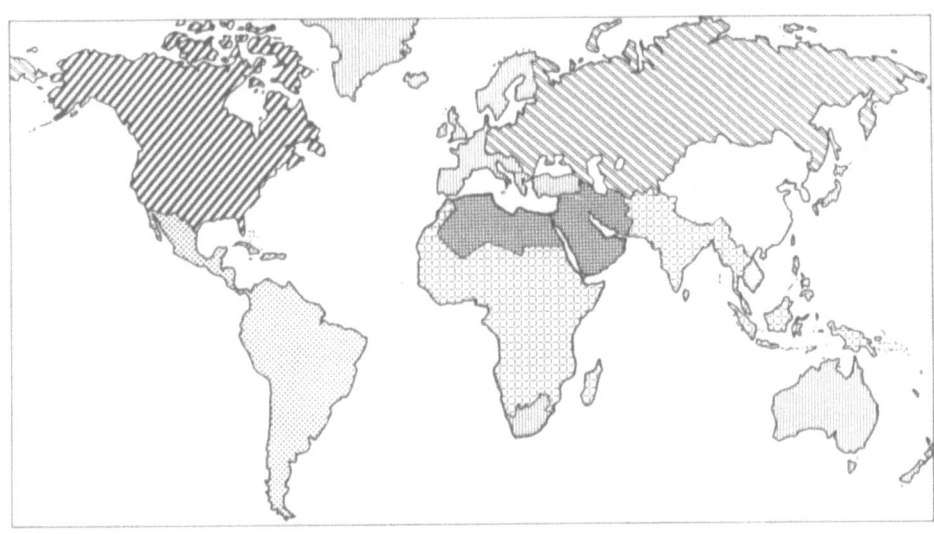

Region	No.	Abbrev.	Name
▨	I	NA	North America
◪	II	SU/EE	The Soviet Union and E. Europe
▥	III	WE/JANZ	W. Europe, Japan, Australia, New Zealand, S. Africa, Israel
▦	IV	LA	Latin America
▣	V	Af/SEA	Africa (except Northern Africa and S. Africa), South and Southeast Asia
▦	VI	ME/NA	Middle East and Northern Africa
☐	VII	C/CPA	China and Centrally Planned Asian Economies

FIGURE 3 - World regions defined for use in IIASA Energy
Systems Program

2. A SUMMARY OF GLOBAL RESULTS

2.1 World Regions and Base Year Data

Seven aggregate world regions have been chosen for study. The
regions have been selected more for the economic and energy systems similarities than for geographical proximity, i.e. one region i

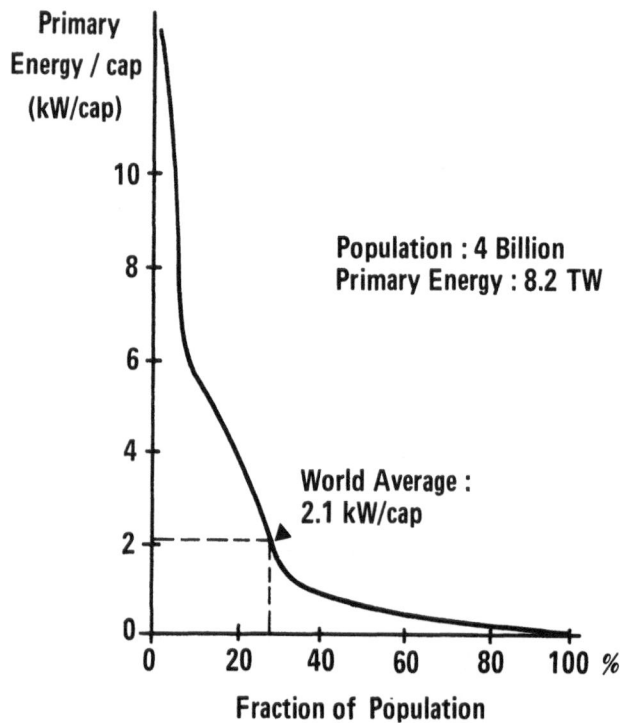

FIGURE 4 - Per capita primary energy consumption, world, 1975

a developed market economy with large resources (Region I, NA) while another is a developing economy with relatively few energy resources (Region V, Af/SEA) while another is a centrally planned economy with large resources (Region II, SU/EE) (Fig. 3).

The year 1975 has been selected as a base year for study. In that year world primary energy consumption was close to 8×10^{12} Wy/y or in short 8TW on average. With the world population of 4 billion average per capita energy use was 2kW. Yet, more than 70% of the world's population lives with less than 2kW/cap and more than 80 countries have a consumption as low as 0.2kW/cap, while only 6% of the world enjoys more than 7kW/cap. Responsible technical planning must assume that the present uneven distribution becomes less uneven- meaning that the global average will increase beyond 2kW/cap (Fig. 4). This has been a starting point for the IIASA energy scenarios.

2.2 Two Scenarios Defined

Two global energy scenarios for the next fifty years have been defined and analyzed: a "High" and a "Low". The High scenario has

an assumed relatively high rate of economic growth and consequent
high energy use – although with strong energy efficiency improvements.
The Low scenario represents relatively low economic growth, but
also strong energy efficiency improvements. Both scenarios assume
a doubling of world population, to 8 billion people by 2030 with
gradually declining growth rates.

Assumed economic growth rates for the several regions of the world
(Table 1) were generated by relying on the general trends of published estimates – with some modifications based on discussions with
experts and on the growth dampening impacts of rising energy prices.

The projected trend of decreasing growth rates is evident in every
region and exemplified by the drop from 4.3%/yr in the first period
High scenario for Regions I (NA) and III (WE/JANZ) to 2.0 in the
final period. The projected growth rates for the developing regions
are generally 1.5 to 2.0% higher than for Regions I and III. This
higher growth in developing regions is due to higher population
growth and industrialization. The economic growth of the developing regions has also been assumed to be tied to growth in developed
regions, which also tends to limit growth in the longer term.

2.3 Energy Demand

World primary energy use is projected to increase more than 5-fold
by 2030 in the High scenario and more than 3-fold in the Low
scenario. Over the 55 year period the average annual energy growth
rate for the High scenario turns out to be about 3% and for the
Low about 2.1%. From the regional breakdown in this projection
(Table 2) we can calculate that the growth rate for Regions I and
III together is lower than the world average, that Region II is
close to the world average and that the developing regions have
much higher energy growth rates than the world average.

The primary energy demand figures of Table 2 result from the full
modelling analysis process within IIASA's Energy Program. This involves detailed analyses of energy consumption by many sectors for
each region of the world, including analyses of useful and final
demand and demands for non-substitutable fuels such as motor fuel,
coking coal, electricity for specific purposes, as well as non-
substitutable fossil fuels for heat needs. No attempt will be made
here to report comprehensively those results. However, two measures
will give a sense of the aggregate character of the final energy
demand calculations as well as an example of the detailed considerations in our energy demand analyses.

The first such measure is the energy-to-GDP elasticity. The energy/
GDP elasticities resulting from our demand analyses are in general
for the developed regions less than 1.0 and for the developing

Real Growth Rate of GDP (% per year)

Region	Historical 1960–1975	High Scenario Projection 1975–1985	1985–2000	2000–2015	2015–2030
I (NA)	3.5	4.3	3.3	2.4	2.0
II (SU/EE)	6.4	5.0	4.0	3.5	3.5
III (WE/JANZ)	5.3	4.3	3.4	2.5	2.0
IV (LA)	5.8	6.2	4.9	3.7	3.3
V (Af/SEA)	5.2	5.8	4.8	3.8	3.4
VI (ME/NA)	9.4	7.2	5.9	4.2	3.8
VII (C/CPA)	6.2	5.0	4.5	4.0	3.6
World		4.7	3.8	2.9	2.6

Region	Historical 1960–1975	Low Scenario Projection 1975–1985	1985–2000	2000–2015	2015–2030
I (NA)	3.5	3.1	2.0	1.1	1.0
II (SU/EE)	6.4	4.5	3.5	2.5	2.0
IiI (WE/JANZ)	5.3	3.2	2.1	1.5	1.2
IV (LA)	5.8	4.7	3.6	3.0	3.0
V (Af/SEA)	5.2	4.8	3.6	2.8	2.4
VI (ME/NA)	9.4	5.6	4.6	2.7	2.1
VII (C/CPA)	6.2	3.3	3.0	2.5	2.0
World		3.6	2.6	2.0	1.9

TABLE 1 – Assumed Growth Rates of GDP for High and Low Scenarios.

Primary Energy Consumption[1] (TW)

Region	Base Year 1975	High Scenario 2000	High Scenario 2030	Low Scenario 2000	Low Scenario 2030
I (NA)	2.66	4.53	7.43	3.78	5.49
II (SU/EE)	1.91	4.21	9.02	3.81	6.50
III (WE/JANZ)	2.26	4.83	8.17	3.73	5.17
IV (LA)	0.34	1.40	4.04	1.02	2.55
V (Af/SEA)	0.33	1.45	4.57	1.11	2.60
VI (ME/NA)	0.13	0.88	3.02	0.58	1.23
VII (C/CPA)	0.46	1.61	5.57	1.03	2.18
Bunkers[2]	0.14	0.15	0.20	0.10	0.15
World	8.22	19.06	42.02	15.16	25.87

1 Totals may not appear to add exactly due to rounding.

2 Energy used in international transport of energy (petroleum and coal).

Note: These, and other energy projections in this paper are preliminary assessments. Revisions are in process, and the formal reports of the studies are in formal review.

TABLE 2 – Primary Energy Projections, by Region, High and Low Scenarios.

regions greater than 1.0 (Table 3). The lower values for developed regions result from several factors including price-induced conservation, efficiency improvements due to technological development and regulation and a changing economic texture towards more services and less energy-intensive industry.

The developing regions have elasticities close to or greater than unity implying a shift towards more energy intensive manufacturing and heavy industry. If the final energy/GDP elasticities is some measure of the rate of improvement of energy productivity increases, then it will be seen that the assumptions are for such increases to be vigorous in the developed regions. Also, the projections do not see the developing countries becoming developed and having the same kind of elasticities as the developed regions in the next 50 years. Possible exceptions to this are Regions IV and VI (LA and ME/NA) which, although their elasticities remain high near the end of the period, have energy per capita and GDP per capita values comparable to these of Europe and Japan today.

Energy projections can also be summarized as the product of income and price elasticities by:

$$\frac{E}{E_o} = \left(\frac{Y}{Y_o}\right)^\gamma \left(\frac{P}{P_o}\right)^\beta ,$$

where E is energy, Y is income and P is energy price in a given year, (($_o$) are initial values), γ is the income elasticity, and β is price elasticity. This measure of energy, economy and prices has been made for our scenarios (Table 4). By assuming energy-income elasticities in the developed regions to be greater than in the developing (i.e., no departure from the recent past), then price elasticities in developed economies are found to exceed, generally, those in developing regions.

2.4 Energy Supply and Conversion

Now a few central conclusions from analyses (using the MESSAGE model) of the energy supply and conversion system for the world over the next five decades can be drawn:
- Coal is required in tremendous quantities: 10-14 billion tons per year in 2030 in our cases. Some 65% of this amount are required for the production of synthetic liquid fuels.
- World conventional oil production peaks shortly after the turn of the century at about 7TW, or two times 1975 output, and declines thereafter. Only large scale exploitation of unconventional shale oil and tar sands seem capable of reversing the trend.
- The level of liquids demands in our cases seems to drive many

TABLE 4— Final Energy-Income and -Price Elasticities, High and Low Scenarios.

Region	High Scenario Income Elasticity[1] γ	High Scenario Price Elasticity[2] β	Low Scenario Income Elasticity[1] γ	Low Scenario Price Elasticity[1] β
I (NA)	0.75	-0.35	0.75	-0.25
II (SU/EE)	0.70	-0.21	0.70	-0.09
III (WE/JANZ)	0.85	-0.29	0.85	-0.21
IV (LA)	1.10	-0.22	1.10	-0.16
V (Af/SEA)	1.15	-0.18	1.20	-0.12
VI (ME/NA)	1.15	-0.08	1.20	-0.13
VII (C/CPA)	1.15	-0.23	1.20	-0.24

1 Average value assumed for 1975-2030 for final energy, based on historic value.

2 Final energy price elasticity, all sector aggregate, for period 1975 to 2030.

TABLE 3— Final Energy-GDP Coefficient, ε_f, High and Low Scenarios.

High Scenario

Region	1975-1985	1985-2000	2000-2015	2015-2030
I (NA)	0.48	0.54	0.54	0.52
II (SU/EE)	0.70	0.64	0.60	0.55
III (WE/JANZ)	0.86	0.71	0.63	0.54
IV (LA)	1.08	1.01	0.99	0.94
V (Af/SEA)	1.24	1.07	1.03	0.98
VI (ME/NA)	1.24	1.18	1.04	0.95
VII (C/CPA)	1.08	1.06	1.02	0.99

Low Scenario

Region	1975-1985	1985-2000	2000-2015	2015-2030
I (NA)	0.43	0.49	0.56	0.48
II (SU/EE)	0.68	0.62	0.55	0.49
III (WE/JANZ)	0.75	0.64	0.61	0.58
IV (LA)	1.11	1.04	0.98	0.93
V (Af/SEA)	1.26	1.11	1.07	0.98
VI (ME/NA)	1.23	1.15	1.07	0.95
VII (C/CPA)	0.98	1.04	1.03	1.02

of the results. In combination with this the assumed oil production ceiling in Region VI (ME/NA) of about 33.6 million
barrels per day has a great impact on the extent and timing of
oil alternatives. As imports become restricted alternatives
appear.
- The mix of sources for electricity generation undergoes a great
  transformation. By 2030 some 60% of oil capacity could be nuclear.
  Coal could decline (globally) as a share from 40% in 1975 to 20%
  by 2030.
- Production of natural gas increases moderately throughout the
  period reaching a range of 2.5 to 5.0TW in 2030 compared to 1.6TW
  in 1975. If the major transportation and the price impediments to
  natural gas could be overcome, world reserves would allow natural
  gas to be a truly global scale fuel in the coming decades.
- Hydropower in our cases meets from 4.0 to 2.7% of primary energy
  by 2030 (or about 10.5% of total electricity demand). Over 60%
  of the world's total hydropotential capacity is in the developing
  regions. In our exploitation of this, the potential supplies 31%
  and 17% of the electricity demand in the developing regions by
  2030.
- Solar, geothermal and other renewable energy sources make important
  but usually local and small scale contributions to the supply mix
  by 2030. Together they add as much as 1.6TW, or about 4% to
  primary energy. This assumes that major technical innovation and
  mass production techniques are not stimulated by government incentives and do not achieve progress unanticipated at present.
  However, if rapid technological advances take place, as many
  expect that they will, the relative contribution of high technology solar devices (e.g. solar photovoltaic systems) could be
  significantly higher than we have estimated.

These few conclusions stem from our energy systems analyses of the
globe's energy prospects in the next 50 years. A great deal of
detailed studies lie behind these figures – very little of which can
be reported or described here. Table 5 summarizes the definitions
of main parameters in our High and Low scenarios and sets the bounds
on the assumptions and central results of our analyses. In order to
indicate the character of the interaction among supplies and demands and to indicate some of the richness of the output of the
tools in use, we consider here world oil supply and demand and the
cumulative uses of fossil fuels.*

In our High and Low scenarios the oil era will not be over by 2030.
We will be moving from the petroleum era to the unconventional
fossil era.

_____

* More complete results are presented in the Supply Scenarios
  chapter in the forthcoming final report of IIASA's Energy Systems
  Program.

TABLE 5 - Two Global Supply Scenarios: Definitions of Main
           Parameters

Parameters	High Scenario	Low Scenario
Population, 2030 (10^6)	7976	7976
World GDP Growth Rate (%/yr)	3.4	2.4
Final Energy Demand (TW)	27.3	16.6
Oil Price* ($/boe)	22.33	16.31
Level reached by	1995	2010
Region VI Oil Production		
Ceiling (TW)	2.38	2.38
(mbd)	33.6	33.6

Note: These parameters include both exogenous assumptions and the
      results of analyses that become inputs to the supply scena-
      rios.

* These prices, of internationally traded crude oil, are the results
  of iterative supply analyses; the figures are 1975 dollars per
  barrel of average crude, fob Persian Gulf.

By 2030 world oil production will have peaked and declined under
the conditions of our scenarios (Fig. 5). At the global oil
production peak around 2010 the annual rate is nearly 7.7TW in the
High case and 5.8TW in the Low compared to 3.8TW today. Much of the
extra oil production by 2030 would have to come from unconventional
oil, tar sands, heavy oils, and oil shales in addition to large
amounts of conventional oil extracted through enhanced recovery
techniques.

The global oil production profile is built up on the basis of liquid
demands in each of the regions as well as estimates of the potential production region by region over time consistent with the
price movements in the scenarios. In order to estimate such maximum production levels detailed assessments must be made region by
region of resource availabilities, discovery rates, and willingness
to produce. In general, the discovery rates used in these analyses are higher than those used in other studies in the initial
period. This allows the supply of oil to meet demands in the coming
two decades _ a result somewhat different from several recent studies.

In addition to this, another critical element in these analyses is
an assumed production ceiling in Region VI (ME/NA). This ceiling

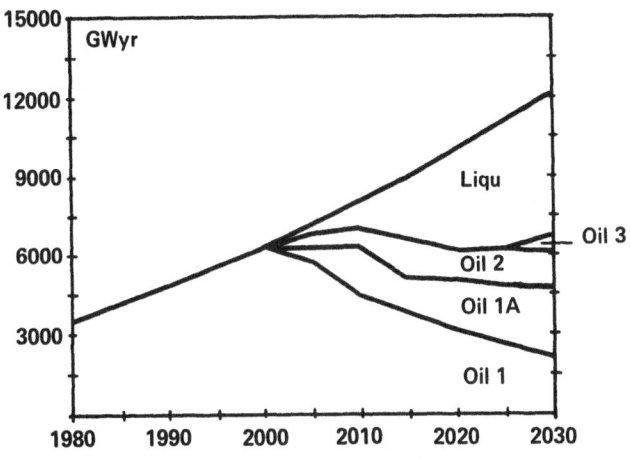

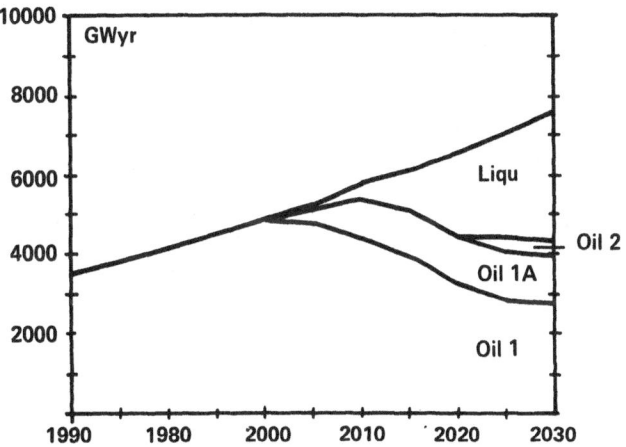

Note: Oil 1 is oil produced at costs of 12$/boe or less, oil 1A is
      at 12-16$/boe, oil 2 at 16-20$/boe, and oil 3 at 20-25$/boe.
      "Liqu" is synthetic liquid fuels from coal.

FIGURE 5- Oil Supply and Demand, 1975 to 2030.

of 33.6 million barrels per day is set arbitrarily at just precisely 50% greater than the 1973 production from this region of 22.4
million barrels per day. Production (and exports) from this region
increase to meet global demands until the imposed ceiling is reached.
This occurs by the year 1995 in the High case and by the year 2010
in the Low.

The production ceiling might be argued. Still, several factors
should be kept in mind. Firstly, it has been widely felt that some

of the major suppliers in this region (e.g. Saudi Arabia and Kuwait, and the United Arab Emirates) will be unable to absorb revenues from extremely high levels of production in the short term. The unbalancing implications for the world financial system are such that many would argue that it is in the best interests of both these countries and the rest of the world to restrain production rates. Secondly, Saudi Arabia has very clearly announced its own intent to maintain a moderate ceiling on production. Much of the world expects the Saudis to increase production to 16 million barrels per day in the 1980's. And the Saudis insist that, at least in the short term, they will restrain their production to less than the 11 million barrels per day recent peak period to about 8.5 million barrels per day. Thus, indications of restraint of production in the region are already announced, acknowledged and felt. Thirdly, it should be remembered that our 50 year projections are surprise-free. They assume neither major technological breakthroughs nor major disruptions or disasters. Within this it is certainly unrealistic to assume that difficulties in sensitive governments in this region (such as recently witnessed in the case of Iran) will not cause some difficulty in expansions of oil production and exports.

Finally, it might be observed that the production ceiling assumption of 33.6 million barrels per day, if maintained through our scenarios to about the year 2030, results in a cumulative oil production which is nearing the bounds of the technical limit for the region, i.e. shortly after 2030 oil production might well have to decline because of reserves-to-production ratio constraints. Therefore the ceiling imposed could well be seen as an attempt to be efficient and to stretch out the lifetime of the reserves - independent of any near term or medium term political or economic arguments.

In any case the global oil results (Fig. 5) are revealing. The cheapest oil category (oil I) is virtually depleted. Other moderately expensive categories are nearly depleted and the unconventional sources of liquid fuels are beginning to be used by the year 2030. Coal liquefaction is used in rather substantial amounts in these cases.

Indeed, the oil supply picture by 2030 is unconventional in this dimension. The large scale use of coal to produce synthetic liquids reaches 5.4TW or a figure that exceeds the present total global oil production by 42%.

What does this mean? Will the coal-rich countries mine their vast resources sufficiently to quench the world's thirst for liquid fuels? Our analyses here point to few other solutions if liquids continue to be required on a large scale - as we think they will, even if only for premium uses.

This is a major observation. Coal will be needed in very great amounts for the production of liquid fuels. Unless the scenario estimates of demand for liquid fuels are greatly in error there seems little margin for other than synthetic liquids production to meet the needs.

Much of our analyses leads to the central observation that liquids supplies is a critical problem in the long term, that meeting essential needs for liquid fuels poses the greatest challenge to future energy supply systems. And in particular it leads to the observation that the world's oil resources are inadequate for the job, unless shale oil and tar sands are developed on a truly global scale and on a globally traded basis. Even then coal is pressed to and possibly beyond its feasible limits of supply. New technologies which stretch out the lifetime of coal resources – by greater efficiencies and coupled with large heat sources such as the breeder or the sun – are described in the final report of IIASA's Energy Systems Program. It may be that these technologies will point to a truly global scale option for mankind in the very long term.

TABLE 6 – Cumulative Uses of Fossil Fuels, 1975-2030, High and Low Scenarios

Oil	Total Resource Available[2] (TWy)	(TWy)	Total Consumed		
			High Case (%of total available)	(TWy)	Low Case (%of total available)
Conventional[1] (Cat. I + II)	473	325	(69)	302	(64)
Unconventional (Cat. III)	365	4	(1)	0	(0)
Natural Gas					
Conventional (Cat. I + II)	400	161	(40)	118	(30)
Unconventional (Cat. III)	120	0	(0)	0	(0)
Coal[3]					
Cat. I	562	402	(72)	303	(54)
Cat. II	1024	24	(2)	0	(0)

1 For definition of terms and categories, see text.
2 Total resources, including to be discovered, as of 1975.
3 Coal use includes coal converted to synthetic liquids and gas.

The cumulative fossil fuel uses in our estimate of total oil resources at costs less than $20/barrel is 473TWy (334 billion tons
or 2437 billion barrels of oil equivalent). This total includes
some small amounts of shale resources, rather large amounts (at
least 90TWy) of heavy crudes, and a large amount of enhanced recovery. Of this total our High scenario uses by 2030 325TWy (1670
million barrels), and our Low uses 302TWy. Of our assumed 365TWy
of unconventional high cost (25$/barrel) oil (mostly shale) only
4TWy are used in the High case and none in the Low (Table 6, which
also includes comparable natural gas and coal figures).

We are not running out of resources but we are running into expensive, difficult unconventional resources. We are running uphill and
remain unsure of what is on the other side. And it is a big hill.

Natural gas is not consumed relative to available resource base
(Table 6) nearly as much as oil. Only 40% of cheap category I gas
is consumed by 2030 in the High scenario, while virtually none of
category II gas is used. This result points to the possibly vast
potential for exploiting gas resources - perhaps to produce liquid
fuels - as well as to the restricted uses and difficult transport
of gas. Ironically enough, in spite of being a highly desirable,
clean and convenient fuel, gas demand is restrained because the
locations of supply are not the same as demand. The problem is
imagining large scale production of relatively higher cost gas from
relatively remote sites and delivering it to consumers. Should this
be doable then the challenge is to greatly expand the possible uses
for gas and so save oil for those purposes for which it is primarily
and almost uniquely suitable. This is at least one major challenge
posed to energy thinkers and planners as the implication of long
term energy analyses.

3. IMPLEMENTATION OF RESULTS

The models described here seem to be rather well-adapted to global
energy projections for the long term. They seem able to deal with
most of the relevant issues - to address the details and maintain
the overview. Still, the iterative process (Fig. 1) is a laborious
one. And the critical interregional, financial and other trade
issues are dealt with only crudely. The imminent addition of an
oil gaming model will help, but more work is in order.

The models have now demonstrated some success at making global
projections. And yet one must observe that no client exists for
such global energy studies. No world-wide decision maker exists.
No one has requested, no one is looking for, no one knows exactly
how to make use of global energy projections such as those produced
by the Energy Systems Program at the International Institute for
Applied Systems Analysis. How then can or should the methods and
results of such studies be implemented?

There is, of course, no single answer to the question. Still, these perceived global trends, these perceived interactions among regions, these perceived binding constraints in various regions, heavily impact or should impact the decisions and plans of each region and country. It is axiomatic that no man and no country is an island in these considerations. Narrow national interests peer into the future with tunnel vison. Our responses to international energy interrelationships have to date been primitive. The creation of the International Energy Agency provides the developed countries with a safety net; it is clearly a step in the right direction. The North-South dialogue was thought to explore (among other things) a common energy ground for producers and consumers. Yet, no one knows exactly what to do if the producers of the Arabian Peninsula cease to meet short-falls stemming from Iranian production cutbacks.

Energy problems are not national at all: they are international problems. No nation would have an energy problem if it had cheap domestic sources of energy. The only nations that do have such sources have precisely the opposite problem: how to produce enough and export it at a fair price. Oil is or should be seen as the world's glue; we are bound in a common, unbreakable bond of oil trade. Still, too many nations approach energy trade issues on the basis of national security, not international peace.

Global energy analyses – as presented here and elsewhere – should find a way into the thinking of national energy policies throughout the world. Government regulations and incentives for energy conservation, energy productivity increases, pricing policies, leasing schedules, incentives or disincentives for fuel A or B, policies regarding the support and/or export of specific technologies etc. should be based as much or more on the perceived energy interactions among regions as on more narrow domestic interests and opportunities. For example in my own country supports and incentives should be expanded toward the production and use of natural gas. Prices should rise so that the percentage of natural gas in total energy use can rise as rapidly as possible, or at least not drop below recent past levels. Incentives should be developed for shifting away from oil. Transport systems should be designed for minimum liquids use. Multinational energy companies should gear investment in energy-intensive industries towards those places of cheapest and most accessible energy resources. Cost-incentive supports for distributed solar technologies, solar cells, and solar photovoltaics should be increased rapidly.

These and other policies in my country and others should be faced in light of international energy perspectives. Mechanisms should exist for so doing. Without these steps we delude ourselves into believing that energy is only our concern and not the common need and challenge of all mankind.

DISCUSSION

Although the IIASA model did not include price elasticities of
fuel demand explicitly, Basile was asked if IIASA had analyzed
the demand elasticities which the engineering parameters in the
model implied. Basile replied that it was no easy task to compute these elasticities - one had to make assumptions about the
relationship between fuel production costs and consumer prices -
i.e. taxes and the like. The group at IIASA had made some assumptions on these add-ons to costs and had calculated that the elasticities in the developed regions were in the order of -0.7 to -0.8
measured on consumer prices. For lesser developed regions, the
elasticities were lower, in the region of -0.1 to -0.2.

Basile was asked on the way in which the oil price trajectory was
calculated in the medium term before the price reflecting back-
stop technology was reached. Basile replied that the calculation
procedure for oil price was complex but basically involved maximizing discounted revenues in region VI. The trajectory which had
been arrived at was then smoothed. In effect, a constant rate of
increase was assumed to a price level reflecting the cost of coal
liquifaction. It was pointed out that in Ervik's presentation,
the price of oil in ca. 1990 had been determined by the availability of alternative energy such that an overshoot in oil price
had been experienced. Basile agreed. This overshoot in oil price
had also been observed at IIASA. He did not feel that this affected the long term results of the IIASA model and was happy to omit
it.

Following up Deam's presentation earlier in the conference, several
participants asked whether the use of methanol was in the model
and whether the team at IIASA had considered the technology required for methanol production. Basile replied that no assessment
of the technology for methanol production from natural gas had
been made at IIASA. The methanol which was included in their
study was based on coal. IIASA had, however, run some scenarios
in which gas supplied a share of the liquid fuels. The result
had been a decrease in coal liquifaction. Based on previous experience of the time delays involved in the introduction of new
technology, Basile did not feel that gas-based methanol would
have a significant impact on the total energy supply picture.
Except for Middle-East and North Africa, natural gas plays little
role in the aggregate. As regards Deam's postulate of the Soviets
and Australia being price-leaders in the methanol market, Basile
was again sceptical: Soviet domestic demand was growing fast and
he doubted that they would export much.

Several speakers were also surprised at the small contribution
which solar and nuclear energy made. Basile replied that the lead-
times in nuclear construction were large. From orders placed one had

a good idea of the probable capacity in 1990 already. IIASA had
assumed a build-up of orders from 5-10 years hence which was
consistent with market penetration observed historically for oil
and coal. As regards solar energy, they had studied the market
potential. It was assumed that 50% of all new buildings in OECD
countries met 70% of their space heating and water heating needs
from this source. He felt personally that solar energy would pene-
trate the market faster than had been the case with oil and coal,
and felt too that their nuclear numbers were too high. A scenario
with 6% of electricity generation based on solar energy had been
tried, but it was not possible to make the model system converge.

It was suggested that the energy efficiency of the transport sector would be enhanced by including electric cars in the scenarios
instead of combustion engines based on liquified coal. Basile agreed. Electric cars had been assumed in developed urban regions.
The growth in the transport sector required to sustain the GNP-
growth assumed in the scenarios led to the projected increase in
demand for liquid fuels, though. He admitted that other studies
such as that of G. Leach in Britain had managed to expand GNP without a significant increase in the transport sector.

MACRO: AN AGGREGATE MACROECONOMIC MODEL FOR THE
IIASA SET OF ENERGY MODELS

H.H. Rogner

International Institute for Applied Systems
Analysis
Laxenburg, Austria

ABSTRACT. An econometric model of macroeconomic development in
global regions is presented. The main purpose of the model is
to enable the prediction of fundamental economic activities for
the next 50-year period, and to provide a check for the consistency of assumptions in the IIASA set of models for world energy
development to the year 2030.

1. THE APPROACH

The long term approach of IIASA's Energy Systems Program covers
the globe entirely. Energy demand and energy supply, however,
are not equally distributed over the globe. Therefore it was
necessary to divide the globe into 7 major regions. The reasoning
as well as the composition of each of the 7 regions can be found
in [1]. A set of models has been developed at IIASA to provide
insight into the complex problems of the dynamics of a transition
from conventional to unconventional energy systems. The purpose,
goals, and the general structure of the set of models have been
described in detail elsewhere [2]. This presentation intends to
illustrate the structure of one of the models – the macroeconomic
component (MACRO) of the IIASA set of energy models. MACRO stands
for a number of highly aggregate one-sector macroeconomic models
designed to reflect the historical structures of various regional
economies.

The dynamics of an energy system in transition is inherently of a
long term nature. Observations in the field of market penetration
[3] for the energy sector show that transitions from an existing
technology to a more advanced one have historically taken about

50 years for the new technology to increase its market share from
1% to 50% (Fig. 1). For this and other reasons the time horizon
for any modelling effort to study transition phenomena has to be
in the magnitude of 50 to 60 years as well. A time span of 50
years and more, however, dictates a comprise among specific features, and some notably differences from traditional economic models which will be elaborated below.

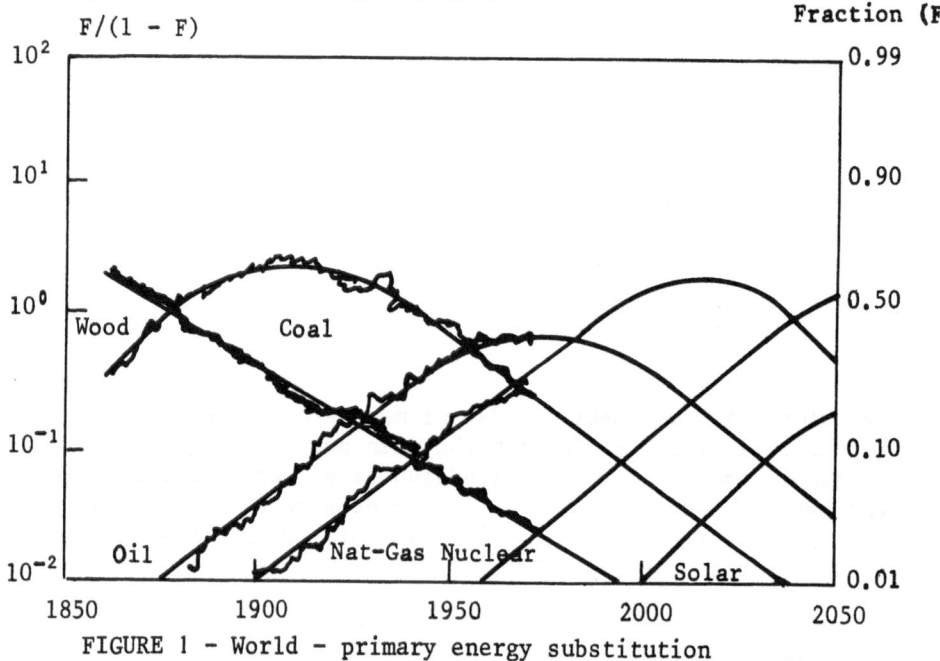

FIGURE 1 - World - primary energy substitution

It is appropriate to have a look at the place MACRO takes in the
overall model chain. Fig. 2 contains a simplified representation
of the information between the models and the modeller's
exogeneous judgement, while in Fig. 3 the detailed interfaces of
MACRO with the other models is given. In this context
MACRO provides a means to check economic consistency of the components, assumptions and results of the energy demand and supply
side. Future energy supply in an aggregate developed market economy, or region such as the European Community (EC) without major
indigenous energy resources of an economically recoverable range
will face increasing capital needs for alternative indigenous
energy production technologies. The present energy import dependency of the EC (>50%) might lead to historically unexperienced
capital requirements if the replacement of existing indigenous
technologies by more capital intensive ones is to take place as
well as a reduction in import dependency. In the case of energy
supply, MACRO interacts with the energy supply model MESSAGE and
the energy investment model IMPACT. IMPACT feeds the investment
requirements of the energy sector into MACRO. The question then

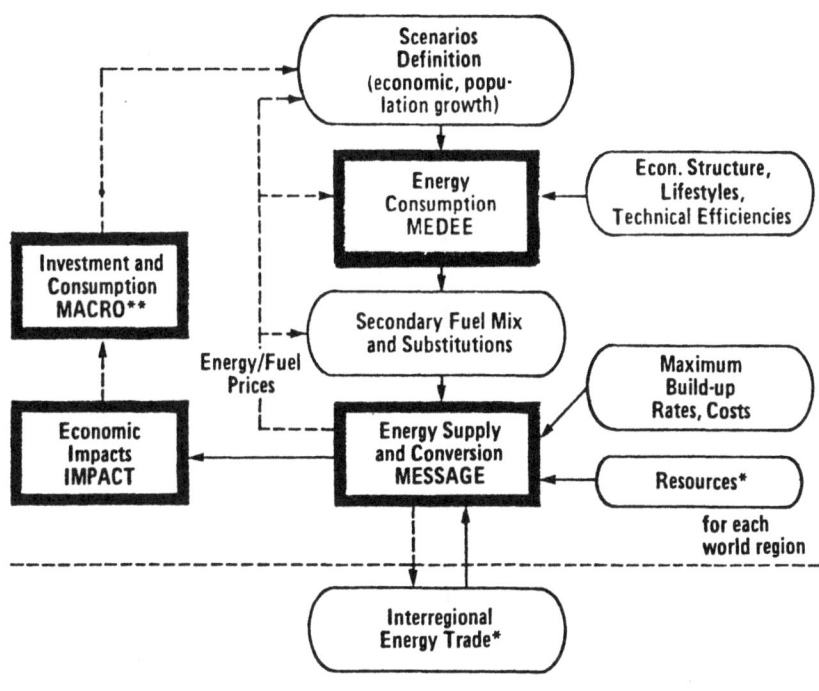

Scenarios Definition (economic, population growth)

Energy Consumption MEDEE

Econ. Structure, Lifestyles, Technical Efficiencies

Investment and Consumption MACRO**

Secondary Fuel Mix and Substitutions

Energy/Fuel Prices

Economic Impacts IMPACT

Energy Supply and Conversion MESSAGE

Maximum Build-up Rates, Costs

Resources*

for each world region

Interregional Energy Trade*

Assumptions, judgments, manual calculations

Formal mathematical models

Direct flow of information (only major flows shown)

Feedback flow of information (only major flows shown)

*Formal mathematical models to replace these judgmental analyses are in process.
**Not yet fully implemented.

FIGURE 2 - The IIASA set of models

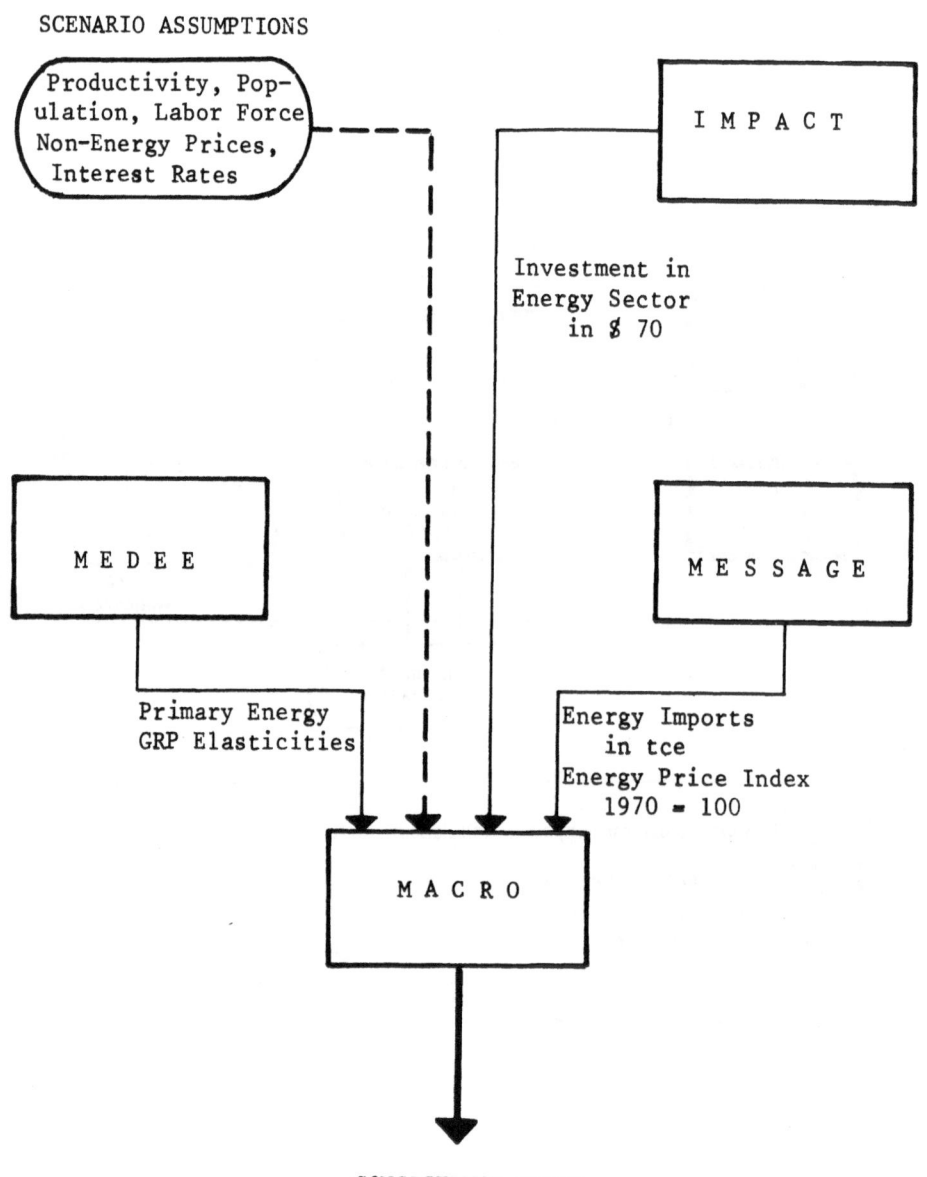

FIGURE 3 - Relation of macro with other IIASA models

asked concerns the relative share of economic product that has to be spent to finance the capital needs in the energy sector. Apart from the questions of capital, energy imports and rising energy world market prices will affect the economy. The transfer of income to oil exporting regions will mitigate domestic economic growth and push inflation. The information concerning energy imports and energy import prices is supplied by MESSAGE.

The output of the energy demand model MEDEE - the final energy demand - reflects certain life style assumptions as well as a conceivable change in the structure of Gross Regional Product (GRP) production. Although the MEDEE output is fed into the energy supply model MESSAGE, MACRO has to check the consistency of underlying assumptions such as the development of labour and capital productivity, future manpower availability or primary energy - GRP elasticities (e_p) as well as the final energy - GRP elasticities (e_f) with the resulting overall growth rates of the components of aggregate demand.

Unlike other models, MEDEE (descriptive), MESSAGE (linear programming) and IMPACT (input-output) whose methodological structures remain unchanged in the application to various world regions and only differ in their initial assumptions, restrictions, boundary conditions, etc., the macroeconomic component has to be conceptualized for each region separately. This procedure necessitates data collection of historical national time series and their aggregation into regional time series, the identification of behavioural relationships, parameter estimation, and finally the construction and validation of the complete model over the sample period. The MACRO model then reflects the historical structure of a specific region, here the European Community.

2. THE STRUCTURE OF MACRO

Despite the long planning horizon of 50 years, the approach used to build MACRO is econometric. Econometric models, in general, are based on large data bases for the sample period containing time series of macroeconomic data as well as cross sectional data of many detailed sectors of an economy. Functional explanations for various economic relationships have been estimated and validated over the time series of the sample period (e.g. demand for certain commodities or consumption functions) which then together with the necessary economic identities form an econometric model (e.g. Gross Regional Product equals the sum of components of aggregate final demand). The planning horizon of such models typically is a small fraction of the sample period; in other words, econometric models traditionally have been applied in short term economic analyses, up to fifteen years. In order to run the set of economic relationships further over the planning horizon, one requires the specification of exogenous or predetermined variables such as demographic data, the overall development of technology, or relative prices.

With a planning horizon of 50 years, however, forward extrapolation of econometric functions estimated over a relatively small sample period is limited, mainly by two factors. First is the user's inability to predict the many exogenous variables necessary for a complete model. Especially, predictions of the development of exogenously determined relative prices for the next 50 years or more immediately lead to serious uncertainties in any model. Second, the structural patterns of behaviour relationships and short-term trends and fluctuations inherent in functions estimated on historic data may not necessarily prevail in the future, or may even be unreasonable in the long run.

With these general considerations in mind MACRO was conceptualized as a simple, highly aggregated, one-sector "potential" GRP model designed to capture the historical structure of the EC economy. The basic structure - rather than the complexities of multi-relationships - is the focus of attention here.

To overcome the above-mentioned deficiencies in applying MACRO for long term economic analysis, the structure of MACRO was based on a number of important macroeconomic relations and variables whose values have been observed to remain fairly stable (within a certain range) for a couple of decades in industrialized countries. Such relations (or better, slow variables) are e.g. the capital-output ratio, investment and consumption rates, population or labour force participation. By concentrating on "slow" variables one obtains a model wherein short run fluctuations of "fast" variables such as Gross Regional Product (GRP), Private Consumption (C) or Gross Private Investment (I) can be essentially avoided. Another advantage of this approach is the possibility of reducing the uncertainty of future exogenous variables by keeping the number of such variables at a minimum.

In addition to slow variables, MACRO contains exogenously determined scenario parameters - parameters which in the long run will balance out short term trends inherent in the estimated parameters (e.g. the downward trend of average working hours per week in the EC for the last 20 years) and which will be the means to simulate evolutionary changes in the economic structure of the EC (e.g. the share of private domestic investment of GRP). Slow variables, together with these scenario parameters enable MACRO to be a useful tool for modelling both historical trends and imposed normative changes in the long run while at the same time guaranteeing consistency in a macroeconomic sense. The conceivable, even likely, substitution of advanced, capital intensive energy technologies may dictate the necessity for certain normative changes.

An important assumption in MACRO is indicated in the term "potential" mentioned above. "Potential" in this sense represents maximum available output of the economy under optimal utilization of all input factors; institutional policy is adequate to maintain aggregate demand under sustained full employment.

In general, one should always keep in mind that MACRO was not developed to predict the future. Its main service is to limit the set of plausible scenarios (represented by scenario parameters and exogenous variables) for the future.

3. CODING THE MODEL

The following equations reflect the historical structure of the EC economy for the years 1960 to 1976 (the sample period). In addition to those equations relevant for a "business as usual" scenario, four important scenario parameters were implemented which are the means to impose exogenously determined scenario changes on the model. With these parameters and the necessary renormalizations of some of the equations one reaches the "potential GRP" version of MACRO able to run this model over the planning horizon from 1977 to 2030.

The yearly observations for the variables over the sample period originate from EUROSTAT "National Accounts ESA", EUROSTAT "Tax Statistics", OECD "Labour Force Statistics", OECD "Main Economic Indicators", ILO "Bulletin of Labour Statistics", and UNCTAD "Handbook of International Trade and Development Statistics". The aggregation of national data whenever this was not provided by EUROSTAT statistics (e.g. GRP, investments, consumption, population) was compiled by applying the weighted average method (e.g. for the calculation of the wage rate index).

The model's parameters were estimated by means of the ordinary least squares method. The validation runs over the sample period achieved an overall outcome of less than 5 percent deviation, or error.

The estimated equations of the model, together with the goodness of fit and autocorrelation statistics, are listed below. The figures in parentheses show the corresponding t-statistic for the estimated parameter. Further, the values for the correlation coefficients R^2 corrected for degrees of freedom, the Durbin-Watson statistic d, and the standard error of estimate se are given below each equation. The relatively high R^2's and low se's indicate that most of the equations provide a good fit to the data. In assessing the goodness of fit, it should be noted that many of the Durbin-Watson statistics are too low. This implies positive autocorrelation and hence upward biases in the R^2's estimates and t-statistics and downward biases in the estimates of the standard errors. However, a model's usefulness is determined by the simultaneous solution of all its equations. The accuracy of the simulation over the sample period is an indication of a model's goodness of fit.

4. CHARACTERISTICS OF AGGREGATE DEMAND AND SUPPLY IN MACRO

4.1 Demand

There are five components of aggregate demand: personal consumption, gross private investment, government purchase of goods and services, import and export of goods and services. The purpose of the demand side of MACRO is to produce a consistent set of estimates for these variables under different assumptions. The estimates should be consistent with assumed behaviour relationships for consumers and producers. Furthermore, they should be consistent with the GRP and the disposable income identities. The government's direct effect on total demand through taxes and transfer function is estimated. These tax and transfer functions provide a means to change the proportion of consumption to GRP (consumption rate) and investment levels, etc. through the specification of scenario parameters. Investment is determined by the difference of desired capital stock for the presented period and the actual capital stock at the beginning of the present period (end of last period) corrected for depreciation.

Definition of real GRP

$$GRP\$70 = C\$70 + I\$70 + G\$70 + X\$70 - M\$70 \qquad (1)$$

The basic equation of a macroeconomic model is the definition of real Gross Regional Product (GRP$70). GRP$70 is the constant 1970 value in European Units of Account of all goods and services produced by labour and property supplied by residents of the European Community member countries. It also equals total purchases of goods and services.

Definition of real national income

$$NI\$70 = GRP\$70 - DEP\$70 - TAXIN\$70 - RES \qquad (2)$$

Real National Income (NI$70) is the total income paid to the factors of production (labour and property). One must substract from GRP$70 all the non-factor charges, e.g. indirect business taxes, current surplus of government enterprises less subsidies, and capital consumption allowances; to finally secure a balance, a statistical discrepancy must be included. In equation (2) the term RES (residual) represents both the statistical discrepancy and convertion error to constant dollars and surplus of government enterprises less subsidies.

Indirect business tax function

$$TAXIN = [0.106 * P * GRP\$70 + 16.85] * \sigma_1 \qquad (3)$$
$$(25.2) \qquad\qquad (5.9)$$

$$R^2 = 0.975, \quad se = 5.16, \quad d = 0.29$$

Indirect business tax and non-tax liability (TAXIN) includes taxes for sales, property, inspection fees, fines, royalties, and donations. This term does not include taxes on corporate income. The estimated indirect business tax function has an average tax rate of 10.6% on GRP. The term σ_1 is the first scenario parameter which together with σ_2 in equation (6) are a means to influence disposable income and therefore the absolute level of personal consumption.

Capital consumption allowance (depreciation)

$$DEP\$70 = 0.041 * K\$72_{-1} - 28.59 \qquad (4)$$
$$(102.9) \qquad\qquad (34.8)$$

$$R^2 = 0.998, \quad se = 0.53, \quad d = 0.38$$

Capital consumption allowance with capital consumption adjustment (DEP\$70) consists of depreciation changes and accidental damage to fixed business capital. Equation (4) estimates the yearly depreciation (DEP\$70) as a function of the total capital stock of the economy at the beginning of each period. The results indicate that 4.1% of the capital stock is worn out per year.

Definition of real disposable income

$$YD\$70 = NI\$70 - TAXDIR/p + GT/p \qquad (5)$$

The disposal income (YD\$70) equals national income corrected for income taxes and government transfer payments to personal accounts. To simplify the model, corporate retained earnings were included with personal disposable income.

Income tax function

$$\text{TAXDIR} = [0.33 * P * \text{NI\$70} - 14.19] * \sigma_2 \qquad (6)$$
$$\phantom{\text{TAXDIR} = [}(66.9) \phantom{* P * \text{NI\$70} -} (5.5)$$

$$R^2 = 0.996, \quad se = 4.76, \quad d = 1.39$$

The income tax function has a surprising result. It is impossible to estimate an incorporate income tax function and a personal income tax without taking account of all the tax rate changes. With the total income tax (TAXDIR) function, however, one obtains a function that works for the entire period. For the 1960 to 1976 period the average income tax rate was 33% of national income. The various changes of the tax structures seem to have changed the proportion in each category and left the total fairly constant.

Government transfer payments to persons

$$\text{GT} = P * (\text{POP65} + \text{UNEMP}) * \text{EXP}(0.035 * \text{TIME} - 0.106) \qquad (7)$$

$$R^2 = 0.945, \quad se = 0.043, \quad d = 1.28$$

The main determinants of government transfer payments are the number of retired persons and the payments for unemployment compensation. The exponential part of this equation represents the improvement of the welfare system over the sample period, while the price deflator p should prevent those concerned from income losses caused by inflation.

Consumption function

$$\text{C\$70} = 0.41 * \text{YD\$70} + 0.28 * \text{YD\$70}_{-1} + 0.14 * \text{YD\$70}_{-2} - 10.51$$
$$\phantom{\text{C\$70} = }(55.8) \phantom{* \text{YD\$70} + }(55.8) \phantom{* \text{YD\$70}_{-1} + }(55.8) \phantom{* \text{YD\$70}_{-2} - }(1.7)$$
$$\tag{8}$$

$$R^2 = 0.995, \quad se = 4.31, \quad d = 0.71$$

Personal consumption is goods and services purchased by individuals and operational expenditures of nonprofit institutions. Personal consumption expenditures are determined by the distributed lagged weights of disposable income – representing the fact that individuals' adjustment on changing income levels will not take place immediately.

Investment function

$$\text{I\$70} = 0.61 * (\text{K\$70*} - \text{K\$70}_{-1}) + 0.44 * E + \text{DEP\$70}$$
$$\phantom{\text{I\$70} = }(10.8) \phantom{* (\text{K\$70*} - \text{K\$70}_{-1}) + }(1.9)$$
$$\phantom{\text{I\$70} = } - I_{\text{ene}}\text{\$70} + I_{\text{ipa}}\text{\$70} + 14.17 \qquad (9)$$
$$\phantom{\text{I\$70} = - I_{\text{ene}}\text{\$70} + }(2.7)$$

$$R^2 = 0.894, \quad se = 3.51, \quad d = 0.94$$

Gross private domestic investment equals desired capital stock minus actual capital stock at the beginning of the period plus depreciation. E is the trade balance, a negative trade balance (caused e.g. by too high and expensive energy imports) therefore will affect the investment rate. The term $I_{ene}\$70$ represents the continuation of the historical share of investments in the energy sector, while $I_{ipa}\$70$ are the "investments" in the energy sector calculated in IMPACT. A transition from cheap fossil fuels towards highly capital intensive energy production technologies influences the overall demand for capital.

Desired capital stock

$$K*\$70 = \left| \sum_{i=0}^{7} w_i * GRP\$70_{-i} \right| * \sigma_3 + \sum_{i=0}^{4} v_i *$$

$$GRP\$70_{-i} * \left| \frac{w - rate_{-i}}{cost - k_{-i}} \right|^{1/2} + \begin{array}{c} 388.6 \\ (89.5) \end{array} \tag{10}$$

$w_o = 0.000$	$w_4 = 0.511$	$v_o = 0.080$
$w_1 = 0.223$	$w_5 = 0.479$	$v_1 = 0.104$
$w_2 = 0.383$	$w_6 = 0.383$	$v_2 = 0.077$
$w_3 = 0.479$	$w_7 = 0.223$	$v_3 = 0.000$

$$R^2 = 0.999, \quad se = 1.47, \quad d = 2.10$$

The desired capital stock $K*\$70$ was estimated by applying the generalized quadratic mean of order rho cost function and Shepard's theorem 4,5. The calculation of desired capital stock requires a longer lagged time horizon since the adjustment for capital takes more time periods than, for example, dereieved demand for man/hours. "Desired" connected with capital stock reflects the fact that private business cannot adjust demand for investment within one time period (see parameter 0.61 in equation 9).

σ_3, the capital-output ratio normalized to 1, is the main scenario parameter in MACRO. Over the sample period σ_3 equals 1. By allowing values for σ_3 to become evolutionarily greater than 1, one forces the model to allocate relatively more capital per unit of output (GRP$70) than in the previous time periods. An increasing demand for investment will be the resulting effect of this manipulation.

Actual capital stock

$$K\$70 = K\$70_{-1} + I\$70 - DEP\$70 \tag{11}$$

Actual capital stock equals capital stock at the beginning of the period plus investment corrected for depreciation (net capital stock).

Cost of capital

$$\text{cost-k} = (BAA * 0.01 + 0.041) * P \tag{12}$$

Cost of capital is determined by the actual rental cost of capital (BAA) and the depreciation on capital (see equation (4) for parameter value 0.041) adjusted for inflation.

Wage rate (nominal)

$$\text{W-RATE} = 57.505 * \left| \frac{GRP_{-1}}{MANHOURS_{-1}} \right| + 0.798 * P_{-1} - 117.754 \tag{13}$$
$$\quad\quad\quad (5.8) \quad\quad\quad\quad\quad\quad\quad (5.3) \quad\quad\quad (9.6)$$

$$R^2 = 0.992, \quad se = 3.68, \quad d = 2.14$$

The wage rate is determined by last year's level of productivity, and the previous year's development of inflation. This behaviour could be observed for many years when trade unions regularly have demanded labour's participation in productivity gains and cost of living increase compensation.

Government purchases - renormalization of equation (1)

$$G\$70 = GRP\$70 - C\$70 - I\$70 - X\$70 + M\$70 \tag{14}$$

The government sector is the residual of the components of aggregate demand.

Derived demand for labour (non-agriculture)

$$\text{MANHOURS} = 0.196 * GRP\$70 + 0.035 * GRP\$70 * \left| \frac{\text{cost-k}}{\text{w-rate}} \right|^{1/2}$$
$$\quad\quad\quad (2.8) \quad\quad\quad\quad\quad (3.1)$$
$$\quad\quad\quad - 6.62 * TIME + 177.78 \tag{15}$$
$$\quad\quad\quad\quad (5.7) \quad\quad\quad\quad (11.2)$$

$$R^2 = 0.929, \quad se = 1.78, \quad d = 1.70$$

In the derived demand function for man/hours to produce total output (GRP$70) prevails the assumption that adjustment for labour will take place within one time period.

Average weekly hours for total private non-agriculture

$$\text{HOURS} = |-0.144 * \text{UNEMPL} - 0.166 * \text{TIME} + 41.2| * \sigma_4 \qquad (16)$$
$$\quad\quad\quad (3.2) \quad\quad\quad\quad (15.7) \quad\quad\quad\quad (182.0)$$

$$R^2 = 0.922, \quad \text{se} = 0.311, \quad d = 0.63$$

The weekly hours worked depend on the level of unemployment and contain a declining trend represented by the negative time parameter. The scenario parameter σ_4 was introduced to eventually convert this short term time trend of decreasing working hours per week in the private non-agricultural sector. One should think of the necessity to weaken this downward trend in order to provide additional man/hours to produce the excess demand for capital caused by an increased capital-output ratio.

Occupied population

$$\text{POPOCC} = (\frac{\text{MANHOURS}}{\text{HOURS}} * 52) * 1000 \qquad (17)$$

Unemployment

$$\text{UNEMPL} = \text{LABOUR} - \text{POPOCC} \qquad (18)$$

Man/hours, physical availability

$$\text{MANHOURS} = \text{HOURS} * \text{LABOUR} * 52/1000 \qquad (19)$$

In case the derived demand for MANHOURS exceeds the physical availability of labour, the above equation replaces equation (15).

4.2 Supply

The demand section of MACRO cannot stand alone; the economy may not be capable of producing the output required by total demand. The limiting factors in an economy can be systematically examined by the fundamental assumption of the existence of an aggregate production function. This production function describes the input-output relations of the economy as a whole. A production function that represents the maximum output obtainable from a set of inputs and assumes cost minimizing producers, is the Cobb-Douglas production function.

Production function

$$GRP\$70 = \exp(-0.58 + 0.301 * \ln K\$70 + 0.699 * \ln MANHOURS$$
$$(11.9) \quad (6.92) \qquad\qquad (6.92)$$

$$+ 0.87 * PROD) \qquad\qquad\qquad\qquad (20)$$
$$(16.9)$$

$$R^2 = 0.998, \quad se = 0.54, \quad d = 1.47$$

4.3 Government Budget and Trade Sector

Government budget (surplus/deficit)

$$SUR\$70 = TAXIN\$70 + TAXES/P - G\$70 - GT/P \qquad (21)$$

The government's budget identity (surplus or deficit) contains the tax revenues on the income side and government expenditures on goods and services plus government transfer payments to persons on the expenditure side. There is no restriction for a balanced budget in MACRO.

Trade balance

$$E = X\$70 - |p_x/p_m| - 1 * M\$70 \qquad\qquad (22)$$

The trade balance is the difference of total exports minus imports corrected for the terms of trade. Terms of trade are defined as export price index over import price index.

Exports

$$X\$70 = 0.287 * p * 0.01 * GRP\$70 - 36.871 \qquad (23)$$
$$(28.0) \qquad\qquad\qquad (5.9)$$

$$R^2 = 0.981, \quad se = 10.28, \quad d = 1.30$$

Exports at present are determined by a simple linear function of nominal Gross Regional Product. In due course this equation will be substituted by a more sophisticated trade submodel.

Energy imports

$$Me\$70 = ENIMP * 0.0142 \qquad (24)$$

The monetary value in constant terms of energy imports is derived by the physical import volume of energy (output of the energy supply model MESSAGE) multiplied with the unit energy import price of 1970.

Non-energy imports

$$Mn\$70 = 0.240 * p * 0.01 \ GRP\$70 - 26.184 \qquad (25)$$
$$\quad (28.1) \qquad\qquad\qquad\qquad (5.0)$$

$$R^2 = 0.981, \quad se = 8.56, \quad d = 1.48$$

Non-energy import demand again is just a linear function of Gross Regional Product (see equation (23)).

Imports

$$M\$70 = Me\$70 + Mn\$70 \qquad (26)$$

Import price index

$$P_m = \frac{Me\$70 * p_{ei} * Mn\$70 * p_{nei}}{Me\$70 + Mn\$70} * 100 \qquad (27)$$

The import price index is highly influenced by the energy import price index (output from MESSAGE) and the energy import volume.

DEFINITION OF VARIABLES

x	:	exogenous
e	:	endogenous
$70	:	European Units of Account
C$70	e :	personal consumption expenditures, 10^9 $70
cost-k	e :	cost of capital
BAA	x :	bond yields and interest rates, percent per annum
DEP$70	e :	capital consumption allowance with cca, 10^9 $70
E	e :	trade balance in 10^9 $70

ENIMP	x :	energy imports in physical units (millions of tce)
G$70	e :	govt. purchases of goods and services, 10^9 $70
GRP$70	e :	gross regional product, 10^9 $70
GT	e :	government transfer payments to persons, 10^9 $
HOURS	e :	average weekly hours total private non-agricultural
I$70	e :	gross private domestic investment, 10^9 $70
$I_{ene}$$70	e :	investments in energy (historical trend) 10^9 $70
$I_{ipa}$$70	e :	investments in energy (IMPACT output) 10^9 $70
K$70	e :	estimated capital stock at end of period, 10^9 $70
K$70*	e :	desired capital stock predicted values from eqn. (10)
LABOUR	e :	civilian labour force, millions of persons
M$70	e :	import of goods and services, 10^9 $70
Me$70	x :	energy imports in 10^9 $70
Mn$70	e :	non-energy import in 10^9 $70
MANHOURS	e :	hours of all persons total private, 10^9 hours
NI$70	e :	national income, 10^9 $70
OCCPOP	e :	occupied population, millions of people
P	x :	implicit price deflators for GRP, 1970 = 100
P_m	e :	import price index, 1970 = 100
P_x	x :	export price index, 1970 = 100
P_{ei}	x :	energy import price index, 1970 = 100
P_{nei}	x :	non-energy import price index, 1970 = 100
POP	x :	population, millions of persons
POP65	x :	population over 65, millions of persons
PROD	x :	productivity development in percent
RES	x :	residual from GRP identity, 10^9 $70
SURPL	e :	governments budget surplus or deficit, 10^9 $
TAXDIR	e :	personal + corp. taxes + social insurance, 10^9 $
TAXIN	e :	indirect tax + govt. surplus, 10^9 $70
TIME60=1	x :	time trend 1960 equals 1
UNEMPL	e :	total employment, millions of persons
w-rate	e :	compensation per hour, 1970 = 100
X$70	e :	exports of goods and services, 10^9 $70
YD$70	e :	spendable income, 10^9 $70

REFERENCES

1. IIASA's Energy Systems Program, "Energy in a Finite World",
    International Institute for Applied Systems Analysis,
    Laxenburg, Austria, forthcoming report.

2. P. Basile, "Global Energy Modelling and Implementation for
    Planning", Presentation at Advanced Study Institute on
    Mathematical Modelling of Energy Systems, Instanbul, Turkey,
    10-22 June 1979.

3. C. Marchetti, and N. Nakicenovic, "The Dynamic of Energy
    Systems and the Logistic Substitution Model", International
    Institute for Applied Systems Analysis, Laxenburg, Austria,
    AR-78-1B, Draft, July 1978.

4. R. Shepard, Cost and Production Functions , Princeton
    University Press, New Jersey, 1953.

5. "Development of Methods for Forecasting the National Indus-
    trial Demand for Energy", EPRI EA-242, Project 432-1 Final
    Report, prepared by Econometrica International, Santa Bar-
    bara, California, July 1976.

DISCUSSION

Rogner showed in the course of his presentation that the statistical significance of several of his model parameters was low.
It was suggested that he should be able to increase the explanatory power of his variables significantly if he extended his data
to pooled time series/cross-sectional sample. At least in the
region covering Europe, data should be reasonably consistent from
one country to the other. Parameter estimates from the data pool
could then be applied to individual countries.

Rogner replied that the reasons for restricting the data to time
series were not statistical, but resulted from the policies of
IIASA.

Several speakers questioned the relevance of the model to the
analysis of long term issues. The model was criticised for focusing attention on short run issues such as unemployment and inflation to the detriment of long term issues such as capital formation and potential GDP. The model basically used aggregate
demand analysis rather than focusing on the supply side of the
markets. The suggestion was that the model was not adequately
formulated for the relevant issue - how will energy system changes influence the growth rates of various economies. It was

argued by participants that it would be more appropriate to estimate aggregate production functions for the various economies and
to derive energy demands from equilibrium conditions for this
demand.

The inclusion of inflation in the model raised doubts in the minds
of several participants on the accuracy of the model in addressing
long term questions. Inflation implied that capital was receiving
more of the economic cake than was labour and this was probably
dependent upon short run economic problems.

Rogner was aware of these drawbacks and the effects of inflation
were usually accounted for in the input to the scenario. This
was, however, an unsatisfactory manner of solving the problem as
it required one to mix what were the desired scenario parameters
with "correction factors".

It was suggested that the inclusion of lagged variables in a model
designed to address long term issues to the year 2030 was unnecessary. The point was made that the model as specified would collapse were the lags to be removed.

The low values of the Durbin-Watson statistics also cast doubt
in several speaker's minds as to the ability of the model to predict
long term effects with a reasonable degree of accuracy.

ENERGY AND ECONOMIC GROWTH: A CONCEPTUAL FRAMEWORK[*]

J.L. Sweeney

Department of Engineering-Economic Systems
Stanford University, Stanford, California
U.S.A.

ABSTRACT. Since the 1973 oil embargo there has been intense interest throughout the world in the economic impacts of changes in the
price or availability of energy. This concern is reflected in
U.S. energy policy debates, OPEC pricing deliberations, and international meetings such as the Conference on International Economic Cooperation.

The relationships between energy and the economy are complex and
multifaceted. Sudden changes in energy price or availability can
lead to structural and aggregate unemployment, to sharp increases
in inflation, and to major changes in the distribution of wealth,
both among nations and within a given country. In the longer run,
when transitional impacts subside and after compensating monetary
and fiscal policies are implemented, changes in the price or availability of energy may influence the long run growth prospects of
a nation.

The purpose of this paper is to provide a simple conceptual framework within which the long run impacts of energy on economic growth
can be considered. Within this framework the question is asked:
What will be the impact on aggregate output, consumption, and welfare of changes in the availability or price of energy?

[*]Many of the ideas presented here were developed jointly with
Tjalling Koopmans. I would like to thank Ernst Berndt, Dale
Jorgenson, William Hogan, Alan Manne, John Weyant, and David Wood
for their helpful comments throughout the research project.
Partial financial support has been provided by the Office of
Conservation, Planning, and Policy, U.S. Department of Energy.
Of course, all errors remain the sole responsibility of the author.

In this long run analysis it will be assumed that the economy
always maintains full employment of all factors of production.
Differences amongst the energy carriers -coal, oil, natural gas -
will be ignored in order to facilitate the conceptual discussion,
although the extension to consider different fuels is possible.

1.   INTRODUCTION

Many economic or policy changes may influence the price or availability of energy. Examples include changes in the OPEC-admin-
istered world price, a moratorium on nuclear construction, a
tariff or quota on imported energy production, a domestic price
control program, a tax on domestic energy production, or a successful energy supply R and D program. Each change can influence
the use of energy in a country and the growth of its economy,
although the ratio of the two magnitudes may vary widely among
programs.

In order to focus on the essential differences and similarities
among the various economic and policy changes, each can be represented analytically as one (or more) of four types of changes:
an import price change, a domestic cost function change, an import
tariff, or a domestic tax. The first two analytical representations are associated with changes in the resource cost structure
of energy in the economy, in the value of other resources that are
used up or foregone within the domestic economy in order to gain
access to a given quantity of energy. The latter two analytical
representations are associated with changes in the price structure
of energy but not with changes in the resource cost of energy.
An import tariff or domestic tax will change the price of energy
facing consumers or producers but will not directly entail changes
in the resources foregone to gain access to a given amount of
energy.

As will be shown in the subsequent analysis, policies that can be
modelled by the first two analytical representations will lead to
greater impacts on the economy than will policies that can be accurately modelled by the latter two analytical representations.

Policies can also be separated into those directly influencing
imported energy and those directly influencing domestically produced energy. The analytical representations maintain this distinction. However, as will be shown, this difference is not fundamental to explaining differences in the aggregate economic
impacts of alternative policies.

Table 1 lists several policy or economic changes influencing the
energy system and their corresponding analytical representations.
Each of the energy system changes listed on the left margin is

equivalent in its impact on the aggregate economy to one or more
of the analytical representations listed across the top margin.
The elements in the table indicate which analytical representations are equivalent to a given energy system change and what are
the correct directions of change. The pluses represent increases
in the specific analytical representation, while the minus signs
represent decreases. For example, a limit on the domestic use of
coal will increase the cost to the U.S. of producing a given quantity of energy. Thus, coal limits can be represented as an increase in the domestic cost function; a plus sign appears in the
appropriate box. An energy conservation tax would influence all
energy, both domestic and imported. This tax would be equivalent
to an import tariff plus a domestic tax; pluses appear in two
boxes. A successful R and D program* to produce inexpensive solar
collecters would reduce domestic cost ; a negative sign appears.
An import quota would limit the amount of energy imported but
would not change the resource costs borne by the economy to gain
access to the imported energy. An import quota would have the
same aggregate impacts as an import tariff. More complex is the
crude oil price control program with entitlements. This program
subsidizes imports and taxes domestic production. Thus, it can
be represented as a combination of a negative import tariff (a
subsidy) and a positive domestic tax.

Many energy system changes are equivalent to one or more of these
four analytical representations. Therefore, subsequent discussions will be couched in terms of the impacts of changes in the
four analytical representations.

2.   EFFECTS ON NET NATIONAL PRODUCT

2.1   The Model

Although much of the current U.S. Congressional energy policy debate is rooted in discussions of wealth redistribution, one of
the most important summary variables for an economy is the gross
national product (gross domestic product) or the net national
product (net domestic product). This section will discuss the
latter aggregate variable, although interpretations in terms of
gross national product follow directly.

The basic model is that of a three-sector economy, consisting of
an energy-importing sector, a domestic energy-producing sector,

---

*Note that although the cost of conducting the R and D program
does not appear in this table, it must be included in an overall
evaluation of the economic impact of such a program.

TABLE 1 - Analytical Representation of Energy System Changes

Energy System Changes	Analytical Representations			
	Import Price	Domestic Cost Function	Import Tariff	Domestic Tax
OPEC Price Increase	+			
Domestic Price Control with Entitlements			-	+
Average Cost Pricing of Electricity				-
Nuclear Moratorium		+		
Coal Limits		+		
Import Tariff			+	
Import Quotas			+	
Domestic Resource Depletion		+		
Conservation Program -- Uneconomic Restrictions			+	+
Energy Conservation Tax			+	+
Successful Energy Supply R&D		-		

and the rest of the economy (ROE). Energy is an intermediate good; output from the ROE may be either a final or an intermediate good.

The first sector imports some quantity of energy (E_I) at a fixed price (P_I), which includes the price paid to the exporting source plus any cost of the import activity. The net resource cost to acquire this energy is simply $P_I E_I$, which becomes one claim on the gross output from the rest of the economy.

The second sector uses output from the rest of the economy in order to produce domestic energy. The quantity of ROE outputs used to produce E_D units of energy will be represented as $H(E_D)$. This is simply the cost function for producing domestic energy.

The total energy used in the economy is the sum of the domestically produced energy plus the imported energy, as represented in equation (2) below.

The rest of the economy uses as inputs a quantity of capital services (K), a quantity of labour services (L), and a quantity of energy (E), and produces as output an aggregate quantity of goods and services (which depends on K, L, and E). This aggregate quantity will be referred to as the gross output of the economy. The relationship between K, L, E, and the resulting gross output can be expressed in terms of an aggregate production function of the economy, F(K, L, E). The net national product (Y) is the difference between the gross output and the costs of importing plus domestically producing energy. This is expressed in equation (1) below. Equation (1) follows directly since, by definition, net national product equals total value of final products produced in the economy or equals the sum over all sectors of the economy of value added.

In a competitive economy* several marginal conditions must also hold.** The price of energy (P_E) will equal its marginal productivity, since competitive firms are assumed to be price-taking profit maximizers. And in a competitive economy the price of energy facing the domestic supplier (P_D) will equal the marginal cost of producing energy. Supply price will equal the energy price plus any tax on domestic production (T_D). These relationships are expressed in equation (3). Similar relationships hold for imported energy: the energy price in competitive equilibrium will equal the import price (P_I) plus the tax on imported energy (T_I). This is represented by equation (4).

Finally, there are necessary marginal conditions for market clearing in the capital and labour markets. The price of labour services (P_L) or of capital services (P_K) is simply equal to the appropriate marginal productivity. These conditions appear as equations (5) and (6).

Six equations thus represent the economy. Equation (1) defines net national product; equation (2) equates energy consumed to domestic plus imported energy; and equations (3) through (6) represent the marginal conditions describing a competitive economy

---

* The results of this paper are also applicable to an efficient centrally planned economy.

**The implicit assumption here is either that there are no market failures in the economy other than those explicitly discussed or that the specific market failures in the economy remain unchanged as energy system changes occur. In particular, this assumption may be violated if an energy conservation program that eliminates market failures is implemented. In that case, the function F(K, L, E) or the marginal conditions (3) and (4) may change as a result of the program.

in equilibrium.* Note that supply conditions for capital and labour have not been written. These will be discussed at a later point.

$$Y = F(K, L, E) - H(E_D) - P_I E_I \qquad (1)$$

$$E = E_D + E_I \qquad (2)$$

$$P_E = \partial F/\partial E = dH/dE_D + T_D \qquad (3)$$

$$P_E = \partial F/\partial E = P_I + T_I \qquad (4)$$

$$P_K = \partial F/\partial K \qquad (5)$$

$$P_L = \partial F/\partial L \qquad (6)$$

Impacts of energy system changes on NNP can be evaluated by use of equations (2) through (6), since the impacts of each of the four analytical representations can be evaluated by means of these equations.

For small changes in any analytical representation, the impact on NNP can be calculated by taking the total differentials of equations (1) and (2) and by using equations (3) through (6). This allows one to derive equation (7), below, which expresses changes in net national product as a sum of weighted changes in capital, labour, domestic energy production, imported energy production, import price, and domestic price. In equation (7) the change in the domestic cost <u>function</u> is expressed as a change in the <u>average</u> cost function (equal to δAC_D) times the quantity of domestic energy (E_D).

$$\Delta Y = - E_I \Delta P_I - E_D \delta AC_D + T_I \Delta E_I + T_D \Delta E_D + P_L \Delta L + P_K \Delta K \qquad (7)$$

Equation (7) will be fundamental to all subsequent analyses of the relationships between energy sector changes and net national product changes. Therefore, it is repeated in the first column of Table 2. The various terms on the right hand side of equation (7) are interpreted as components of NNP change. These components are: direct effects of cost changes, consisting of changes in the resource cost function; "welfare costs" of price changes stemming from divergences between prices and costs; and induced effects via labour quantity and capital quantity. The change in the net national product equals the sum of the components.

---

*While this paper discusses E as a scalar variable under the implicit assumption that energy can be viewed as a single aggregate quantity, E can also be viewed as a vector, thereby allowing a disaggregation by energy type. The interpretations in each equation will be straightforward.

To use equation (7) (or equivalently, Table 2), it is necessary to specify a complete model that describes capital, labour, and energy quantity changes in response to changes in the underlying environment. This can be done by more completely specifying the production functions and cost functions described previously and modelling the supply functions for capital and labour, or else simpler models can be postulated in order to illustrate the fundamental interactions. The latter procedure will be followed in this paper.

TABLE 2 - Components of NNP and Welfare Changes in
Response to Energy System Changes

Effects \ Gauges	Changes in NNP	Monetary Equivalent Welfare Change
Direct Effects of Cost Changes:		
Import Price	$-E_I \Delta P_I$	$-\sum_t \alpha_t E_{It} \Delta P_{It}$
Shift of Domestic Cost Function	$-E_D \delta A_{CD}$	$-\sum_t \alpha_t E_{Dt} \delta AC_{Dt}$
"Welfare Cost" of Price Changes:		
Import Tax	$T_I \Delta E_I$	$\sum_t \alpha_t T_{It} \Delta E_{It}$
Domestic Tax	$T_D \Delta E_D$	$\sum_t \alpha_t T_{Dt} \Delta E_{Dt}$
Effects via:		
Labour Input		
Efficient Allocation	$\left.\begin{array}{c} \\ \\ \end{array}\right\} P_L \, \Delta L \left\{\begin{array}{c} \\ \\ \end{array}\right.$	0
Two Labour Prices		$\sum_t \alpha_t T_{Lt} \Delta L_t$
Capital Input		
Efficient Allocation	$\left.\begin{array}{c} \\ \\ \end{array}\right\} P_K \, \Delta K \left\{\begin{array}{c} \\ \\ \end{array}\right.$	0
Two Interest Rates		$\sum_t \alpha_t T_{Kt} \Delta K_t$
	$\alpha_t = \sum_{\tau=1}^{t} \dfrac{1}{1 + r_\tau}$	

2.2 Taxes vs. Cost Increases

As a first illustration, equation (7) can be used to examine the differences in NNP impacts stemming from the resource cost increases in those motivated by tax increases. It will be assumed that capital and labour are supplied perfectly inelastically and that there are no changes in the cost function for domestically produced energy. Therefore, the first two terms and the last term on the right hand side of equation (7) will be identically zero.

If all taxes remain zero while imported energy prices change, then equation (7) is reduced to the simple differential equation:

$$\frac{dY}{dP_I} = - E_I \tag{8}$$

If the import price remains constant while the tariff on all energy (imported plus domestic) changes, equation (7) is reduced to the differential equation:

$$\frac{dY}{dE} = T, \tag{9}$$

where T is the tax rate on all energy.

Note that in the case of import price changes, the impact on NNP of a price change is proportional to energy imports. However, in the case of a tariff on all energy, the impact is independent of the quantities imported, but depends on the change in the quantity of energy consumed.

Upper and lower bounds on the NNP impacts of tariffs and of changing import prices can be evaluated very simply by use of equations (8) and (9). If imported energy prices were to increase from P_{10} to P_{IF}, then the change in NNP (Y) could be calculated by integrating equation (8):

$$\Delta Y = \int_{P_{10}}^{P_{IF}} (- E_I)\, dP_I.$$

Since E_I is decreasing in P_I and increasing in Y, this allows simple limits to be set:

$$E_{10}\Delta P_I > - \Delta Y > E_{IF}\Delta P_I, \tag{10}$$

where E_{10} and E_{IF} are the initial and final import levels, respectively, and $\overset{\bullet}{P}_I$ is the change in energy price:

$$\Delta P_I = P_{IF} - P_{10}.$$

Inequality (10) provides tight limits on the NNP reductions of an import price increase whenever the level of imports occurring with the higher prices is near to the level occurring with the lower prices. However, when the price change leads to large changes in imports, the bounds are very loose.

Inequality (10) allows an unambiguous comparison between the NNP reduction and the revenue gains accruing to the exporter of energy. Initial revenue is $P_{10}E_{10}$, while final revenues is $P_{IF}E_{IF}$. Then the revenue gain for the exporter (ΔR) is:

$$\Delta R = P_{10}(E_{IF} - E_{10}) + E_{IF}\Delta P_I. \tag{11}$$

This equation provides an upper limit to the revenue gain as prices increase:

$$\Delta R < E_{IF}\Delta P_I.$$

Comparing this to inequality (10) shows that for price increases:

$$-\Delta Y > \Delta R. \tag{12}$$

The revenue gain to the exporting country from increasing energy price **must** be strictly smaller than the resulting NNP loss borne by the importing country. Furthermore, if the exporter's marginal cost of energy production is no greater than the initial price, then the gain in revenues minus costs (net profits) must also be strictly smaller than the resulting loss in NNP borne by the importing country.

A similar analysis can be conducted for a tax rate increase. Integrating equation (9) provides the inequality:

$$T_0 \Delta E > Y > T_F \Delta E, \tag{13}$$

where T_0 and T_F are the initial and final tax rates on all energy, respectively.

Inequality (13) can be used for any change in tax. For example, assume the initial tax is zero and that either a positive or a negative tax (a subsidy) is imposed. In either case, inequality

(13) shows that ΔY must be negative: any tariff or subsidy will reduce NNP. The larger the tariff or subsidy and the larger the change in energy consumption, the greater will be the possible NNP reduction.

In order to use equations (8) and (9) to obtain more precise estimates of NNP change than are provided by the above inequalities, it is necessary to describe the model more fully. In particular, the aggregate production function must be specified more completely in order to relate energy demand or energy imports to energy price and NNP. For this discussion it will be assumed that all energy is imported, so that $E_I = E$. And it will be assumed that the aggregate production function is such that in a competitive equilibrium the quantity of energy demanded is the following constant elasticity function of NNP and energy prices: Equation (14) will be somewhat loosely referred to as a demand function for energy in this economy, and ε will be referred to as the price elasticity of demand for energy.

$$E = A\, Y\, P_E^{-\varepsilon}. \tag{14}$$

To calibrate equation (14) it will be assumed that in the year 2010, if the OPEC price increase had not occurred, the energy price would be ¢ 0.80/MMBtu,* the energy demand would be 220 quads, and the NNP would be $ 4,400 billion. These assumptions correspond to those of Hogan and Manne [1], except that those authors assume a constant elasticity of substitution production function rather than a constant elasticity demand function.

What then are the impacts on NNP of an import price change? Equation (14) can be used to eliminate energy consumption in equation (8), giving a simple differential equation relating Y and P_I:

$$\frac{dY}{dP_I} = -\,A\, Y\, P_I^{-\varepsilon}.$$

This differential equation can be solved to give the following equation relating NNP to imported energy price:

$$\frac{Y}{Y_0} = \exp\left\{ \frac{S_0}{1-\varepsilon}\left[1 - \left(\frac{P_I}{P_{I0}}\right)^{1-\varepsilon}\right]\right\}, \tag{15}$$

---

*MMBtu = million Btu

where

$$S_0 = \frac{P_{I0}E_0}{Y_0} \qquad (16)$$

S_0 is the expenditure on energy in the base case as a fraction of base case NNP, and will be referred to as the value share of energy in the economy. The symbols E_0, Y_0, and P_{I0} are the base case values of E, Y, and P_I, as indicated previously.

Equation (15) shows that the impact of increasing energy import prices on NNP is increasing in S_0: the greater the base case expenditure on energy (as a fraction of NNP), the greater will be the economic impact of an import price change. Also, the greater the elasticity of demand, the smaller the impact of a given import price increase on NNP.

Equation (15) has been solved under the base case assumptions indicated earlier. These assumptions give a value of S_0 of 4 percent, which corresponds to the pre-embargo experience. The results appear in Figs. 1 and 2 for several different values of demand elasticity (ε).

Fig. 1 shows net national product as a function of imported energy price for increases or decreases in energy price for increases or decreases in energy price from the 1972 levels. Note that increasing imported energy price always reduces NNP and decreasing price always increases net national product. The impacts depend significantly upon the elasticity of energy demand. For low elasticities an increase in energy price has a larger impact on the economy that would occur with a high elasticity. But even with relatively high elasticities, increases in energy price can lead to significant reductions in net national product. For example, an increase in the energy price from $ 0.80/MMBtu to $ 3.00/MMBtu reduces net national product by almost 10 percent if the elasticity of demand is .1 and by 6 percent if the elasticity of demand is .7.

Fig. 2 shows the values of net national product and energy use as the energy prices change. The greater the elasticity of demand, the smaller will be the reduction in net national product for a given reduction in energy use. A 30 percent reduction in energy use (from 220 quads*to 154 quads) reduces net national product by billion (2.2 percent) if the elasticity of demand is .7, but reduces net national product by $ 70 billion (16.3 percent) if elasticity of demand is .1. Note, however, that the former reduction is motivated by an import price increase of roughly 33 percent, while the latter is motivated by a 370 percent increase in imported energy price.

---

*quads = 10^{15} quadrillon Btu

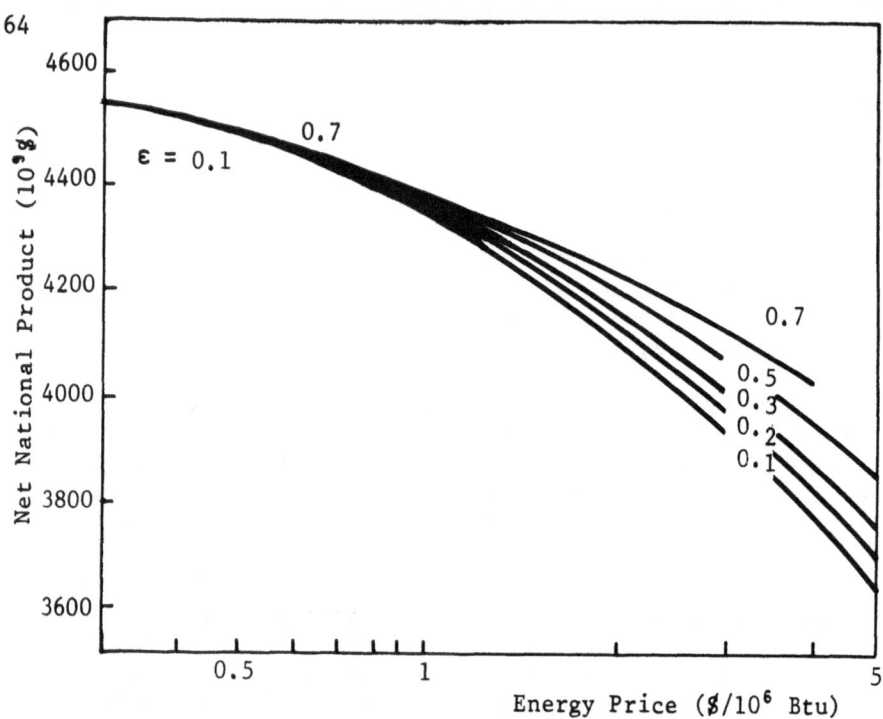

FIGURE 1 - Import price change case

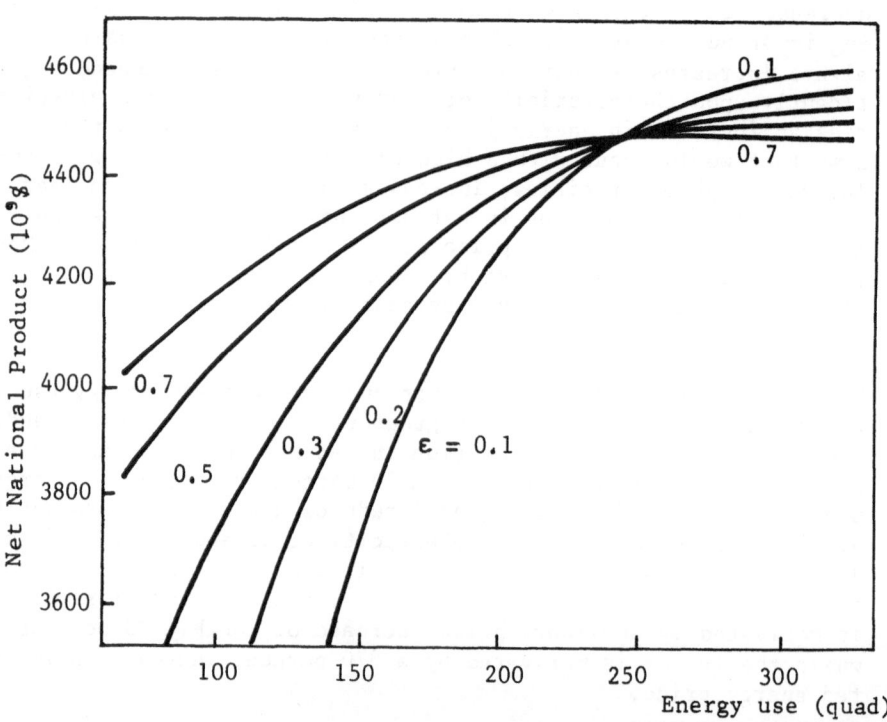

FIGURE 2 - Import price change case

This analysis has focused upon increases in imported energy price when there is no domestic production. If some energy is domestically produced, then the impact of import price changes on NNP is reduced correspondingly, as shown by equation (8).

Increases in the domestic cost function would have impacts on NNP analogous to those stemming from increasing import price, as can be shown by equation (7). In fact, an analysis could be conducted for all energy domestically produced with an infinitely elastic domestic supply curve. In this case, exogenous shifts in the supply price would have NNP effects identical to those analyzed here for an import price increase.

The key issues for evaluating NNP impacts of cost increases are (1) the magnitude of the cost increase, (2) the expenditure on energy as a fraction of NNP in the base case, and (3) the elasticity of demand for energy. Whether the cost increase stems from domestic cost function increases or import price increases is irrelevant for evaluating impacts on NNP.

What are the impacts on NNP of a tax change? Under an assumption that a tax T is imposed on all energy, whether imported or domestically produced, equation (9) is the fundamental differential equation. It will be assumed that the domestic supply price of energy is determined by the fixed import price of P_I. Then, from equation (14), demand for energy can be expressed as follows:

$$E = A\, Y(P_I + T)^{-\varepsilon} . \qquad (17)$$

Equations (9) and (17) are sufficient to determine NNP and energy consumption as functions of the tariff. These equations are independent of the fraction of energy imported and thus the impact of a tariff on NNP will be independent as well.

No closed form solutions to equation (9) and (17) have been found. However, solutions can be obtained numerically. These results are provided in Figs. 3 and 4.

Fig. 3 shows NNP as a function of energy price (P_E) as the tariff changes. When energy price is $ 0.80/MMBtu, the tariff is zero. NNP will decline for increases or decreases in energy price away from the import price. Net national product is maximized at zero tariff for energy price equal to the import price. Small changes in tax around zero lead to virtually no loss in NNP, while equivalent tariff changes beginning at non-zero levels lead to greater NNP impacts. These results all occur because, as shown in equation (9), the change in NNP associated with a given change in energy use is proportional to the tariff, which in turn is the difference between the marginal value of energy (P_E) and the marginal cost (P_I) of attaining energy.

66

Fig. 3 also casts light on the role of the demand elasticity.
For a given tariff, the higher the elasticity, the <u>greater</u> the
impact on NNP. For example, an increase in energy price to
$ 3/MMBtu reduces net national product by $ 23 billion if the
elasticity of demand is .1 and by $ 95 billion if the elasticity
of demand is .7. This is precisely opposite from the result obtained for a change in import price. The explanation is that the
higher the elasticity, the greater the change in energy consumption for a given tariff, and thus the greater the NNP change [see
equation (9)]. The consumption of energy is reduced by 28 quads
(13 percent) in the former case and by 135 quads (61 percent) in
the latter.

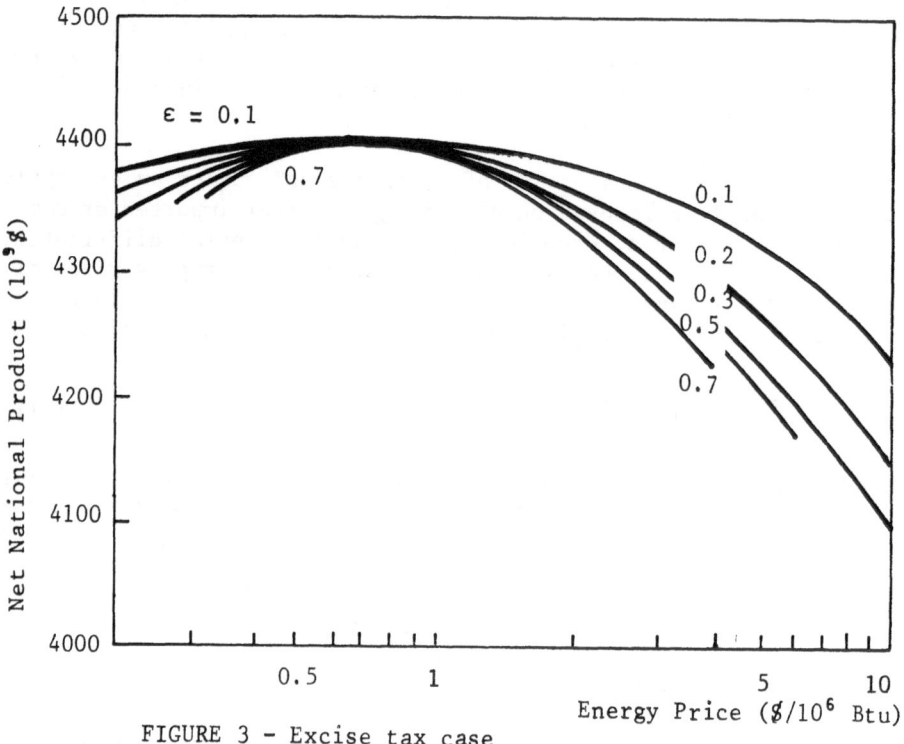

FIGURE 3 - Excise tax case

Fig. 4 plots changes in net national product against changes in
energy consumption motivated by changing excise taxes. For relatively high elasticities, large percentage reductions of energy
use occur with relatively small percentage reductions in net national product. For example, a 30 percent reduction in energy
consumption leads to a $ 14 billion reduction (0.3 percent) in
net national product if elasticity is .7, but a $ 270 billion re-
duction (6 percent) if elasticity is .1.

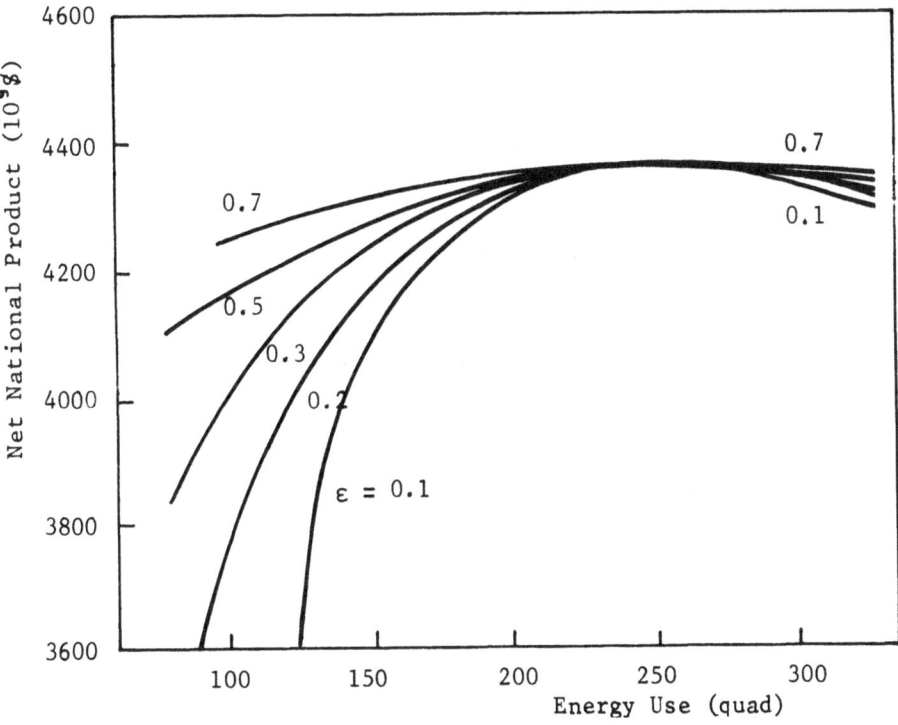

FIGURE 4 - Excise tax case

Note that an increase in energy use motivated by a negative excise tax (a subsidy) will also reduce net national product. Therefore, attempts to control energy prices below marginal costs or otherwise subsidize energy consumption will reduce NNP. However, the result is not symmetrical. A 50 percent increase in energy use motivated by a subsidy leads to a much smaller reduction in net national product than would the same proportionate reduction in energy use motivated by a tax.

Fig. 5 compares the NNP impacts of import price increases to those of excise tax increases for two different demand elasticities (.2 and .5). The numbers under the curves are the demand elasticities, while the letters I or R denote import price change or excise tax change, respectively. A given energy consumption reduction will be associated with a greater impact on NNP if motivated by an import price increase than if motivated by an excise tax increase.

The effects of simultaneous shifts of import price and tariff are illustrated in Fig. 6. The downward sloping curve, labeled "Tariff = 0, P_I Changing" shows the NNP as a function of import price

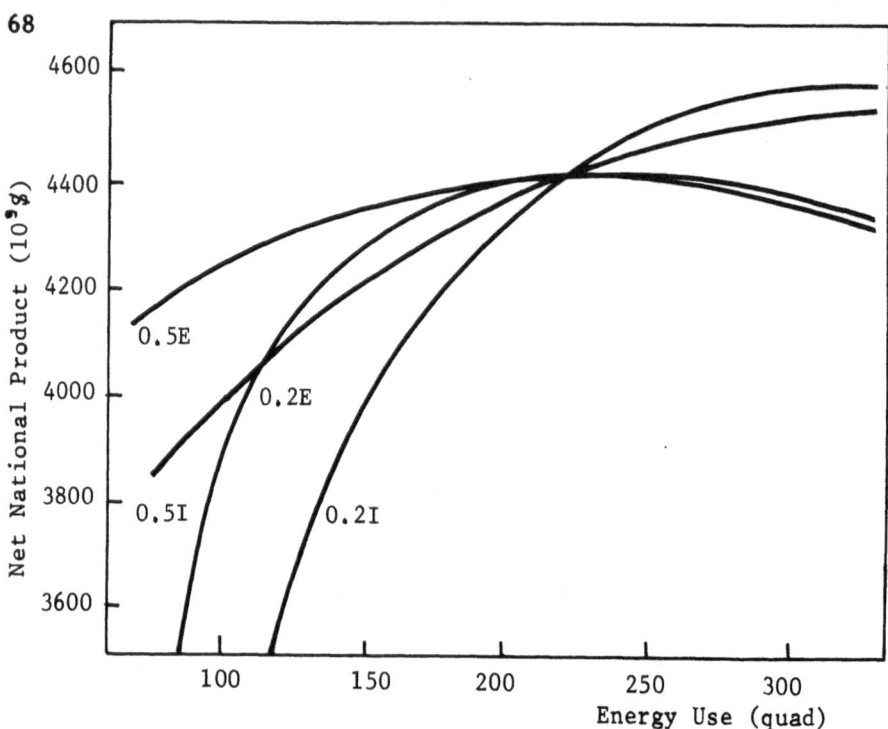

FIGURE 5 - Excise tax vs. import price change

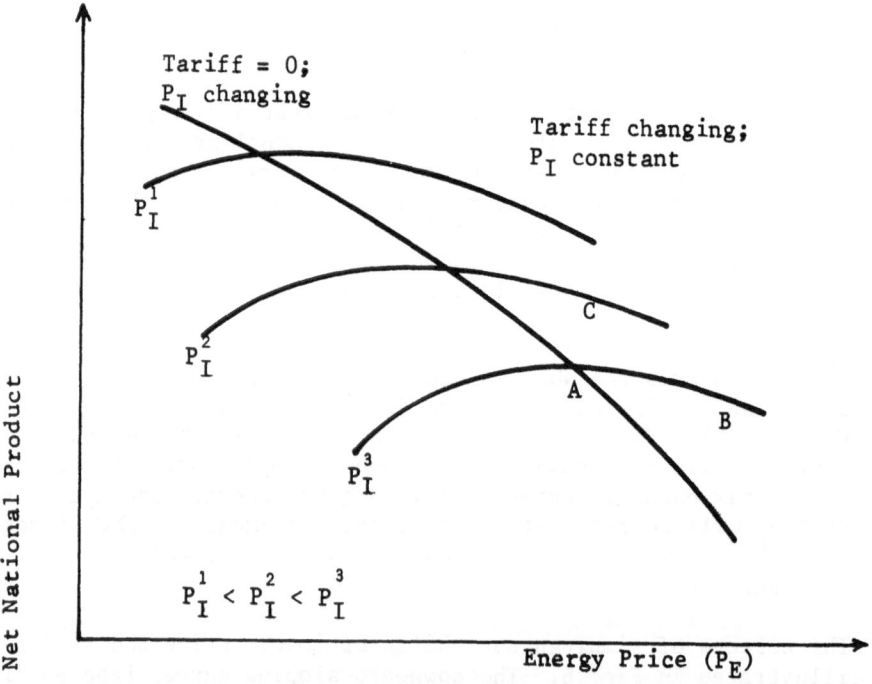

FIGURE 6 - Simultaneous changes in tariff and import change

for zero tariff. Each of the three curves intersecting this curve illustrates NNP for import price constant but tariffs changing. At the point of intersection the tariff equals zero. As shown, a given energy price may be associated with many possible values of NNP. Higher values of NNP occur when the import price is low but the tariff is high.

The differences between the NNP impacts of excise taxes and of import price changes cast some light on the potential value of a tariff on energy, if the tariff influences import prices. Suppose that importing countries were to impose a tax on oil. If the β import price of energy were to remain unchanged, this tariff would reduce the net national product of each importing country, as is illustrated by the move from point A to point B in Fig. 6. However, since the tariff would reduce demand for OPEC oil, it might induce the OPEC nations to reduce the oil price from what it would have been otherwise. This price reduction would increase net national product, corresponding to a move from point B to point C. Whether the consuming nations would be made better or worse off by the combination of changes would depend upon which of these effects were to dominate. For a relatively small excise tax, however, such a strategy would increase net national product even if the reduction in import price were significantly smaller than the excise tax.

This result can be examined mathematically in the context of a tariff on imported oil. Assume that for each unit reduction in the rate of oil imports the world price of oil is reduced by β dollars per barrel. Consider a tariff of T_0 dollars per barrel on imported oil, and assume that an additional unit increase in this tariff results in a change in imports of ΔI (where $\Delta I < 0$). In this case the change in national product associated with the tariff can be simply determined by means of equation (7), which is rewritten below assuming that the quantities of capital and labour remain constant:

$$\Delta Y = T_0 \Delta I + I \, \Delta P_I. \tag{18}$$

Including the impact of import quantity changes on import prices, this equation becomes:

$$\Delta Y = (T_0 - I \, \beta) \, \Delta I. \tag{19}$$

By equation (19) national product increases as the tariff increases as long as the tariff is less than $I \beta$. Further increases in the tariff above $I \beta$ would reduce the national product. That tariff which maximizes GNP is often referred to as the optimal tariff, T^*, and is equal to:

$$T^* = \beta\, I. \tag{20}$$

As an example assume that in the year 1985 the U.S. were importing oil at the rate of 10 million barrels per day and that every 1 million barrels per day reduction in the rate of imports were to reduce the world oil price by 15 ¢ per barrel. Then, by equation (20) the optimal tariff on imported oil would be $ 1.50 per barrel. Note that the optimal tariff does not depend upon the elasticity of demand for energy but depends only upon the level of imports and the impact on the world oil price of changes in imports.

The "optimal tariff" indicated above is optimal from one country's individual perspective, given its impacts on the world oil price. However, the tariff set by one country will benefit all other net importing countries and will impose costs on all net exporting countries. Thus, a coalition of net importing countries would find that the optimal tariffs from their collective perspective will be greater than the "optimal tariff" derived above.

The analysis suggests the value to the energy importing countries of collective strategy to reduce imports of energy. However, the value of a collective tariff strategy depends upon the fraction of energy imported. Without the world price decrease, imposing an excise tax which is identical for each country would have the same proportional negative impact on NNP for a nation virtually self-sufficient in energy as it would for one importing all energy. The gains from the import price reduction, however, would vary greatly between the two, with the virtually self-sufficient nation gaining little while the nation totally dependent upon imports would gain much. The latter may see a net gain in NNP and, therefore, may be motivated to support the plan while the former may see a net loss and may not be so motivated. Therefore, the tariff level acceptable to a given importing country can be expected to be larger, the greater is the proportion of energy which is imported.

The example may become more relevant when it is remembered that many domestic programs can be analytically represented as excise tax on energy. Therefore, the comments above on tariffs are also applicable to other actions such as incentives to "conserve" energy or to produce more domestic energy. Thus, nations that are heavily dependent upon imports would be more willing to engage in domestic actions to force a world price decrease than would nations that were virtually self-sufficient.

2.3   NNP Effects Via Capital and Labour

The preceding section incorporates an assumption that changes in
energy availability will not influence the quantities of other
input factors. However, equation (7) is more general. Any induced increase in capital or labour services will increase NNP by
an amount equal to the factor price times the quantity increase.
Conversely, energy sector changes that reduce the capital formation or the labour supply would reduce NNP. The magnitude and
direction of labour or capital service changes depend upon characteristics of the aggregate production function and of the supply
functions for these factors.

Characteristics of the production function are important because
changes in the energy price, with other prices constant, may either increase or decrease the demand for capital or labour. If an
increase in the price of energy leads to an increase in the demand for labour, then labour and energy are said to be supplementary commodities. Conversely, if labour demand decreases, then
energy and labour are said to be complementary commodities.*
Whether energy and capital or energy and labour are complementary
or supplementary depends upon properties of the production function. Empirical research is necessary to determine which relationship is correct.

Empirical studies have indicated that energy and labour tend to
be supplementary products: an increase in the price of energy
tends to increase the demand for labour. This occurs fundamentally because increases in energy price cause firms and consumers
to use more labour-intensive and less energy-intensive processes
and products as energy prices increase.

While the result is still far from conclusive, empirical evidence
tends to support the hypothesis that capital and energy are complementary. Berndt and Wood [2], in their excellent review and
reconciliation of the evidence, conclude that capital and energy
are complementary products: increases in energy price probably
reduce the demand for capital. This conclusion, however, is not
accepted by all researchers in the field.

Supply functions for capital and labour are also critical in determining changes in capital and labour quantities. If these
supplies are independent of prices, then the issue of complementarity vs. supplementarity is irrelevant, for changes in energy

---

*There are several different, but related, definitions of comp-
lementarity and supplementarity. A complete discussion is beyond the scope of this paper, but is discussed in Hogan [3].

or capital demand functions will not influence the equilibrium
capital or labour quantities, but only their prices. If, on the
other hand, capital or labour supply is very responsive to price,
then the change in equilibrium quantity roughly equals the constant price demand change.

Labour supply depends both upon labour price and upon NNP, with
either increasing wage rates or decreasing NNP leading towards
increases in the labour supply. While the current empirical evidence is inconclusive, it suggests that the elasticity of supply
of labour is small. Therefore, it can be expected that increases
in energy price will only slightly change NNP via changes in labour
quantity. Rather, increases in energy prices will **tend** to increase
wage rates.

The capital stock of the economy increases slowly over time whenever net investment (the net rate of capital formation) is positive and decreases when net investment is negative. Thus, any
factor that increases the rate of capital formation will, in the
long run, increase the capital stock, although the effects would
not be felt rapidly.

The rate of capital formation depends upon both NNP and upon the
interest rate, which is the relevant price of capital services.

While empirical evidence is inconclusive, the rate of capital formation seems to be a significantly increasing function of the interest rate. Under the assumption that capital and energy are
complementary, increases in energy price will decrease capital
demand. In the short run, the supply of capital is relatively
inelastic, so this will simply tend to depress interest rates.
However, in the long run, reductions in capital demand would induce significant reductions in the equilibrium capital stock and
thus would reduce NNP as energy prices increase.

For fixed interest rates, capital formation is roughly proportional to NNP. Therefore, reductions in NNP reduce the rate of capital formation and, after some time, reduce the capital stock.
This further reduces NNP. The net result of this feedback loop
is a multiplied impact on NNP.

The various effects via capital and labour are summarized in Fig.
7 under an assumption of capital/energy complementarity and la-
bour/energy supplementarity. This illustrates the positive feedback loop through capital and NNP and the negative feedback loop
through labour and NNP.

In summary, in response to increases in the price of energy, several effects tend to reduce NNP. First are the direct effects
of cost changes or "welfare costs" of price changes, as described

in the previous section. Second, if capital and energy are complementary, then price changes induce a reduction in the demand for
capital and the equilibrium quantity of capital, which in turn
reduces NNP. Finally, the first two effects, by reducing NNP,
reduce the supply of capital, further reducing its equilibrium
quantity, and thereby lead to further NNP reductions.

CAPITAL/ENERGY COMPLEMENTARITY ASSUMED:

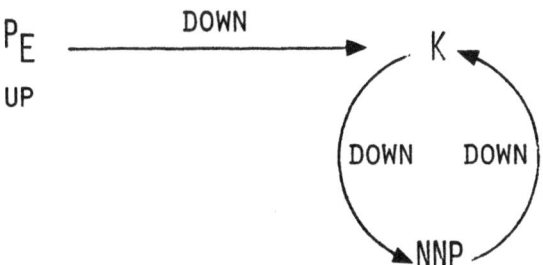

LABOUR/ENERGY SUPPLEMENTARITY ASSUMED:

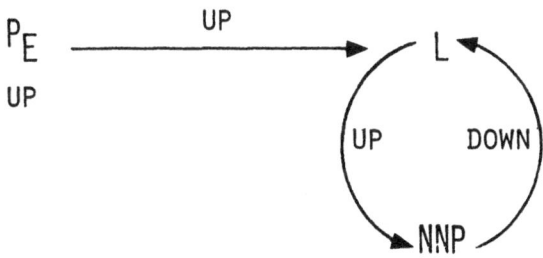

FIGURE 7 - Effects via capital and labour an
increase in energy price

3.   EFFECTS ON WELFARE

The previous section has been focused on net national product.
However, there are several difficulties with the use of NNP as a
measure of well-being, two of which are particularly relevant here.
First, increases in labour use expands output but decreases leisure time. The increase in output is included in NNP but the
value of the foregone leisure time is not deducted. Second, increases in the rate of capital formation (for fixed NNP) imply decreases in the current output available for consumption. Thus,
current well-being is not uniquely determined by NNP whenever the

investment rate changes. Finally, since people make tradeoffs over time, even if NNP were a good static measure it would be necessary to find some method of aggregating NNP changes over time to obtain a welfare criterion. This section focuses attention on a better measure of welfare than NNP.

It will be assumed that both individuals and the economy as a whole can preferentially rank consumption trajectories and alternative consumption/leisure time (or labour time) combinations and that the individual and the societal rankings are identical.

Under this assumption, one can define, as a measure of performance, some welfare functions, W:

$$W = \sum_{t=0}^{\infty} U(C_t, L_t) \, e^{-\rho t} , \tag{18}$$

where C_t is consumption at time t, L_t is labour used at time t, $U(\cdot)$ is a utility function (which may itself depend on time), and $e^{-\rho t}$ (for $\rho > 0$) is the discount factor on future utilities.* Welfare, under this well-known utilitarian formulation, is represented by a weighted sum of individual utilities of consumption and labour. This formulation incorporates the notions that well-being depends on the goods and services available to be consumed, rather than simply upon the net output of the economy, and that well-being also depends upon leisure time available. Additionally, it allows for inter-temporal tradeoffs.

Equation (18) provides the welfare criterion to be used in the subsequent analysis. However, it could easily be generalized if necessary. In order to evaluate how welfare changes in response to changes in the energy sector, it is useful to have a more intuitive measure than that provided by the welfare function itself. For small changes in welfare, define the monetary equivalent welfare change (ΔMW) as the ratio

$$\Delta MW = \frac{\Delta W}{\partial U / \partial C_0} . \tag{19}$$

The monetary equivalent welfare change equals the change in welfare divided by the welfare change that would occur with a unit increase of consumption at time 0. Thus, the monetary equivalent welfare change can be interpreted as the number of dollars' worth

---

*Not to be confused with a discount factor on future consumption streams.

of consumption in time period 0 that would provide the same change
in welfare as that stemming from the arbitrary change in the consumption and labour trajectories. Thus, the monetary equivalent
welfare change is a conceptually simple, workable measure of welfare changes.

3.1   The Model

All elements of the model of NNP determination from the previous
section will remain valid in this section. However, in addition,
it is necessary to specify the relationships between NNP, consumption, and capital formation. The NNP can be allocated between
two purposes: consumption or capital investment. The capital
change from one period to the next is just equal to the net investment in the economy. Therefore, the change in capital stock
plus consumption must equal NNP at each time:

$$C_t + [K_{t+1} - K_t] = Y_t, \tag{20}$$

where K_t and Y_t are capital stock and NNP at time t, respectively.

Marginal conditions of market equilibrium, in addition to those
specified previously for the energy sector, must be specified
here.

On the demand side of labour markets, labour price (P_{Lt}) must
equal the marginal productivity of labour:

$$\frac{\partial F}{\partial L_t} = P_{Lt} . \tag{21}$$

On the supply side, consumers in the economy face a labour/leisure
choice. If consumers maximize well-being under a budget constraint,
each must satisfy a marginal condition relating labour price (net
of taxes) to a ratio of marginal utilities of labour and consumption. If the utility function U(·) in the welfare function adequately represents individual preferences, then the marginal conditions describing labour supply can be written as follows:

$$- \frac{\partial U/\partial L_t}{\partial U/\partial C_t} = P_{Lt} - T_{Lt}, \tag{22}$$

where T_{Lt} is the tax on labour.

Similar relationships hold for the capital stock. On the supply
side, consumers choose the allocation of their income between

consumption and capital formation. A unit reduction in consumption at period of time t allows the consumer to increase consumption at the next period of time by an amount equal to $(1 + r_t)$, where r_t is the interest rate facing the consumer (the consumption interest rate). An optimizing consumer, choosing consumption and rate of capital formation, will select that consumption rate for which the ratio of discounted marginal utilities of consumption between the two time periods is equal to $1(1 + r_t)$:

$$\frac{\partial U/\partial C_{t+1}}{\partial U/\partial C_t} \, e^{-\rho} = \frac{1}{1 + r_t} \, . \tag{23}$$

On the demand side of capital markets, firms choose the capital to use based upon its price and productivity. The price facing the user of capital (P_{Kt}) is the supply price (the interest rate) plus the tax on capital (or on profits earned using capital), denoted by T_{Kt}. This gives:

$$\frac{\partial F}{\partial K_t} = P_{Kt} = r_t + T_{Kt}. \tag{24}$$

The entire market over time is now represented by equations (1), (2) and (20), which relate physical flows, and by equations (3) through (6) and (21) through (24), which express necessary conditions for a competitive equilibrium. The measure of welfare is provided by equations (18) and (19).

In order to evaluate the effects of energy system changes on welfare, equations (1), (2), and (20) can be differentiated totally, the marginal conditions (3) through (6) and (21) through (24) can be inserted, and the results can be combined with equation (19). This process provides the following equation for monetary equivalent welfare changes:

$$\Delta MW = \sum_{t=0}^{\infty} \alpha_t \left\{ - E_{It} \, \Delta P_{It} - E_{Dt} \, \delta AC_{Dt} + T_{It} \, \Delta E_{It} \right. \tag{25}$$

$$\left. + T_{Dt} \, \Delta E_{Dt} + T_{Lt} \, \Delta L_t + T_{Kt} \, \Delta K_T \right\} ,$$

where

$$\alpha_t = \prod_{\tau=1}^{t} \left(\frac{1}{1 + r_\tau} \right).$$ (26)

$_t$ is simply the discount factor on monetary flows obtained using the <u>consumption</u> interest rate. This factor is quite different from $e^{-\rho t}$, the discount factor on <u>utilities</u>.

Equation (25) allows the monetary equivalent welfare change to be obtained as simply a discounted sum of individual components are presented in the second column of Table 2.

The components of the monetary equivalent welfare change, for the most part, are simply discounted sums of the components of NNP change, as shown in this table. The direct effects of import price changes or shifts in the domestic cost function and the "welfare costs" of an import tax or a domestic tax correspond precisely. However, the effects via the labour input and the input of capital are not discounted sums of changes in the corresponding NNP components.

A comparison of the two columns of Table 2 shows that if a given policy leads to no changes in capital or labour quantities, then the monetary equivalent welfare change associated with that policy equals the discounted sum of NNP changes. If a given energy system change were to induce capital or labour to increase, however, then the monetary equivalent welfare change would be strictly less than the discounted sum of NNP changes.

The monetary equivalent welfare change via capital or labour always has the same sign, but a smaller magnitude* than the discounted sum of the corresponding component of NNP change. Therefore, discounted sums of NNP changes will not correctly measure changes in monetary equivalent welfare whenever there are induced changes in capital or labour.

In evaluating the welfare changes via labour and capital input, it is necessary to differentiate between economics that are perfectly efficient and those that are not. In an efficient allocation, the price of labour facing a consumer and that facing a firm are identical; the tax on labour is zero. Similarly, the tax on capital is zero.

---

*This occurs because the tax on capital or labour must always be smaller than the capital and labour price facing consumers (unless the supply prices are zero or negative).

In an efficient world, induced changes in labour or capital input do not change welfare even though they do change NNP. In the case of labour markets, this occurs because the consumption benefits obtained through an increase in labour input are just equal to the losses due to the decreased leisure time. In the case of capital markets, an increase in capital formation at one time period leads to a stream of additional consumption at later times. In an efficient allocation, the present value of this stream of future consumption changes is just equal to the reduction in current consumption required to increase the capital stock. Thus, if there is an efficient allocation of capital, changes in the capital stock have benefits just equal to their cost and, therefore, lead to a zero change in welfare.

Whether energy is complementary or supplementary with capital or labour is totally irrelevant for evaluating welfare in an efficient allocation. Only in the case of taxes on labour or capital will these relationships be significant for welfare evaluations.

The situation is different if the allocation is not efficient or if there are taxes on labour or capital. In this case, the monetary equivalent welfare changes via capital or labour inputs can be important. There will be a welfare gain from increasing labour supply whenever there is a tax on labour because the marginal productivity of labour will be greater than the marginal value of leisure foregone, the difference (per unit) being equal to the tax. Similarly, there will be a welfare gain from increasing the rate of capital formation whenever there is a tax on capital because the discounted sum of future consumption gains will be greater than the requisite current consumption reduction.*

An example is helpful in showing that with two different interest rates a change in capital formation may have significant impacts on welfare. Assume that the consumer interest rate is r for all time, the tax is r and, therefore, that the price of capital facing a firm is 2r.** Suppose that in the year 0, investment is increased by $ 1.00 and that, therefore, consumption is decreased by $ 1.00. The capital stock at all subsequent times will be increased by $ 1.00 as long as the additional output produced each year is simply consumed. This assumption will be made.

---

* A tax on labour would be associated with too little labour use. A tax on capital would be associated with too little capital formation.

**This perhaps corresponds to the situation in the U.S. with a 50% corporate income tax coupled with a personal income tax on the earnings.

Under these assumptions, the additional consumption in years 1 and beyond will be 2r dollars, the marginal productivity of capital. This sum of 2r dollars per year, discounted at an annual interest rate of r is $ 2.00. Thus, the discounted sum of NNP increases stemming from the $ 1.00 increase in capital formation is $ 2.00. The discounted sum of <u>consumption changes</u> is $ 2.00 minus the $ 1.00 reduction in year 0. Thus, the <u>monetary equivalent welfare change</u> of investing $ 1.00 more is $ 1.00, if the interest rate facing firms is twice the consumption interest rate! The same result could be obtained directly from the corresponding component in Table 2: the tax of r, discounted at a rate r, gives a monetary equivalent welfare change of $ 1.00. Induced changes in capital stock can have proportionately large welfare impacts.

The divergence between the consumption interest rate and that facing firms may have implications for energy policy. A simple example will illustrate. To the assumptions of the preceding example, add the assumptions that all energy is imported and, further, that there is no concern for international vulnerability (a clearly difficult assumption!). Assume that capital and energy are complementary. The question then is, what tariff on imported energy would maximize welfare?

To illustrate the point, particularly simple mathematical forms will be specified, as indicated:

$$\Delta K_t = - \gamma \ \Delta P_E = - \gamma T,$$

$$\Delta E_t = - \theta \ \Delta P_E = - \theta T.$$

What is the monetary equivalent welfare change associated with the imposition of a tariff? In an efficient world the change in monetary welfare is equal to simply:

$$\frac{-\theta T^2}{2r} \ .$$

In this situation, any positive or negative tariff reduces welfare: the optimal tariff is zero. In this case of no externalities, no import vulnerability costs, and in an efficient world, the optimal tariff on energy is zero.

The answer is different, however, in the situation of two interest rates. In the case assumed, the monetary equivalent welfare change is:

$$- \gamma T - \frac{\theta T^2}{2r} \ .$$

The optimal tariff is no longer zero. All positive tariffs reduce welfare by reducing capital formation and by reducing energy
consumption below its optimal level. However, a small negative
tariff (a subsidy on energy consumption) would increase welfare
by increasing capital formation. The welfare gain from the increase in capital formation would dominate the welfare loss from
excessive energy consumption. A negative tariff having a value
of $-r\gamma/\theta$ would minimize welfare. Thus, in the situation of no
externalities or vulnerability costs, capital/energy complementarity, and too little capital formation, welfare can be increased
by subsidizing the use of energy.

This example, though unrealistic, does suggest that the existence
of taxes on labour and capital that lead to non-optimal rates of
capital formation or labour supply, may imply different policy
prescriptions from those appropriate in a fully efficient world.
The magnitude of these effects is unknown. And even the sign of
the optimal tariff in the simple model is open to question if
there is uncertainty whether capital and energy are complementary.
However, the example does suggest that the interactions between
energy and capital need to be considered before simple policy
prescriptions are adopted.

4. SUMMARY AND CONCLUSIONS

This paper has presented a conceptual framework and a few simple
examples of the long run relationships between energy sector changes and economic growth. Both a simple descriptive measure - NNP -
and a more complex normative measure - welfare - have been considered.

It has been argued that many energy system changes could be represented by combinations of four analytical representations - tariffs and taxes, import price changes and domestic cost function
changes. Thus, the conceptual framework was developed considering
impacts of only the four analytical representations, rather than
separately examining each of the many possible energy system changes. However, the conclusions developed here can be applied to
the many basic changes in the energy system.

Changes in welfare and in NNP can be decomposed into components
representing the direct effects of import prices or cost increases,
the "welfare" effects of tariff or tax increases, and the effects
via induced changes in the quantities of other factors of production, particularly capital and labour. The first two classes of
components influence both net national product and welfare, with
the components of monetary equivalent welfare change being equal
to a discounted sum of the corresponding components of NNP change.
However, the welfare change components via induced shifts in capital and labour are not discounted sums of the corresponding NNP

components. In an efficient world, changes in capital and labour will influence net national product but will have no impact on welfare. In a world of two different interest rates, changes that induce an increased rate of capital formation will increase welfare. The monetary equivalent welfare change via capital or labour is of the same sign but smaller magnitude than the corresponding discounted sum of NNP change components.

Policies that change costs of energy (import price changes, domestic cost function changes) have far larger impacts on the economy than policies that simply change energy prices (tariffs and taxes). The relative differences are particularly pronounced for small changes in energy price or cost. For equivalent cost increases, whether the cost increase is imposed on domestic energy or on imported energy is not relevant to the NNP or welfare impact.

The impact of a tariff or a cost increase depends upon the elasticity of demand for energy, a measure of the ability of the energy-using sectors to adjust to changing prices. For a given import price increase, the greater the demand elasticity, the smaller the economic impact. For a given tariff, the greater the demand elasticity, the larger the economic impact (and the energy impact). However, for a fixed reduction in energy consumption, the greater the demand elasticity, the smaller the economic impact of either an import price change or a tariff.

For increases in the world price of energy, the NNP loss borne by importing countries will always be greater than the revenue gain by the exporting country. And if the world price exceeds the exporting country's marginal cost of energy supply, then the NNP loss will also exceed the net revenue gain (net of costs).

Simple examples were produced that suggested conflicting policy advice. A tariff on all energy consumption (or an energy conservation programm) would increase NNP and welfare if the tariff, by reducing energy demand, motivated a decrease in imported energy price. On the other hand, if capital and energy were complementary, if there were less than optimal capital formation, and if the import price were fixed, then an energy subsidy would increase both NNP and welfare. In the face of the conflicting advice, only more carefully articulated quantitative analysis (to say nothing about a consideration of other interactions) could be relied upon to develop the appropriate recommendation.

This paper has presented only a simple conceptual structure for thinking through the relationships between energy and domestic growth. More complete analysis depends upon the use of carefully articulated models of the energy system and at least of capital and labour markets. Such studies have been initiated through the CONAES project [4] and the Energy Modelling Forum [5], but many issues are still far from resolution.

REFERENCES

1. W.W. Hogan, and A.S. Manne, "Energy-Economic Interactions: The Fable of the Elephant and the Rabbit?" in Energy and the Economy, Vol. 2, EMF Report 1, September 1977, Stanford University, Stanford, California. Also in C. Hitch (ed.), Modelling Energy-Economy Interactions: Five Approaches, Resources for the Future, Washington, D.C., 1977.

2. D.O. Wood, and E.R. Berndt, "Engineering and Econometric Approaches to Industrial Energy Conservation and Capital Formation: A Reconciliation", M.I.T. Energy Laboratory Working Paper (77-040WP), November 1977.

3. W.W. Hogan, "Capital-Energy Complementarity in Aggregate Energy-Economic Analysis", in Energy and the Economy, Vol.2, EMF Report 1, Stanford University, Stanford, California, September 1977.

4. T.C. Koopmans, et al.,"Energy Modelling for an Uncertain Future," Report of the CONAES Modelling Resources Group, National Research Council of the National Academy of Sciences, January 1978.

5. Energy Modelling Forum,"Energy and the Economy," EMF Report 1, Stanford University, Stanford, California, September 1977.

DISCUSSION

The ensuing discussion was concentrated on the following points:

- the effect of an excise tax on oil on net national product

- the consequences for international trade

- the benefits of studying the long term as opposed to the short term effects of oil price change

- the applicability of the model to "the real world".

Effects of tax on net national product: The point was made that by placing an excise tax on oil imports one is in effect forcing OPEC to pay taxes to the oil importing country. From this, it follows that the amount of revenue which the government requires to raise from domestic sources in order to balance its budget is reduced and net national product (NNP) should increase.

In his reply, Sweeney maintained that the effect on NNP need not be positive. The way in which the tax benefits were ploughed back into the economy was decisive. Were the benefits given in the form of reduced corporate income taxes, capital formation would be bolstered and NNP would increase. However, if the benefits were reinjected as e.g. subsidies to farmers, distortions already present in the economy would be increased and NNP would be negatively affected.

Consequences for international trade: A change in oil price would affect the international pattern of trade in several ways. Since oil is used as an input for the production of other commodities, the prices of these other goods would change, but not necessarily in the same proportions. Because of substitution possibilities between the different commodities, trade patterns would alter to the detriment of some countries and gain of others. These effects were complex and not included in the model. The important issue was, however, real price change, a concept which was difficult to quantify.

The second major point was that countries faced with a higher oil import bill would have less foreign exchange to buy other import goods with. This could lead, for example, to decreased demand for U.S. exports, a glut of goods on the U.S. domestic market, and economic depression.

The last effect was a short term effect and as such not included in Sweeney's model. His model was a long term model in which it was implicitly assumed that fiscal policies were utilised to maintain full employment.

Short vs. long term: Sweeney's analysis showed in fact that the long run effects of both changes in tariffs and import prices were relatively small. Assuming an aggregate demand elasticity of between 0.4-0.6, the long run effect of tariffs was less than the yearly variations in NNP observed at the present. Even assuming an import price increase from $ 0.8-$2.0 per MMBtu only led to a 5% decrease in NNP.

On this background, the point was made that it would be more beneficial to the economy to study short term (5-10 year) phenomena which can have quite severe repercussions on the economy. Much of the debate on energy policy was centred on immediate problems (i.e. employment effects in the coming 1-2 years). Some policy issues could, however, be addressed by the model; for example whether or not one should subsidize oil extraction from tar sands and shales. The real impediment to studying short term dynamics was the lack of an accepted theory on the dynamics of the economy in the short run.

The idea of extending the model to include OPEC pricing as a function of elasticity of demand in oil importing countries was also proposed as a refinement to the model. This would highlight the need for a good theory for OPEC's behaviour.

Applicability of the model: Doubt was cast on the idea of extending a marginal (differential) analysis - as Sweeney's model was in fact based on - to the analysis of large changes in the long run. In the long term (Sweeney's model was run to zero) both the technology, the nature of the production functions, and the composition of the economy would change. The relationships in Sweeney's model were really only true for small changes in energy price etc. and these could easily be lost in the "noise".

The model assumed perfect competition and profit maximization. These assumptions were not always relevant in the "real world" and there was therefore a danger that decision makers could be misled by an oversimplified hypothesis.

Sweeney's model was static and did not address the dynamic problems implied in part of the question. Even though demand elasticities change as a function of price, one could get a good idea of the effects from Sweeney's model as demand elasticity was a parameter in the results presented.

As regards the criticism of assumptions on perfect competition, Sweeney asserted that the relevant question which the model should address is whether or not changes in the energy system make the economy optimize better.

THE NORWEGIAN ELECTRO-INDUSTRY: A STUDY OF THE ECONOMIC IMPACTS OF ENERGY USE

L.K. Ervik and D.W. Nunn

The Chr. Michelsen Institute
Department of Science and Technology
Bergen, Norway

ABSTRACT. Historically, the Norwegian electro-industry has consumed about one third of the country's annual electricity production. Rising electricity prices and growing public pressure to conserve many of the rivers not already a part of hydroelectric schemes has brought this industry's future role in Norway's industrial picture into question. This paper describes an analysis of the future development of the electro-industry under different assumptions about Norway's economic and industrial policy.

1. INTRODUCTION

Globally, there are in principle two ways of reducing the ratio between energy use and economic growth measured as GNP:

- the energy productivity of individual production processes can be improved

- consumption patterns can be progressively altered toward products using fewer joules per dollar sales value

For a single country, a further option exists:

- produce less energy intensive products domestically

In the Norwegian energy debate this last option has been at the fore. The reasons are to be found in Norway's peculiar pattern of energy use and a vocal conservation lobby.

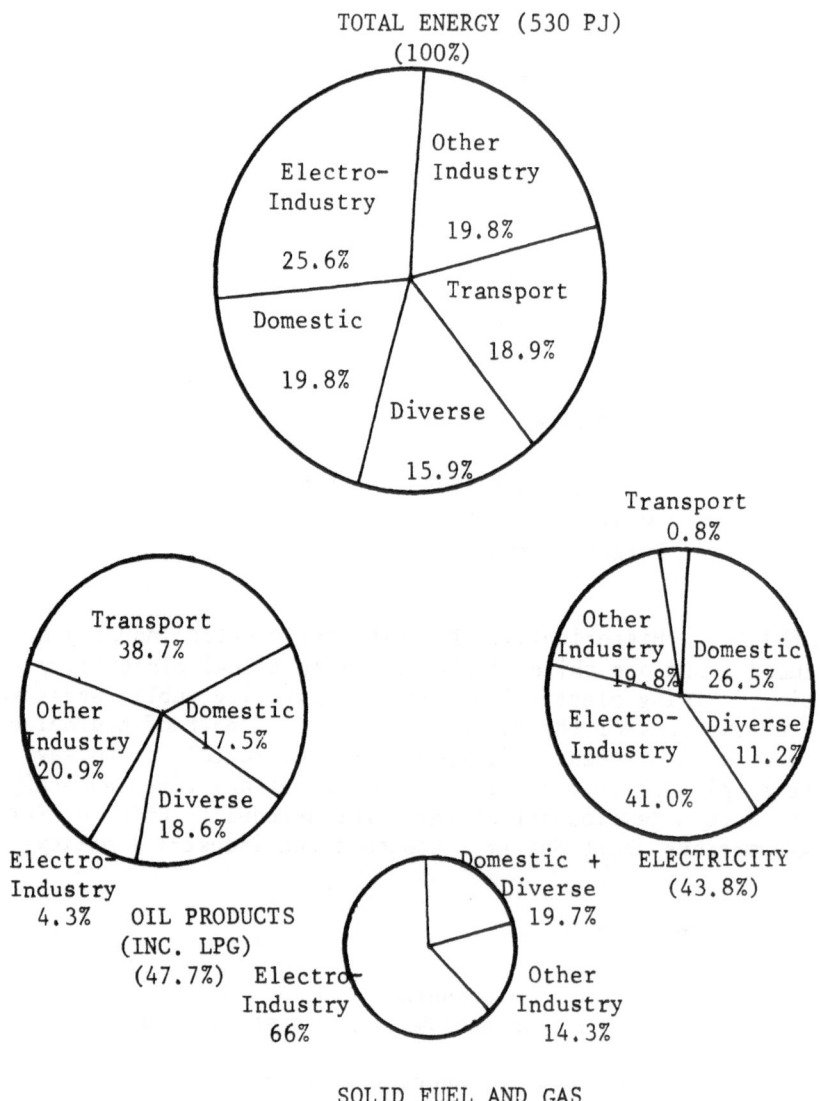

FIGURE 1 - Breakdown of Norwegian energy consumption in
1975 based on the theoretical energy content of
the energy carriers. "Diverse" encompasses services, commerce, public use, agriculture, and
fisheries.

Source: Energy balance sheets, Central Bureau
of Statistics.

Norway meets her primary energy needs from solid fuels and gas 8.5%, oil 47.7%, and hydroelectricity 43.8%. 41.3% of the national net electricity consumption is used within industries which are especially energy intensive's production of primary aluminium, ferro alloys, iron and steel, magnesium, zinc, copper, nickel, silicon, calcium and silicon carbides, and production of ammonia based on electrolysis.

The environmental lobby is very vocal in its desire to hold future environmental costs to a minimum. The aims of this group are to oppose the regulation of all river systems which are not already a part of a hydroelectric scheme, and to prevent the construction of thermal power stations, both fossil-based and nuclear. This group argues that Norwegian electricity consumption is already high enough (at almost 19 000 kwh/capita, Norway has the highest annual electricity consumption in the world) and that by improving the use made of the existing electricity supply, more generating capacity is unnecessary. Because the electro-industry is a large electricity user and because rising power prices have cast doubt in some people's minds on the long term profitability of this industry, the future role which the electro-industry should play in the Norwegian economy has been brought into question.

The electro-industry has long been the flagship of Norwegian industrialisation, so to shift the industrial emphasis away from this industry towards others has understandably aroused sharp reaction from many Norwegians. Historically, hydroelectricity provided Norway with a springboard into the industrial era. Just after the turn of the century, the technology became available to harness its indigenous waterfalls and to utilise the power so generated in electro-thermal, and later, electro-chemical processes. In the 1920's Norway produced 12% of the world's aluminium, twice its present market share.

The electro-industry was also established as a foreign exchange earner. In 1975 it still accounted for 28% of Norway's industrial exports and 16% of the country's foreign currency income. Its percentage of total export earnings has been relatively constant during the past twenty-five years, but this will change in the future due to the boom in oil exports.

The electro-industry's gross domestic product can be used as an indicator of its contribution to the generation of national wealth. The electro-industry's gross product has tripled between 1950 and 1975 from 1.2 to 3.5 thousand million kroner (constant 1978-kroner). It has maintained about a 10% share of total gross industrial product over the twenty-five year period. This can be seen in relation to a share of total industrial employment which has increased from 4% to 5% and a share of total industrial electricity consumption which has increased from 33% to 60% over the same period.

Thus far in the seventies, the electro-industry has contributed 2-3% to the total Norwegian GNP.

The industry has in fact been actively used by earlier governments in their regional development policy. The industry's high growth rate during the past 25 years has meant that although productivity has risen annually on an average by 5.7%, employment has also been able to increase in absolute terms. This trend has, however, changed over the last five years because of lower industrial growth. Employment has now stagnated.

TABLE 1 - Growth of production in the electro-industry and Norwegian industry as a whole 1950-1976. Source: [1].

	1950	1955	1960	1965	1970	1973	1976
ALUMINIUM T/YEAR	47.000	75.000	168.000	280.000	522.000	623.000	618.000
INDEX	100	160	357	596	1.110	1.325	1.314
FERRO-ALLOYS T/YEAR	138.000	189.000	276.000	455.000	580.000	692.000	858.000
INDEX	100	137	200	330	420	501	622
IRON AND STEEL T/YEAR	167.000	294.000	863.000	1200.000	1505.000	1545.000	1565.000
INDEX	100	176	517	719	901	925	937
OTHER NON-FERROUS METALS T/YEAR	62.000	85.000	106.000	128.000	165.000	194.000	158.000
INDEX	100	137	171	206	266	313	254
CARBIDES T/YEAR	34.000	52.000	74.000	175.000	228.000	167.000	163.000
INDEX	100	153	218	515	671	491	479
NORWEGIAN INDUSTRIAL INDEX	100	124	152	201	250	282	290

Limitations in the technology for electricity transmission before the sixties meant that the electro-industry had mainly to be built up in company towns in regional Norway where the hydroelectric potential was located.

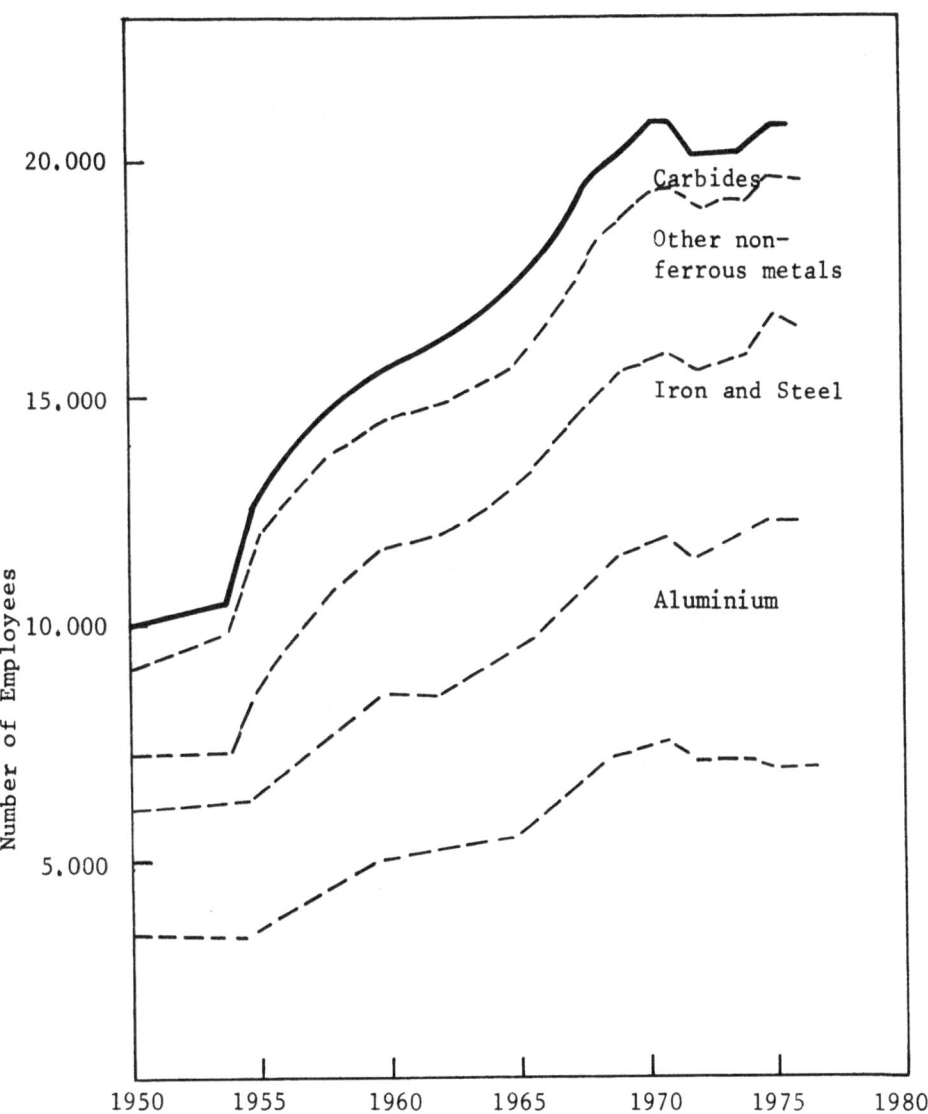

FIGURE 2 — Employment in the electro-industry 1950-1976.
Source: [1].

It was against this background that The Chr. Michelsen Institute, together with the Resource Policy Group in Oslo, were given a contract by The Royal Norwegian Council for Scientific and Industrial Research to look into the future of the Norwegian electro-industry.

2.   METHOD OF ANALYSIS

Although the Research Council provided the funding for the project,
the "clients" most likely to make use of the research results were
not other research workers, but decision makers within the public
sphere. Research results had therefore to be presented in such a
way that they were accessible to people active in the decision
making process governing industrial and energy policy, i.e. members of parliament, government departmental staff and members of
the various factions involved - industrial interests, trade unions,
conservationist groups, regional interests. It should be mentioned
at the outset that the variables governing the amount and price of
electricity which is available to the electro-industry are under
parliamentary control. We are thus dealing with a partly regulated
and not a free market situation. At an early stage in the project,
the decision was taken to present the results in the form of a
popularised book rather than a more traditional research report
[1].

The aim of the research effort was to develop a set of scenarios
each of which gave a consistent picture of the electro-industry
in the future were the policies of a given interest group to be
followed. The goal of the work was thus not to recommend what
course of action ought to be followed, but to analyse the consequences of a set of possible paths which could be followed. This
approach is more in line with the desires of Norwegian politicians
who have often complained about the lack of broad analyses which
also look at the long term consequences of political decisions.
The project was thus designed to structure the information base
for a part of the forthcoming debate on Norwegian energy policy
which is scheduled for the autumn 1979 session of parliament.
The future role of the electro-industry will play a major part
in this debate.

2.1   Project Reference Group

The first rule for any research effort of this sort which hopes
to be of use to decision makers, is to involve the client at the
earliest possible stage in the work and to continue the dialogue
until the work is over - see [2] . The problem when dealing with
macro social systems is that there is no clearly defined client
to converse with, just the spectrum of interest groups and decision makers described above. This problem was solved by defining
as our user a reference group comprising of 14 representatives
of the different interest groups that affect or are affected by
the development of the electro-industry. The group was drawn from
persons of stature with long practical experience and included two
former cabinet ministers.

During the early stages of the project (the first six months)
meetings were held once a month. These stages of the project
covered problem definition, identification of the strategies to
be explored and the primary simulation of consequences.

The agenda for the meetings was deliberately wide ranging with the
purpose of generating unprepared discussions with the following
objectives:

- The reference group discussions serve to focus the
  simulation study on the main issues of relevance,
  on which possible strategies are the most interesting,
  and on which variables are the most important indicators of the success or failure of a proposed strategy.

- The reference group serves as a readily available and
  versatile information source supplying both quantitative
  and qualitative information about how the real world
  works. In addition, the group facilitates access to
  other information sources.

- The reference group serves as a discriminating sounding
  board for the tentative hypotheses, syntheses, models
  and reports of the analysts.

- The reference group members serve as an effective distribution channel for study results into their respective interest groups.

- The reference group serves to give the results of the
  study enhanced credibility which is of help in the
  early dissemination stage.

2.2   Use of Simulation Modelling

Parallel to the discussions in the reference group, a simulation
model of the aggregate electro-industry was developed. The model
dealt with production of metals and carbides, but excluded ammonia
production based on electrolysis. Apart from other things, the
market situation for fertilizer which ammonia is used to produce
is very different from that of primary metals. (Carbides are used
for welding and production of abrasives so their market outlook
is similar to that of metals.)

The simulation model was not the primary focus of the research
effort, but an aid to be used to check the consistency in assumptions on market price, profitability in the industry, the number
of jobs in the industry as well as its power requirements, and
the extent and cost of hydro power development. The main elements

in the model are outlined in Fig. 3. The model is documented in a technical report [3].

The model proved to be useful when the consequences of following the different strategies were discussed in group meetings.

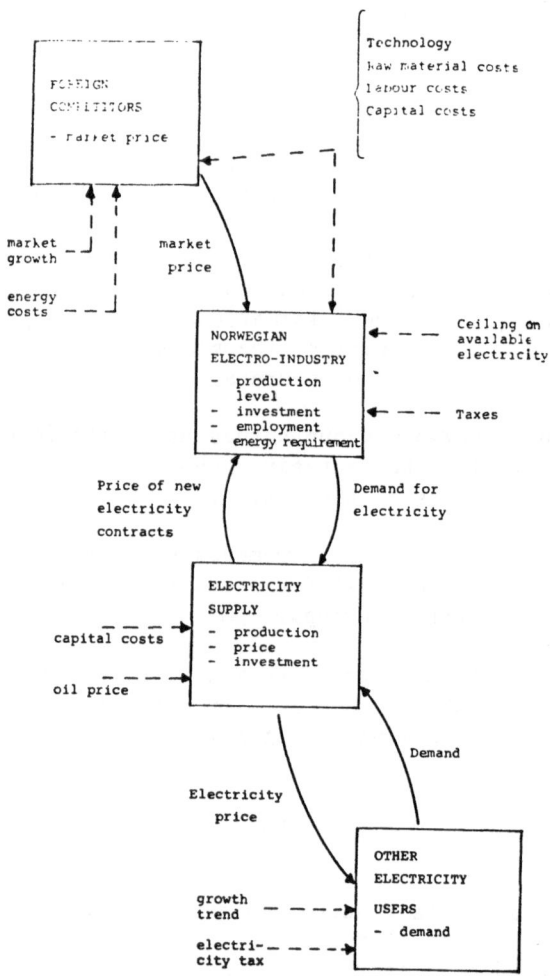

FIGURE 3 - Outline of the model structure. Dashed lines indicate exogenous input.

2.3   Choice of Scenarios

The aim of the research effort was to identify a set of scenarios
each of which could be said to be representative for the goal of
a particular interest group. A list of the scenarios which were
finally included in the analysis is given in Table 2. After identifying the interest groups' main goals, the next stage was to
identify the policies which could realize these goals. There are
often several policy options available to achieve a stated aim.
The consequences of achieving this aim will differ according to
which path is chosen.

For example, the employees within the electro-industry wish to
maintain the number of industrial jobs in the areas in which the
electro-industry is already located. There are three ways of
accomplishing this goal:

- increase primary metal production capacity at the
  same rate as productivity increases

- increase production of semi-finished or finished
  metal products

- diversify and begin production of other industrial
  products.

It is apparent that each of these strategy alternatives will have
a markedly different effect on future electricity requirements.

Six variables were chosen as principal indicators of the consequences of following a given strategy:

- primary metal production capacity
- number of jobs
- average electricity price paid by the electro-industry
- profitability (gross return on investment)
- remaining fraction of hydroelectric potential
- installed thermal power generating capacity

The development of these variables was simulated until the year
2020. The reason for choosing such a long time span was the long
term contract structure which the industry has for supply of electricity. Historically, the industry has been able to obtain long
term electricity contracts, usually for a period between 20-40
years. Concessions which allowed companies to produce hydroelectricity themselves were often given for a 60-year period, although
some of the first hydropower schemes were built before the law
requiring concessions for hydropower production was passed. Few
contracts expire before the middle of the 1990's, and the industry is in fact guaranteed 20 TWh/yr in the year 2000 - 76% of

its consumption in 1976. By the year 2020, only the production
rights which the industry itself owns will remain available to
it unless contracts are renewed as they expire - see Fig. 4.

TABLE 2 - Outline of the scenarios explored in the
study

GOAL	STRATEGIES	CHANGES IN MODEL
II Better the electro-industry's operating conditions	- reduction of tax burden	- tax on labour and investment eliminated - new power contracts to marginal production cost (7% interest-rate)
IIA Maintain electro-industry as a leading export earner	- increase primary production	- contracts renewed to original price on expiry, new contracts to marginal cost based on 5% interest rate
III Maintain employment in regions with electro-industry today	- increase primary production - increase production of semifinished products - establish other industries	- all power priced to average production cost
IV Maximize value of the country's hydro resources	- price elctricity to marginal pro-duction cost	- price all electricity consumed to marginal production cost by gradually increasing electricity tax until 1990
V Stabilize electricity consumption	- place ceiling on amount of power available to electro-industry	- ceiling on power availability; present level of power prices frozen
VI Protect remaining rivers from hydroelectricity development	- reduce power consumption in electro-industry to cover growth in other electricity users - sell surplus power to other electricity users. Income is invested in alternative industry	- all remaining non-developed rivers conserved - no further investment in electro-industry - the industry maintains ownership of its power contracts and sells surplus power to the grid

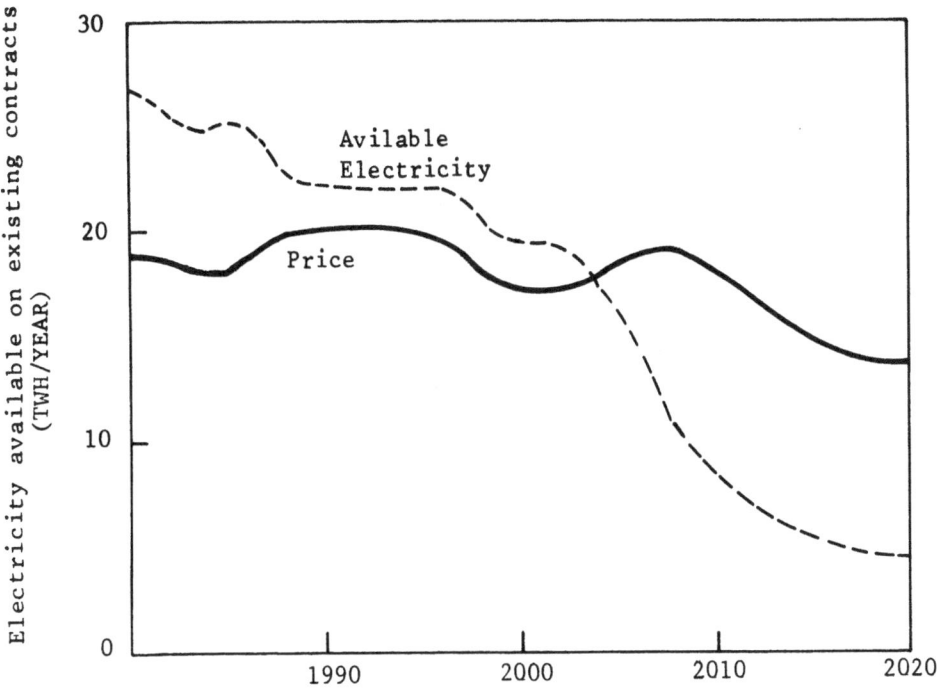

FIGURE 4 – Electricity available to the electro-industry
on contracts signed before 1978
Source: [1].

The contract structure which the electro-industry has already se-
cured can exert a very stabilizing influence on future dynamic
behaviour unless measures are taken to adjust the price which the
industry must pay for its power. Such a measure might be to increase the electricity duty which is levied on each kilowatt-hour
used.

3. ANALYSIS OF THE RESULTS

The reference scenario (scenario I in Fig. 5) shows a gentle increase in the industry's production capacity until the middle of
the first decade in the next century when capacity is drastically
reduced. The buffer which existing power contracts afford the
industry means that it is able to pay relatively high prices for
marginal increases in its total power consumption in the short
term. Although this development leads to gradually falling profits, margins are still high enough for a modest increase in production capacity to be attractive. After the turn of the century,

many existing contracts expire and the industry loses much of its buffer against higher energy costs. The resulting lack of profitability causes the industry to close down.

The reduction in the electro-industry's electricity demand comes too late to postpone the development of the remaining hydroelectric schemes or the construction of thermal power stations. This is because demand from other electricity users has grown in the meantime. In the model, we assume the demand from other users follows the forecast from the Electricity Board modified by a price elasticity.

Attempts to ameliorate the situation by reducing the taxes on labour and investments allow the industry to expand in the short term (scenario II in Fig. 5). The industry can easily afford to pay the relatively high prices for new power contracts and is able to continue along its historical pattern of expansion until the end of the century. The total abolition of labour and investment taxes cannot, however, from that point counterbalance increased energy costs which result from so many power contracts expiring.

3.1   Electricity Pricing is the Crucial Policy Decision

An analysis of the results of the different scenarios shows that the strategy which the government adopts as regards pricing of electricity to the electro-industry is of cardinal importance. Were the government to adopt a marginal pricing policy in which all the electricity which the industry consumed was continually priced to the average cost of producing more electricity from a new power station, the industry's international competitive edge would soon be eroded (scenario IV in Fig. 5).

The reason for this is that the remaining hydropower projects which have not yet been constructed have increasing capital costs. As capital costs amount to between 80 and 90% of all production costs, the cost at which hydropower can be produced approaches the cost of thermal (oil) power. The development in marginal production cost assuming that the cheapest hydro schemes are constructed first is shown in Fig. 6. In the scenario shown here, we have assumed that the alternative to producing hydropower is to generate electricity in an oilfired power station with oil priced at $ 20/bbl (1978 dollars).

Were the government, however, to give highest priority to regional employment and sell power to the electro-industry at a rate reflecting the mean production cost in the whole electricity supply system, the industry would be able to continue a strong competitive position (scenario III in Fig. 5). The fundamental

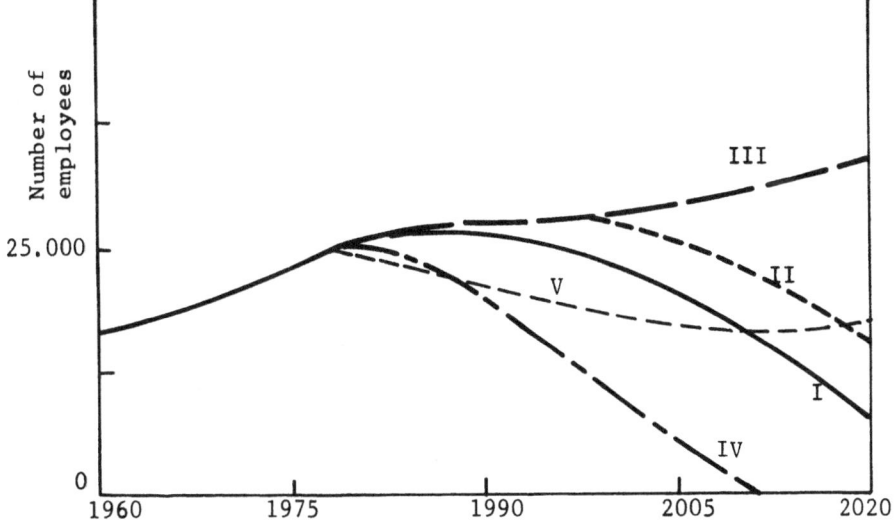

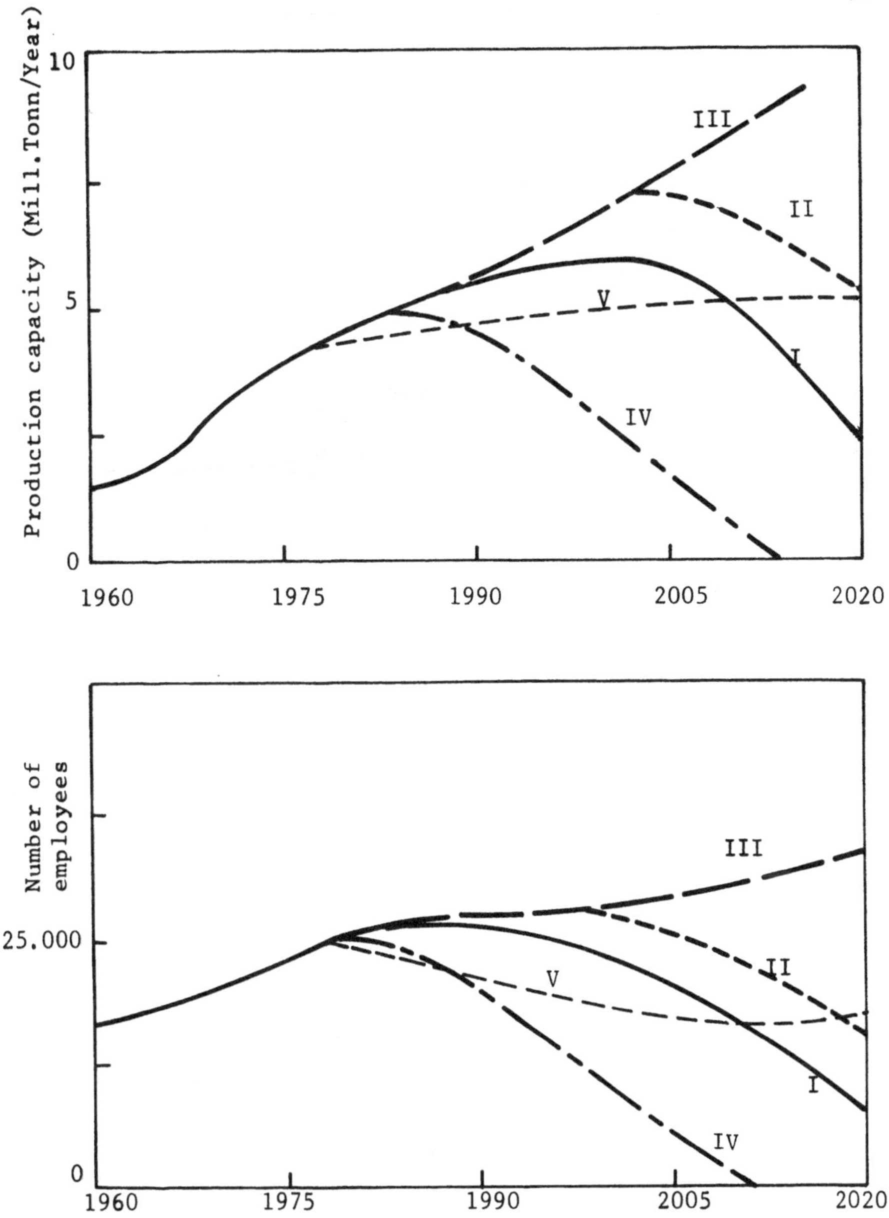

FIGURE 5 - Extract of results from scenario analysis.
The scenario numbers refer to Table 2.

I Reference scenario
II Better economic conditions - elimination of taxes on labour and investments
III Maintain employment - plentiful supply of cheap power assured
IV All electricity priced to marginal cost after 1990
V Limit placed on future electricity consumption
VI All remaining hydro potential conserved

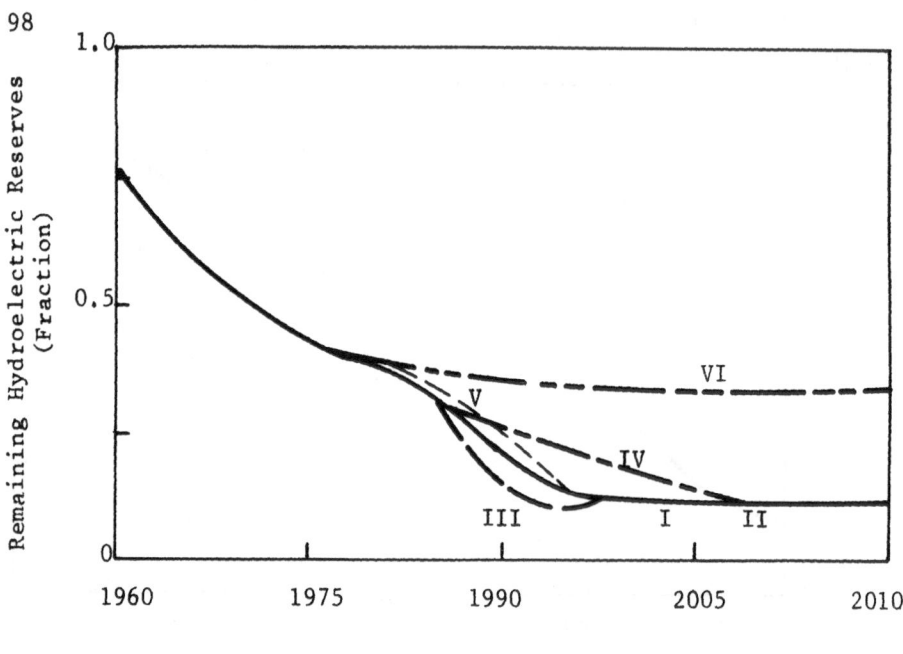

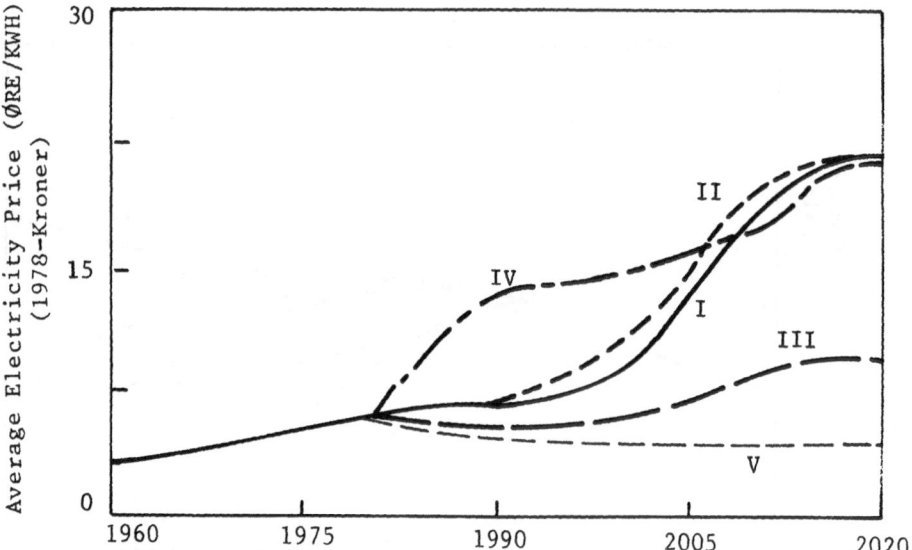

Figure 5 continued....

I   Reference scenario
II   Better economic conditions – elimination of taxes on
        labour and investments
III   Maintain employment – plentiful supply of cheap power
        assured
IV   All electricity priced to marginal cost after 1990
V   Limit placed on future electricity consumption
VI   All remaining hydro potential conserved

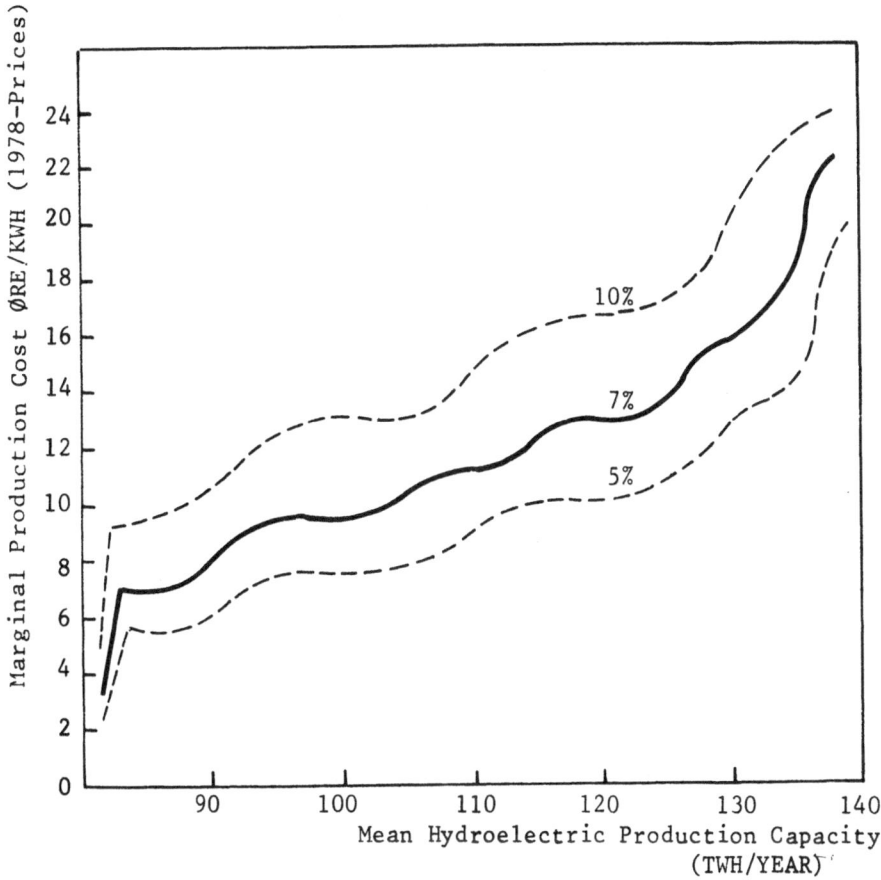

FIGURE 6 - Marginal production cost of remaining hydroelectric schemes as a function of cost of
capital (interest rate) and mean production
capacity.
Source: [4].

assumption in our model is that the Norwegian electro-industry
sells its products at a price fixed on the world market. The
level of the Norwegian industry does not affect price fixation,
but the ratio between the market price and production costs in
Norway determines capacity utilisation and, in the longer term,
the level of investment.

This last point brings us to the fundamental issue in the entire
debate. The cost of producing power from existing hydropower
plants falls over time because the decline in capital costs more
than compensates for increasing variable costs. This means that

although the marginal cost of producing electricity may be high
in the year 2000, the average cost will be low. Compared with a
country which produces electricity solely from oil-fired generating plants, the hydro cost advantage will annually amount to ca.
$ 800 per capita in the year 2000 (1978 dollars).

The crucial issue is thus, who should reap the benefits of this
cost advantage: the regions where the hydropower is located, the
electro-industry (which also provides substantial employment in
these regions) or the country at large? The days when the only
way of utilizing hydropower on a large scale was to smelt primary
metals are gone. Advances in the technology of high voltage electricity transmission have made it possible to sell electricity
to other users in the cities or even abroad who are willing to
pay more for their electricity than the electro-industry can possibly pay and remain solvent.

3.2   Jobs Versus Free-flowing Rivers

There is a clear conflict between the aims of both the unions and
local interests of maintaining employment in the electro-industry,
and the aims of environmentalists who wish to limit increases in
future power consumption. The scenario involving maintaining
employment levels in primary production leads to total development
of the remaining hydro potential within the early 1990's and
a comprehensive construction programme of thermal power plants
if also the projected demand of other electricity (scenario III
in Fig. 5) users is to be met.

Conservation can be achieved by a stringent use of marginal pricing which would cause the electro-industry to close down as well
as damp demand from other users. Conservation can also be achieved
by freezing the electro-industry's electricity consumption and by
a comprehensive programme of energy saving in other sectors. The
aim for the other sectors should then be to limit the increase
in their electricity consumption to 2% annually until the year
2000 with no increase after that. We will not expound on the
plausibility of following this last course of action now.

4.   CONCLUDING REMARKS - THE RELATIONSHIP BETWEEN GNP AND
     ENERGY USE

The simulation model used in this analysis did not include a calculation of the Norwegian GNP consistent with the assumptions on
which the different scenarios were based. This will be done at
a later stage using a multi-sector national economic model which
is at present under development. However, it is possible to make
some general comments as to how the different development paths
outlined in Fig. 5 will be reflected in the GNP.

The first point is that in all the scenarios involving development of the remaining hydropotential, the need for investment in
the electricity supply sector will rise from its present level
of 2% of GNP (6% of total national investments), to around 4% in
1990. This is clear from an inspection of the estimated capital
costs of the remaining hydro schemes. After the introduction of
thermal power plants, the combined yearly expenditure on plant
investment and fuel costs will remain at this level.

The electro-industry is relatively capital intensive itself, so
any reduction in its growth would potentially liberate quite large
amounts of capital which could then be productively employed in
other less energy intensive sectors of the economy.

The challenge which the country's industrial policy then has to
face is to find growth sectors which can productively employ this
capital and which ideally provide job opportunities in the regions
where the contraction of the electro-industry will be felt most.
As many of Norway's other traditional industries such as ship-
building and manufacturing of pulp and paper are in economic difficulties at the present, the solution to this problem is by no
means apparent.

At present, the political consensus is centred around a scenario
involving establishment of no new electro-industrial plants, but
providing existing firms with the electricity necessary to carry
out modernization plans. This will necessitate a moderate reduction in employment, because productivity is expected to rise faster than capacity, but no greater than can be covered by natural
wastage (scenario V in Fig. 5). To further accommodate environmental interests, emphasis is being focused on saving electricity
amongst other users.

This paper has examplified the use of a combination of scenario
and simulation techniques to analyze the economic impact of energy
use. A country can, if it so desires, reduce the ratio between
energy growth and GNP growth by systematically changing the composition of its industrial structure. Whether or not a country
decides to opt for this course of action will depend on how much
emphasis policy makers place on the various national objectives
which comprise their decision function. Energy consumption is
after all not an end in itself, but a means of achieving national
political ends.

REFERENCES

1.   L.K. Ervik, C. Tank-Nielsen, D. Nunn and J. Randers, "Smel-
      teverkene i Smeltedigelen", Cappelens Forlag, Oslo, 1979.

2.   J. Randers, "The Potential in Simulation of Macro-Social
      Processes or How to be a Useful Builder of Simulation Models",
      GRS 111, Resource Policy Group, Oslo, 1977.

3.   D. Nunn, E. Moxnes and L.K. Ervik, "Den kraftintensive
      industri i framtiden: Beskrivelse av simuleringsmodell SMELT",
      Chr. Michelsens Institute, Bergen, 1979.

4.   J. Owens, "Hydroelectric Energy: The Cost of Conservation
      in Norway", GRS 121, Resource Policy Group, Oslo, 1978.

DISCUSSION

The point was made that a scenario in which a transition to marginal electricity pricing was assumed did not reflect a relevant
strategy. Rather it was better to price electricity to its shadow
price. Nunn replied that the use of shadow prices would give the
same result as the use of marginal prices. If the Norwegians wanted to maximize the present value of their hydro resources, they
should place a moritorium on the construction of more hydro schemes
until the market price had risen to the cost of producing more
hydroelectricity. In the long run the shadow price and the marginal price would be the same. In the short term they would differ, of course, and it would be correct to use the shadow price.
This was done in the model.

There were dangers in analyzing one sector of the economy separate
from the rest. The investment requirements in the electro- and
electricity supply industries could have a significant impact on
the rest of the economy, just as a phasing out of the industry
could provide problems for the balance of payments and for employment. These dangers were, however, recognized and it was
planned to link the sector model with a model of the complete
Norwegian economy later in the year. In the meantime, the effects
were discussed qualitatively and found not to be excessive in
comparison with other changes which have taken place in the Norwegian economy of the past few years. In the most "drastic" scenarios, the industry was phased out over a 25 year period. The
employment effects were not great in macro though they could of
course be severe locally.

There was no accord to put extra emphasis on the electro-industry
because it was a foreign exchange earner. The relevant criterion
was whether or not an industry was profitable in a macroeconomic
sense. The proviso was that resources which were "released" by
the electro-industry should it be run-down could be productively
employed in other sectors of the economy.

The question of the international division of labour was also raised. Would it not be better to let developing countries develop their own bauxite reserves and produce aluminium? Arguments both for and against this attitude had been put forward in Norway and there was no consensus on whether or not it "helped" the rest of the world by phasing out the electro-industry there.

PART II

NATIONAL ENERGY SYSTEMS

Simulation Techniques in Energy Analysis

A Dynamic Optimization Model for Energy
Policy Analysis

About the Use of an Energy Optimization
Model

Modelling Energy Demand for a Fast Growing
Economy

Integration of Optimization and Simulation
Models

SIMULATION TECHNIQUES IN ENERGY ANALYSIS

K. Schmitz, W. Terhorst, A. Voss

Programmgruppe Systemforschung und Technologische
Entwicklung (STE), Kernforschungsanlage Jülich
Federal Republic of Germany[*]

ABSTRACT. Simulation is one of the most frequently used techniques
in energy modelling. After some general remarks on the nature of
simulation models, a more detailed description of a large scale
dynamic energy simulation model for the Federal Republic of Germany
is given. The paper continues with a discussion of some model results and concludes with some brief remarks on the limitations of
the simulation approach.

1. INTRODUCTION

The last two decades have seen the emergence of what has come to
be known as the "systems approach". The systems approach is a
methodology and a practical philosophy of how best to aid a decision maker with complex problems of choice under uncertainty.
This approach, at first successful in military and managerial
contexts, has now become widely used in many fields, and energy
policy and planning are no exception. An important step in the
systems approach is the development of models, using them as an
appropriate framework for searching out objectives and alternatives, and comparing them in the light of their consequences. Model building is one way to understand complex relationships within
a system. A model is always a simplified reflection of reality
and can be conceptually regarded as a substitute for the real system. It is used to capture the functional essence of the complex
problem under investigation, but not necessarily the detail of the

---

[*]Programme Group of Systems Analysis and Technological Development of the Nuclear Research Center of Jülich.

whole real system. A model permits experimentation among alternative policy strategies and can illuminate their consequences. To the extent that the model is an appropriate representation of the system and problem to be analysed, it can be a valuable aid to policy analysis and policy making.

In view of these potential benefits, it is not surprising that during recent years there has been growing interest in using models to help plan our way out of the energy problem facing mankind today. The International Institute of Applied Systems Analysis has published a set of very useful review reports of energy models [1].

The 144 models analysed and classified so far range in scope from models for a single fuel to those of the whole energy supply system. They include models related to the energy and economy interactions, on a national as well as on an international scale.

A variety of methodologies are used in the different models. Most employed are econometric methods, simulation, linear programming, and I/O techniques.

Today, there is no common agreed definition of a simulation model. In the following, simulation models are referred to as a special class of mathematical models which express the dynamic relationships among the variables and parameters of the system modelled. Running a simulation model results in the calculation of changes in the state of the system through time.

Simulation models are dynamic models. Simulation models may be classified as predictive models; while they help to answer questions of the type, "What will happen, if...?", whereas optimization models belong to the class of normative models answering questions of the type, "What should be done, in order to achieve a desired goal?". Within the class of simulation models usually a distinction is made between "deterministic" and "probabilistic" models. In deterministic models it is assumed that the exact values of all variables can be computed, whereas in a probabilistic model, at least some variables have an unpredictable randomness, and must be represented by a probability distribution.

During the last few years a number of special purpose simulation languages have been developed. These languages are generally thought of as easier to learn and apply, by simplifying the programming of simulation models. However, special purpose simulation languages have somewhat limited flexibility and range of application compared with general programming languages such as FORTRAN.

After these more general and methodological remarks on simulation models, we will in the following describe in some detail a simulation model of the energy system of the Federal Republic of Germany.

2.   A DYNAMIC ENERGY SIMULATION FOR THE FEDERAL REPUBLIC
      OF GERMANY

During the last years the Programme Group of Systems Analysis and
Technological Development of the Nuclear Research Center in Jülich
(FRG) has developed a dynamic energy simulation model called LESS[*]
[2-7] to be used as a flexible tool in analysing important issues
for the development of the energy system of the FRG. The long
term energy simulation system (LESS) is part of a large energy
model system (JES: Jülich Energy-model System) (Fig. 1) which consists of a set of different energy and energy related models, a
data base, and a method base for linear and nonlinear regression
and correlation analyses [8,9].

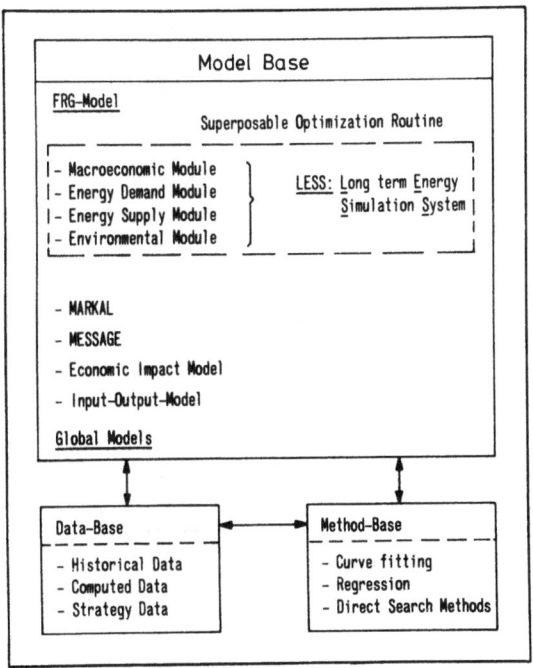

FIGURE 1 - JES-Jülich energy-model system

LESS consists of four modules

      -    a macroeconomic module,
      -    an energy demand module,
      -    an energy supply module,   and
      -    an environmental module.

[*]LESS - Long term Energy Simulation System

which are interconnected by input and output flows as outlined in
Fig. 2. The structure of the four modules will now be described
in more detail.

2.1   The Macroeconomic Module

The reasons for developing a macroeconomic module derive from the
fact that the production and utilization of energy is very closely
connected with the economic development of a nation. Consequently,
future options of the energy demand and supply system cannot be
analyzed and modelled independently from the economic forces within the system. They are always based either explicitly or implicitly upon certain economic assumptions, such as GDP, income, or
capital allocation. Generally there are two ways of covering the
economic impacts within an energy model:

1.   by selection of economic scenarios which provide straight-
     forward input to the energy sector alone;

2.   by utilization of complete models in which the energy
     sector interacts with the economic sector.

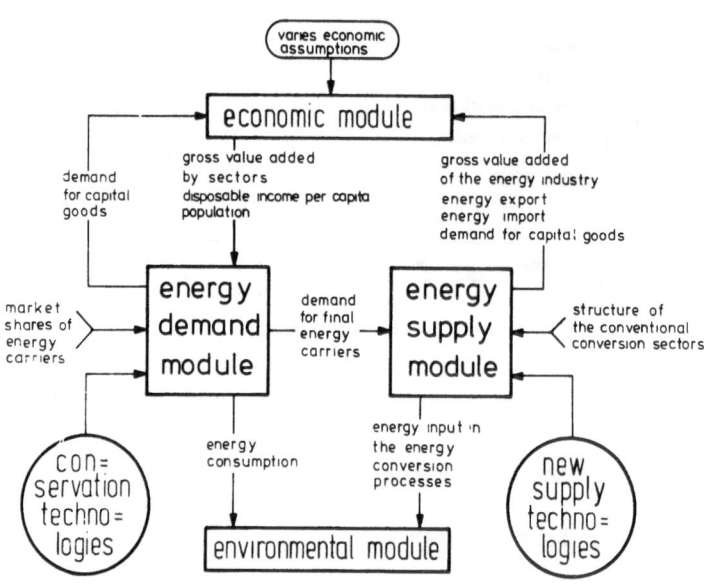

FIGURE 2 - LESS - Long term energy simulation system

LESS makes use of the second approach. A macroeconomic model has been developed which generates the growth in different economic sectors. The sectors have been selected with respect to the energy intensity. They are the four industries (iron and steel production, chemical industries, stone, clay and construction materials, and other industries), the commercial sector and the energy branch. For each branch individually, the growth rate is calculated by means of the allocated production factors (labour and capital), the production outputs and the gross value added. The allocation of production factors is demand driven, i.e. determined endogeneously by intermediate and final demand, but is limited since it depends on a number of constraints such as labour force participation, capital allocation, and intermediate inputs.

In each economic branch, the input of goods which is needed in the production process has to be calculated, as most of these goods have to be produced by the other economic branches. These necessary input goods of a branch are dependent on the production output and on the production structure. In the macroeconomic module, the production structure of a branch is given by technical coefficients, which are the input coefficients of an input-output matrix. As the production structures vary with time, the technical coefficients are also time dependent. With the exception of the coefficients describing the energy input into the branches, these values are estimated exogeneously using input-output matrices and their projections which are available to us but calculated by other institutions. By summing up all input goods, the intermediate demand for the products of each economic branch is obtained.

The components of final demand are determined in the following way. The disposable income of private households is calculated as a function of the gross domestic product, which is the sum of the gross value added of all economic branches. Depending on the development of this disposable income, the consumption of private households is estimated. The collective consumption, that is, the consumption of general government and ¬nonprofit institutions, is given as a function of the gross domestic product. Depending on the development of the production factor capital, the demand for capital goods can be calculated by using a gross-investment matrix. The export of each branch is given as a fraction of its production output. This fraction is an exogeneous value.

Summing up the consumption of private households, the collective consumption, the demand for capital goods and the export, the final demand for the products of each branch is received. The estimation of the primary production factors as mentioned above is dependent on the development of the total demand for goods of the single economic branches, which are calculated as the sum of the intermediate and the final demand. Thus, the economic growth loop is closed.

2.2   The Energy Demand Module

The consumer has a direct requirement for:

- heat,
- light,
- entertainment,
- comfort,
- food,
- transportation,

under the influence of

- economic,
- social, and
- political constraints.

These requirements are transformed to a demand for

- products,
- services, and
- energy.

Therefore it is necessary to distinguish between a direct and an indirect energy demand. From the viewpoint of the final consumer, for example, the energy for space heating marks the direct energy demand, while energy to build the radiators, the boilers etc., is considered as an indirect energy demand.

In a long term view of the industrial sector, the indirect energy demand for building the radiators and boilers equals the direct energy demand for industry.

Starting from the needs of the consumers, we can differentiate between:

- basic energy demand with respect to an applied technical system,
- final energy demand,
- secondary energy demand,
- primary energy demand.

Within the energy demand module, the final energy demand in different sectors is determined by the demand of energy services, e.g. the hot water demand in the residential sector, the persons or goods transport volume in the transport sector, or the production of steel in the iron and steel industry. This is achieved via economic indicators such as the personal disposable income, and the gross value added of the different economic branches. The final energy is determined for the following sectors:

-   the industry (with four industrial sectors corresponding to those in the macroeconomic module),

-   the transport sector, in which the four transportation media - road, rail, water, and air - are distinguished.

-   the residential sector, in which the energy demand for the three purposes - space heating, water heating, and others - are considered.

-   the petrochemical sector (with its socalled nonenergetic energy consumption).

Fig. 3 shows in some more detail the structure of the transport sector as represented in the model. A distinction is made between goods and passenger transport and the different transport modes. The specific energy consumption of the different transportation systems together with their share of the overall transport volume is used to determine the final energy demand by energy carrier of the transport sector.

It should be mentioned that the module allows the computation of the energy savings by introducing energy conservation technologies like:

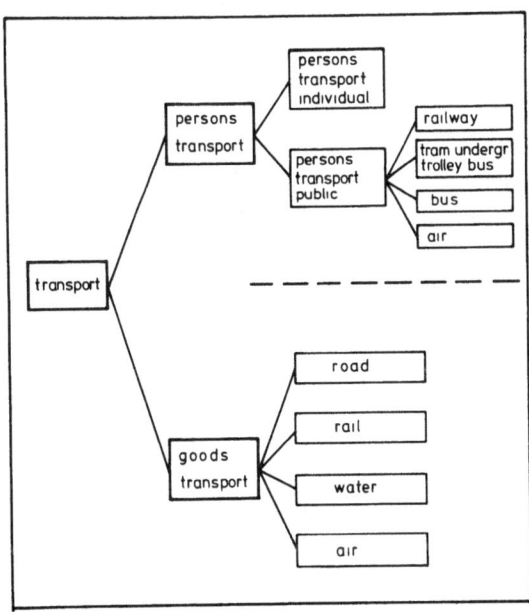

FIGURE 3 - Structure of the transport sector

- better insulation of buildings,
- heat pumps, and
- solar room heating systems.

The output of the energy demand module, i.e. the final energy demand by energy carriers, provides the input for the energy supply module.

2.3. The Energy Supply Module

The energy supply module has two tasks:

1. to calculate the primary energy used to meet the final energy demand,

2. to feedback the employees, the net production and the investments within the energy sector to the macroeconomic module.

For calculating the primary energy consumption, a demand orientated flow model of the mining and conversion processes has been built up. Fourteen energy carriers are balanced with the equation:

$$IP + IM - EX - BU + ST = PC = FC + NC + DL + IT + OT + CT$$

From left to right the following items are taken into account:

Indigenous production (IP) e.g. mining or pit gas, imports (IM), exports (EX), bunkering (BU), and changes in stockpiling (ST), on the left hand side of the primary energy consumption (PC), final energy consumption (FC), nonenergetic fuel consumption (NC), i.e. petrochemical consumption, distribution losses (DL), inputs (IT), and outputs (OT) of all conversion processes and consumption by the energy branch itself (CT), i.e. the self consumption, on the right hand side.

Four mining processes (hard coal, lignite, crude oil, natural gas) and 21 conventional and new conversion processes are considered. The conventional ones are cokeries, gasworks, blast furnaces, conventional steam power plants, light water reactors, and heat production and refineries.

The following new technologies are taken into account: high temperature reactors, fast breeder reactors, and windpower stations for electricity generation; methanol production, gasification of lignite and hard coal in each case, using the conventional methods as well as the processes based on nuclear process heat from HTRs; autothermal coal liquefaction, hard coal combined cycle plants; and, finally, electrolytic and nuclear thermochemical hydrogen production, primarily to meet the hydrogen demand of the iron and steel industry when shifting from conventional steel production to direct reduction of iron.

Each process is mainly characterized by its energy output broken down to the different energy carriers, the inputs, and the self consumption. For the primary energy carriers, the total demand, that means the final demand plus input into transformation processes, plus consumption of the energy sector plus distribution losses plus bunkers, determines the mining up to an upper limit and the net imports as a remainder. For the secondary energy carriers, the total demand and the net imports determine the output of one single, or by market shares the output of two or more, alternative transformation processes, the outputs from other transformation processes being substracted.

The reserve situation, the maximum of production capacity and a share of primary energy consumption which should be covered by indigenous production, determine the gross production. After substracting the self consumption, the available quantity of crude oil and natural gas are computed. Fig. 4 describes the production of crude oil and natural gas.

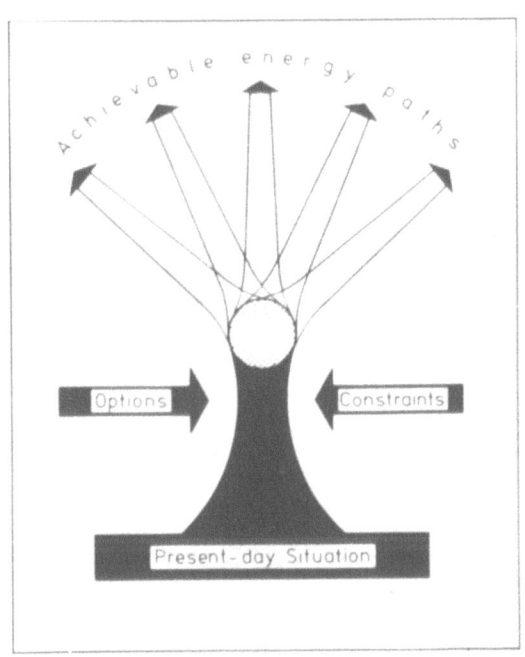

FIGURE 4 - Working out "hard" decisions

As mentioned before, the second task of the energy supply module
is the calculation of the number of employees, the gross value
added, and the investments in order to close the loop economic
growth - energy demand - energy supply - economic growth. The
value of these three factors are coupled in general with the energy
output of the different conversion technologies or with the indigenous production in the mining sector.

2.4   The Environmental Module

The environmental module is an emission model, calculating total
emissions due to energy consumption and conversion by multiplying
the energy inputs with the relevant specific coefficients of emission. The calculation of the annual energy consumption is made in
the energy demand and energy supply module. The specific coefficients are exogeneous variables which for the past are derived
from statistics, and for the future are either kept constant or
changed - in most cases decreased according to different environmental abatement technologies and policies.

3.   OPERATION OF THE MODEL

The model system can be operated in two ways:

-    in an event-oriented way
-    in a decision-oriented way.

Using the first method, a set of reasonable and plausible assumptions have to be defined in order to observe consequences after a
model run in the second, a desired goal has to be defined and
one has to find the conditions to achieve that goal. The first
way consists of a straight-forward computation, the second implies an interactive approach in which normally certain constraints
have to be observed, often caused by large scale global linkages.

Both ways allow the investigation of alternative energy scenarios
for the future so that the energy system in its complex structure
becomes transparent with respect to simulated external events and/
or interferences caused by defined energy policy targets. Even
the next decision to be taken can be evaluated in its future consequences and a minimum number of necessary actions and decisions
which will fit into a flexible energy future can be outlined. What
is practicable and what is controllable can be made apparent and
an incentive is given to necessary structural changes and innovations. Fig. 4 makes this procedure evident. Within a spectrum
of lines of reasonable developments of the energy system under
various conditions and assumptions - economical, technological,
environmental - it is possible to define a set of decision steps

which have to be taken to keep open as many options as possible, especially with regard to the long term security of energy supply and economic welfare. This set of decisions can be seen as located within the circular area of Fig. 4 and represents the minimum of decisions to be taken which will fit into each decision chain symbolised by the branching paths defining alternative paths of development of the energy system.

In contrast to traditional outcomes of energy economic planning which usually provide, as a decision aid for future planning, a so-called energy prognosis and which consists of relatively clear statements presenting a future pattern of development, this new way of producing decision aids for energy planning does not rely exclusively on one future development line, but takes into consideration the possibility and probability of alternative developments. In this way, a fixed decision sequence which will preclude a secure energy supply for different courses of events can be avoided. The certainty that at least a minimum of decisions which have to be taken are reliable and "hard" will reduce some of the uncertainties which are always inherent in the planning of the future.

4.   SOME RESULTS

The model system has recently been operated in the ways mentioned above making possible the elaboration of future, feasible energy paths for the Federal Republic of Germany. The conditions for realization of these paths revealed for the national energy policy some important facts which up to now had not been recognized. Policies for conservation of energy and the necessary development and market penetration of new technologies could be formulated. Besides the projections of future requirements of energy in the various sectors of the whole economy, it was possible to identify priorities for R&D, in particular, by making apparent "hard" decisions - in the sense described above - which have to be taken. In the following, a summary outline of the main results [5] is given.

-    The total economic growth in Germany is and will remain - at least in the short and medium term - the main determinant of energy consumption. Since the current dominant political goals assume a desire to solve problems of unemployment and other social and political problems by means of economic growth, a further increase in energy demand is to be expected. The present "glut" in the energy is an effect of the present economic situation and therefore should not lead to any wrong conclusions regarding long-term developments.

118

- The future increase in energy demand depends not
only on economic development but also on the patterns of consumption in energy demand and supply.

Fig. 5 shows the expected development in primary energy consumption under the assumption of medium economic growth - defined as
3-2-1 case, that means an average growth of 3% per year from 1975
to 1985, 2% per year from 1985 to 2000 and 1% per year from 2000
to 2010 - and extrapolation of the historical structures of energy
demand and supply.

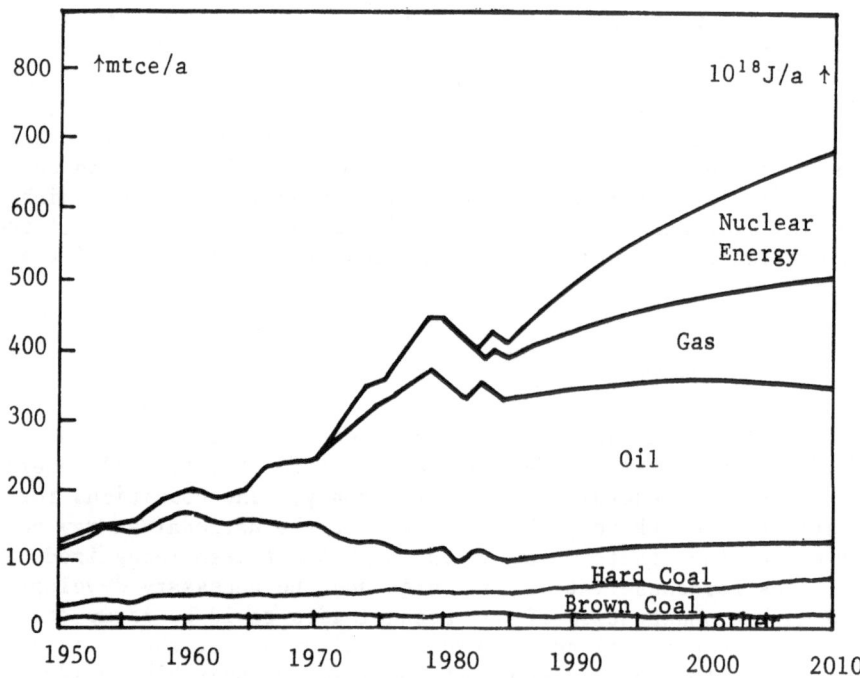

FIGURE 5 - Development of the primary energy consumption
(Trends on continued scenario: 3-2-1 case)

Under these assumptions, the results indicate an energy demand
which in 2000 is about 60% and in 2010 about 75% more than that
of today.

- The primary energy demand structure is impossible
to realize, because of the bounds for the possible
availability of individual energy carriers, as seen
in Table 1. That is very much the case for mineral
oil as Fig. 6 shows. The lower curve of mineral
oil imports up to 2010 would correspond to the development of primary energy consumption under medium
economic growth shown in Fig. 5.

TABLE 1 - Possible bounds to future energy availability
in the Federal Republic of Germany (10^6 tce/a)*

Energy carriers or sources	Of which	Year 1976	Year 2000	Up to the year 2010
Hard coal	Domestic production Net import	89.6 - 12.4	< 100 < 40	→ →
Brown coal	Domestic production Net import	36.3 1.2	< 45 -	→ —
Mineral oil	Domestic production Net import	8 194.9	- < 200	↑ ↓
Natural gas	Domestic production Net import	20.1 31.0	< 15 < 100	↓ ↘
Uranium/Thorium	Net import	7.9	< 250	→
Renewable energy sources(incl. heat pumps)	Indigenous	-	< 50	↑

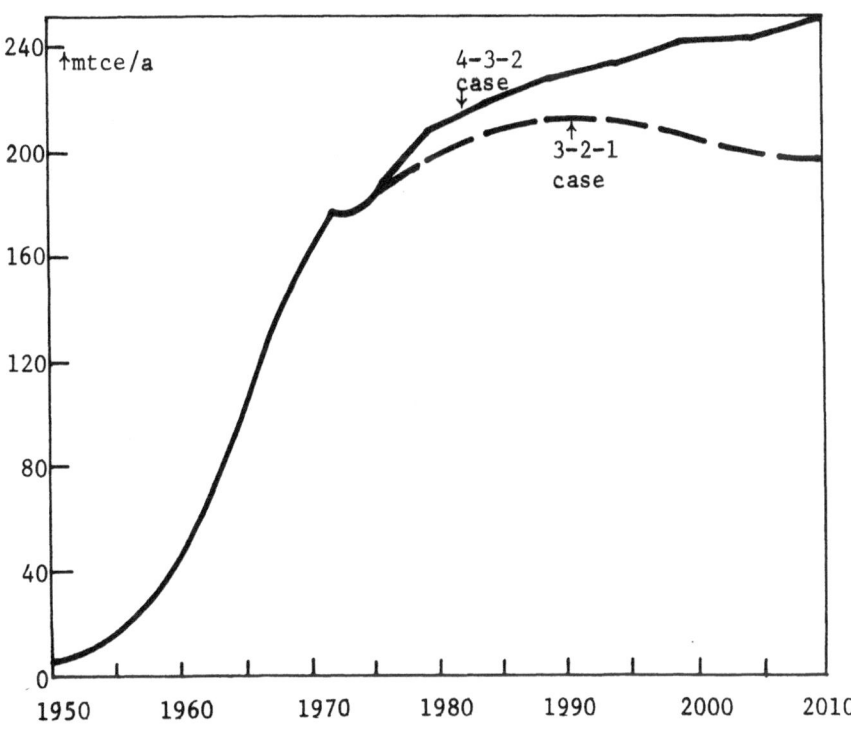

FIGURE 6 - Development of the net mineral oil import

*Tons coal equivalent / annum

- Higher economic growth rates, meaning one more percentage point per year (defined as 4-3-2 case), would
make this situation even more serious as the upper
curve in Fig. 6 shows.

- To minimize such a risk, immediate action must be
taken. Otherwise the race against time could be
lost to the disadvantage of the economy as a whole.

- The measures to be taken should be limited to single
sectors. Thus neither energy conservation nor coal
can solve the problem alone. A long term set of
planned measures is therefore required which will
combine to ensure a secure "energy future".

- To keep open the option for a politically and energe-
tically secure rate of economic growth in the long-
term, the following particular steps, e.g. "hard"
decisions, must be taken:

i. Energy saving must be supported and carried out
with more urgency if an effective contribution
is to be expected. This effort must extend to
all sectors of the economy. These last statements should become clear as the results of calculations are presented shortly, which were based
on the assumptions shown in Fig. 7.

In the residential and commercial sector, all
buildings should be fitted with better insulation
by the year 2010, such that in 2010 50% of all
buildings for human habitation would be equipped
with improved means of heat insulations. Solar
collectors and heat pumps would be implemented
to the lower levels but only in suitable buildings,
mainly detached and semi-detached houses.

It was further assumed that in the industrial and
transport sectors energy savings are made which
are shown in Fig. 8.

In Fig. 8 we see the assumed development of the
conservation factor in industry (separately for
fuel and electricity), and in the transport sector.
The values are based on an estimate of realizable
and mutually supporting measures.

Based on such a list of desired measures which,
although of considerable scope, are still realizable,
the following results were attained in the various

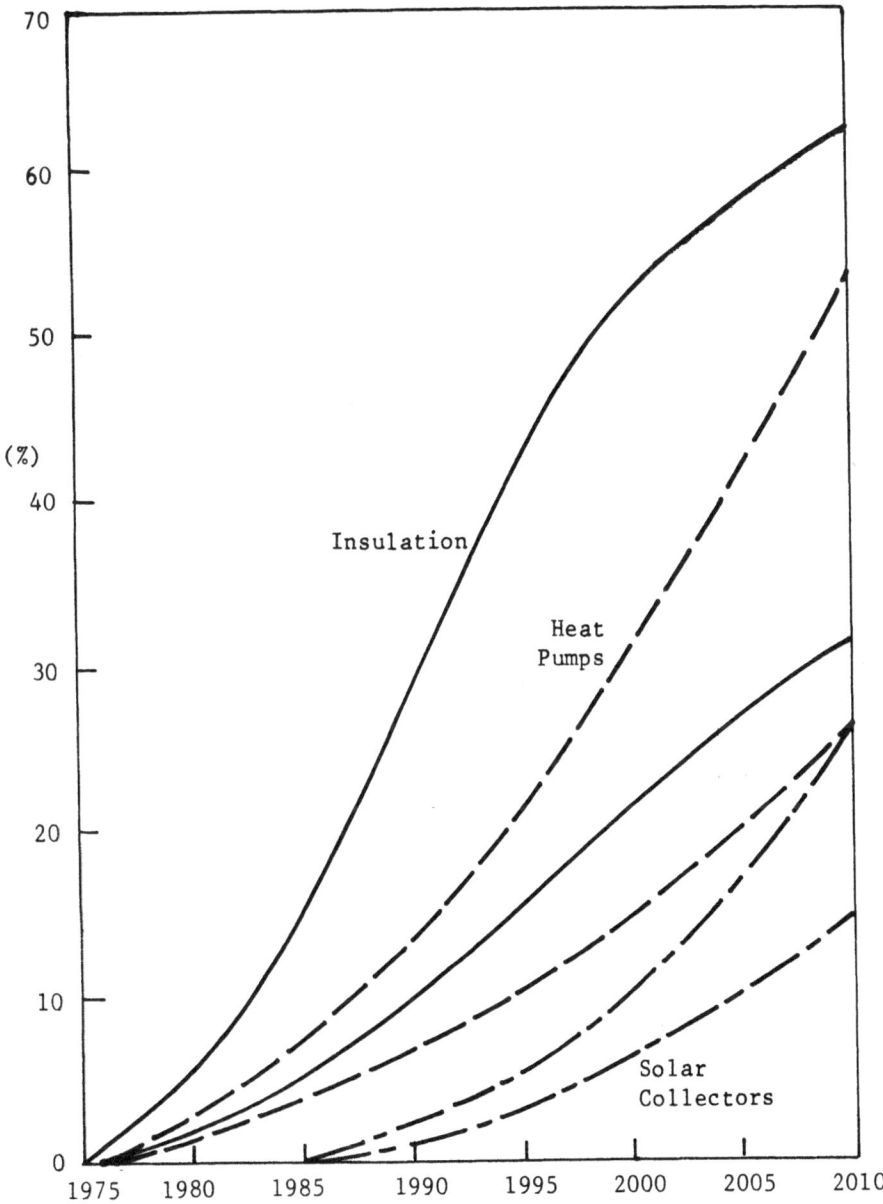

FIGURE 7 - Upper and lower levels of application of
alternative measures of energy conservation

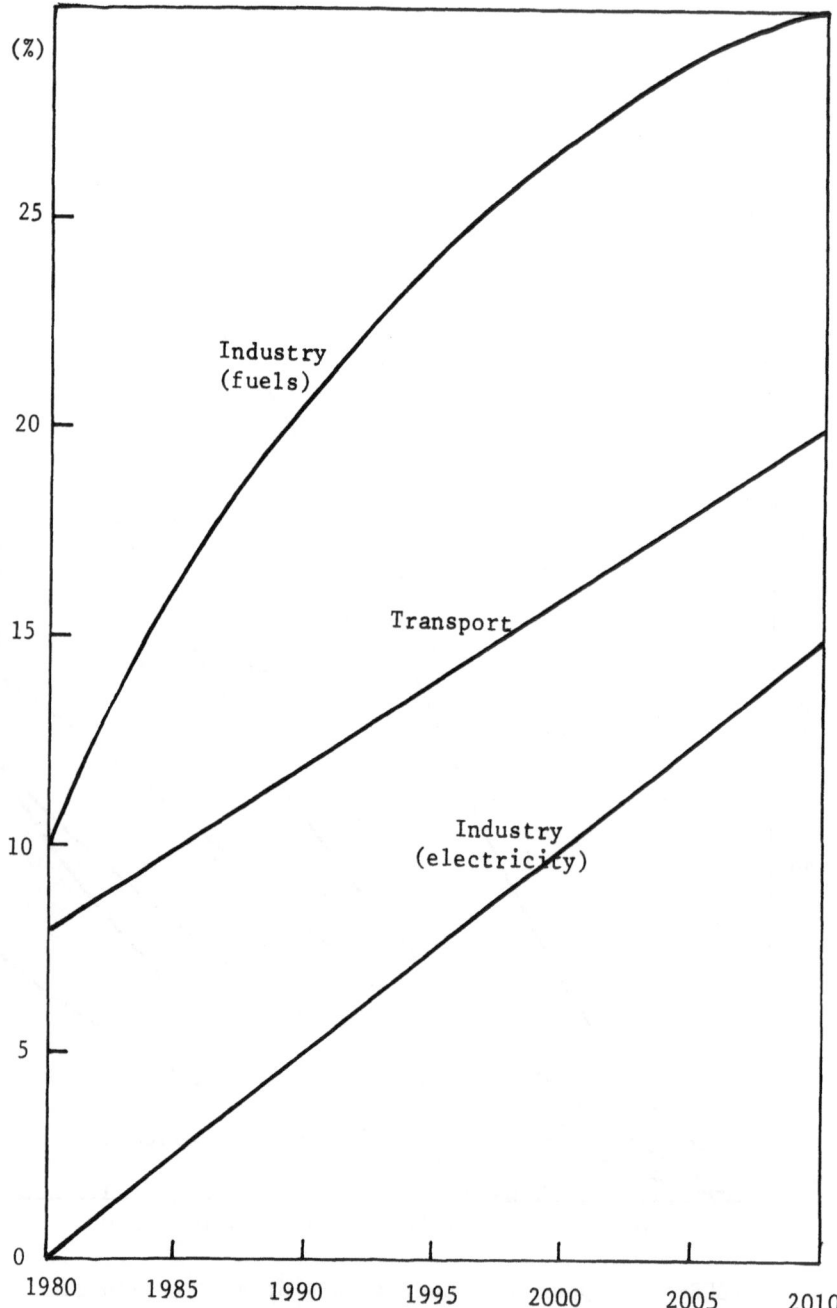

FIGURE 8 – Development in the conservation factors in
industry and transport

branches of the economy and society (See Table 2).
Considering only private households, 2% energy
savings in end energy or 1.5% in primary energy
could be achieved. This is due to the fact that
in the historical trend case (Fig.5) energy conservation measures due to present regulations are
already taken into account. If we include commercial users, which is the sector which provides
services, 6.5% end energy or 6% primary energy
savings would be possible. However, the figures
20% and 16% for the case of a comprehensive energy
saving strategy in private households, commercial
users, industry, and transport, make it clear that
all sectors of the economy and society must be
involved in order to make a significant total contribution to energy saving. Nevertheless it is
obvious that this contribution is not sufficient
in the context of secure energy planning if we consider that an underestimate in the long term economic growth of just one percent (compared to medium growth) would be enough for the effect of the
energy savings achieved to be cancelled out.

TABLE 2 - Possible energy savings with alternative
conservation measures in the year 2000.

SECTORS INCLUDED	ENERGY SAVING END ENERGY (%)	ENERGY SAVING PRIMARY ENERGY (%)
PRIVATE HOUSEHOLDS	2.0	1.5
PRIVATE HOUSEHOLDS AND COMMERCIAL USERS	6.5	6.0
ALL SECTORS OF THE ECONOMY	20	16

TABLE 3 - Development of the primary energy consumption
in the 3-2-1 case (4-3-2 case)

			PRIMARY ENERGY $(10^6$ TCE)					
Time	Hard coal	Brown coal	Mineral Oil	Gas	Nuclear Energy	Hydro and other	Consumption Σ	
1985	81 (85)	40 (40)	210 (225)	83 (87)	30 (42)	10 (11)	454 (490)	
1990	88 (92)	41 (41)	215 (233)	97 (105)	56 (75)	12 (12)	509 (558)	
2000	92 (101)	42 (42)	208 (241)	118 (138)	109 (160)	14 (15)	583 (698)	
2010	96 (110)	42 (43)	204 (245)	125 (152)	165 (248)	15 (18)	647 (816)	

In Table 3 once again are the results for primary energy consumption in those scenarios which in a sense are an extrapolation of
historical trends and which differ in their economic growth rate
by 1%. We see that for the year 2000 the primary energy consumption turned out to be 583 and 698 mtce. 16% savings as compared
with now would almost exactly correspond to this difference.

-   Further steps, e.g. "hard" decisions, to maintain a
    politically and energetically secure long term rate of
    economic growth can be formulated as follows:

ii. The present production capacity for coal must be
    maintained at all costs because large quantities
    of coal must be available in the long term.

iii. A long term build-up of coal fired power stations
    should only be allowed if substitution of nuclear
    electricity is necessary. Otherwise coal should
    be used for conversion to other products to be
    able to solve the problems in the oil and gas mar-
    ket. So it seems necessary around the end of the
    1980's not to undertake a further construction
    of coal for other purposes than electricity pro-
    duction.

    In a programme of coal gasification and liquefaction
    plant construction, care should be taken that the
    coal power stations still in operation are supplied
    with fuel for the remainder of their lifetimes from
    the available coal. This point is illustrated for
    the case of brown coal in Fig. 9. A halt to the

125

building of brown coal power stations from 1988
would mean that the fuel use in the remaining
power stations would have to fall off according
to the curves shown. An agreed programme of allo-
thermal (nuclear) or autothermal brown coal gasification plant construction takes account of this,
as Fig. 9 shows. This means that the building
of brown coal gasification plants is such that the
difference between the available brown coal, represented by the curve for the trends continued
scenario, and the coal used in power stations remaining after the halt in construction, can be taken
for gasification.

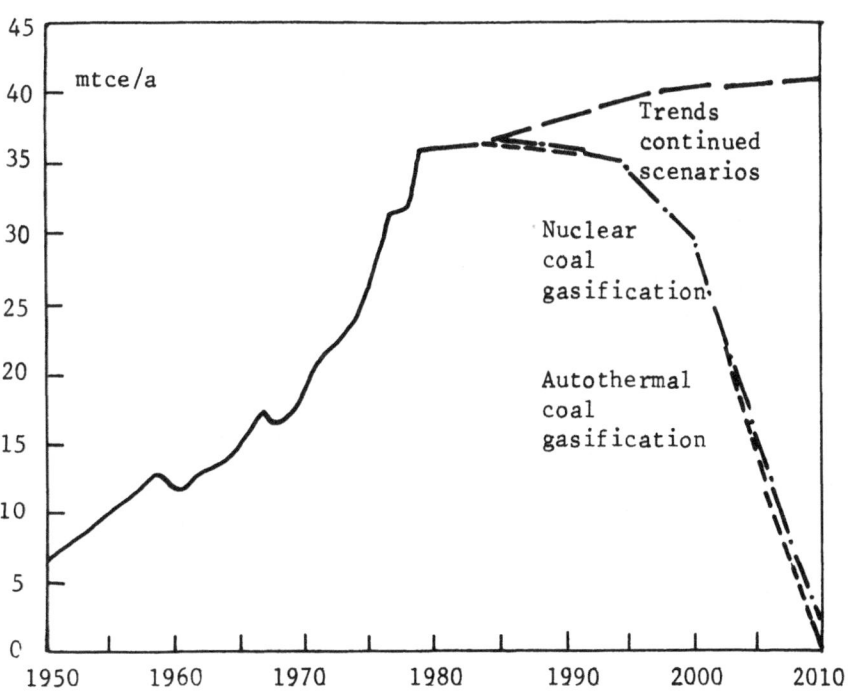

FIGURE 9 - Brown coal input into power stations

iv. Nuclear power must have a firm place in the energy
supply of Germany, both for electricity production,
and for supplying process heat. Without nuclear
energy, economic growth could be reduced even in
the medium term. In Fig. 10, we see the possible
economic development based on the energy availability

in the case of a limited amount of 30 GW nuclear
from the year 1992. A strong deceleration of
economic growth could be the consequence. One
might even expect negative growth rates in the
long term, in spite of the fact that accelerated
energy conservation measures and processes to
substitute for nuclear electricity were simulated
in the calculation.

v. Finally:
The steps referred to should be further supported
by other measures which permanently displace mineral oil from the end energy sector. Stronger im-
plemention of electricity and district heat is one
possibility, but also the products of coal conversion such as gas, methanol and heat are suitable
substitutes.

By simulating this set of measures, a path of development for the
energy economy could be worked out which may be seen as realizable. Fig. 11 shows the corresponding primary energy consumption.

5. SIMULATION MODELS: VALIDATION AND LIMITATIONS

Finally, it seems necessary to make some remarks regarding the
validation of simulation models, and the frontiers of model application.

Regarding the structure of a model, the question immediately arises whether or not the model is a reliable representation of the
real world system in its behaviour. This problem of model validation is an important task within model development. There is
no doubt that a procedure to obtain a complete validation does
not exist. But there are possibilities to analyse at least partially the validity of a model by use of special tests based on
plausibility considerations. Reasonable test procedures are as
follows:

-   The rational and logical inquiry into the model struc-
    ture with respect to relevant influence factors and
    reasonable relations between variables.

-   The reproduction of system behaviour in the past. A
    positive outcome of this test, however, does not prove
    the reliability of the model. Moreover, this test re-
    quires a great amount of time and data.

-   Investigation of the model behaviour in exceptional
    or extreme environments. Incorrect model relationships
    may lead in such cases to illogical and unexplainable
    results.

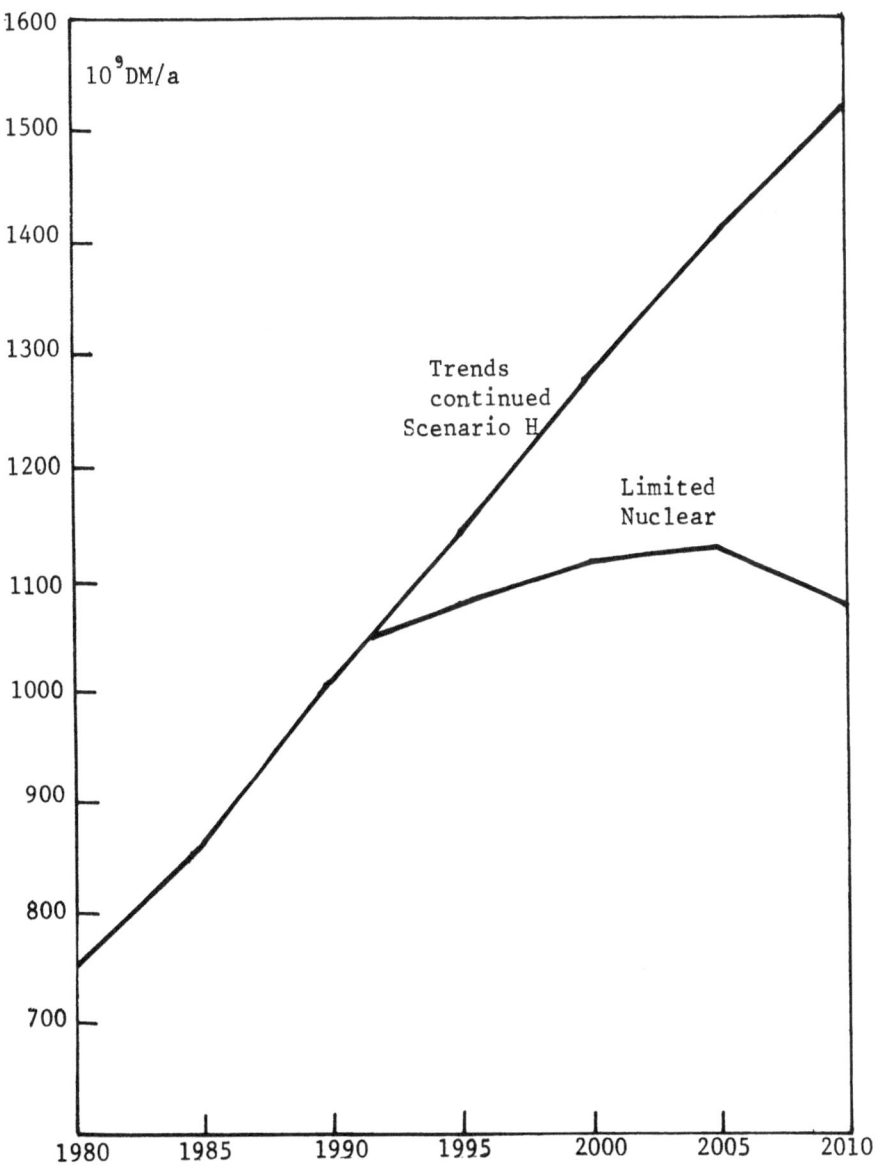

FIGURE 10 - Development of the gross national product

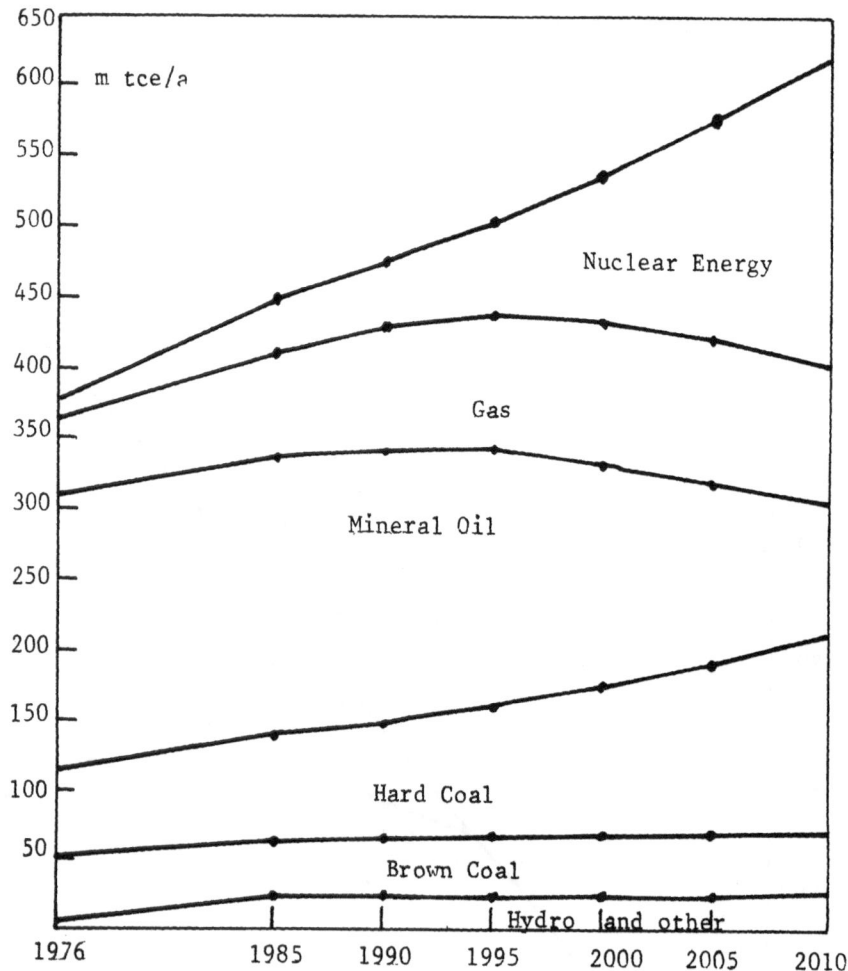

FIGURE 11 - Development of primary energy consumption

Despite these limitations, at least a partial validation is a necessary and essential part of any model development task.

Energy models offer energy decision makers a promising means to achieve a better understanding of the problems and choices before them. To develop the potential of this decision aid and to take advantage of it, it is extremely important to be aware of its limitations. It is true that a mathematical model:

  — forces a precise statement of the problem and objectives and requires an in-depth study of the system being described, as a rule resulting in a better understanding of the system,

- offers a framework within which experiments can be conducted and the consequences of alternative decisions and actions can be analyzed, and

- is able to handle a mass of data.

But it should be kept in mind that a model is not reality; it is always only a simplification of the real system it represents.

REFERENCES

1. J.M. Beaujean and J.P. Charpentier,"A Review of Energy Models No.4", July 1978, RR-78-12, International Institute for Applied Systems Analysis, Laxenburg, Austria, 1976.

2. St. Rath-Nagel, "Alternative Entwicklungsmöglichkeiten der Energiewirtschaft in der Bundesrepublik Deutschland", ISR 28, Birkhauser Verlag, Basel und Stuttgart, 1977.

3. A. Voss, "Ansatze zur Gesamtanalyse des Systems Mensch-Energie-Umwelt", ISR 30, Birkhauser Verlag, Basel und Stuttgart, 1977.

4. A. Voss, et.al., "Dynamische Energiemodelle als Planungs - und Entscheidungshilfe dargestellt an einem Energiemodell für die Bundesrepublik Deutschland", in: Energiemodelle für die BRD, Ed. by Ch. König, ISR 42, Birkhauser Verlag, Basel und Stuttgart, 1977.

5. K. Schmitz, "Langfristplanung in der Energiewirtschaft", ISR 65, Birkhauser Verlag, Basel und Stuttgart, 1979.

6. K. Schimtz, H.P. Schwefel, "Finding Reasonable Energy Policies by Means of a Dynamic Simulation Model", in Proceedings of the International Symposium Simulation '77, ACTA Press, Zürich, 1977.

7. U. Schöler, et.al., "A Dynamic Energy Model for the Countries of the European Communities", EUR 5953, Commission of the European Communities, Brussels, Luxembourgh, 1978.

8. F. Drepper, "A Data-Model Interface for Modular Dynamic Simulation", Winter Simulation Conference, Miami Beach, Florida, 1978.

9. H.P. Schwefel, R. Heckler, "Superposing Direct Search Methods for Parameter Optimization onto Dynamic Simulation Models", Winter Simulation Conference, Miami Beach, Florida, 1978.

DISCUSSION

Voss was questioned on his assumption of fixed market shares for
the different supply technologies in the simulation model. He
replied that although the market shares were fixed for a given
simulation, they were obtained from an optimization model and
manually transferred to the simulation model, so that in effect
an iteration process was occurring.

Several participants raised points on the feedback between the
energy supply model and the national economic model. If consumers' energy bills were reduced through greater energy conservation they would have more money to spend on other goods, how was
their increased spending power accounted for in the model? Voss
replied that increased demand for better insulation etc. was taken
care of in the model via the input-output matrix, which was exogeneous and static. There did not, however, appear to be a consumption function which accounted for the secondary effects eluded
to in the question. Voss was asked if they had computed the price
elasticity implicit in the model. This had not been done.

Another question centred on how increased investments and labour
requirements in the energy supply sector affected the rest of the
economy. Voss replied that if more labour were required in the
energy sector, less would be available to other production activities. In the case of investments, increased demand for investment goods from the energy sector would cause other sectors of
the economy to produce more to meet this demand. The input-output
matrix provided the check on consistency. There was, no capital
constraint, but a capital feed-back in the model. Neither was
there any feed-back to wage-rates of rate of interest as a result
of increasing scarcity.

Voss was asked about his criteria for model validation. These were
not statistical, he replied, but were based on subjective judgement of goodness of fit for the past 15 years of data. Much of
the work that had been put into the model (about 80%) had gone
into validation. There, the main problem was sudden changes and
fluctuations in the past.

In response to other questions, Voss explained that methanol was
produced from coal; the environmental module was simply a pollution model; and that no depreciation of the capital stock was
included.

A DYNAMIC OPTIMIZATION MODEL FOR ENERGY POLICY ANALYSIS

İ. Kavrakoğlu

Boğaziçi University
Bebek, İstanbul, Turkey

ABSTRACT. An optimisation model has been developed primarily for
the planning of the national energy system. It is also shown,
however, that the same model can be used as a tool for decision
analysis.

The dynamic linear programming mini-model includes four primary
energy sources and three forms of secondary energy. Decision
variables include investment as well as production activities in
the refining, conversion, and transmission of energy. The model
contains 186 columns and 226 rows. The objective function is the
discounted costs of fuels, investments, operating costs, and under-
utilised capacity. The model has been used to analyse decisions
in the Turkish energy sector for the years 1960-1975, and possible
developments for the period between 1980-1995.

Through the use of this model, the underlying assumptions concerning the import prices of petroleum have been analysed, as well as
the impact of a delay in the construction of a large hydro project.
Furthermore, the effect of the shadow costs of foreign currency,
the consequences of rapid industrialisation, rapid economic growth,
and substitution among fuels have been investigated. The model is
currently being used for the analysis of possible nuclear strategies for Turkey.

1. INTRODUCTION

The energy sector has certain characteristics and dimensions that
makes its study as part of the national economy an interesting
undertaking. The size and complexity of the national energy system

also calls for tools that are suitable for such a study. Mathematical models of the energy system find increasingly more use as tools for policy analysis and decision-making.

The primary objective of this paper is to demonstrate the use of a mathematical model for decision analysis in the energy sector. The model, which has been applied to Turkey for the fifteen year period between 1960-1975, also sheds light on the underlying assumptions that shaped energy policy for these years. Specifically, the questions to which answers have been sought are as follows:

    i.   What were the predictions by the planning authorities concerning petroleum prices?

    ii.   How did the 4-year delay in a large hydro project influence decisions?

    iii.   Would it have been desirable to have idle capacity in the supply subsectors?

    iv.   How would the shadow price of foreign currency have affected energy policy decisions?

    v.   What would have been the consequences of faster economic growth?

    vi.   What would have been the effect of substitution among petroleum and coal?

As shown in the following, other questions concerning the future development of the energy sector can also be answered through the use of the model.

2. MODELLING THE NATIONAL ENERGY SYSTEM

It is possible to model the national energy system in many different ways. The important factors that will influence the final choice; i.e., the level of aggregation, the method of approach, type of mathematical representation of the relations, etc. are largely dependent on the characteristics of the system as well as on the type of policy decisions that are to be analysed. The approach taken here is to keep the model as small as possible while retaining all the elements that are needed for a meaningful analysis of the system as a whole. Furthermore, since the model will be applied to Turkey, it should reflect all likely developments in the energy sector of the country.

The energy flow diagram of the model is shown in Fig. 1. The primary energy sources include only oil, coal, nuclear and hydro potential. Other sources such as natural gas, geothermal, and solar energy are used in negligible quantities. Noncommercial sources like fire wood and cow dung are utilized locally and are not really influenced by policy decisions.

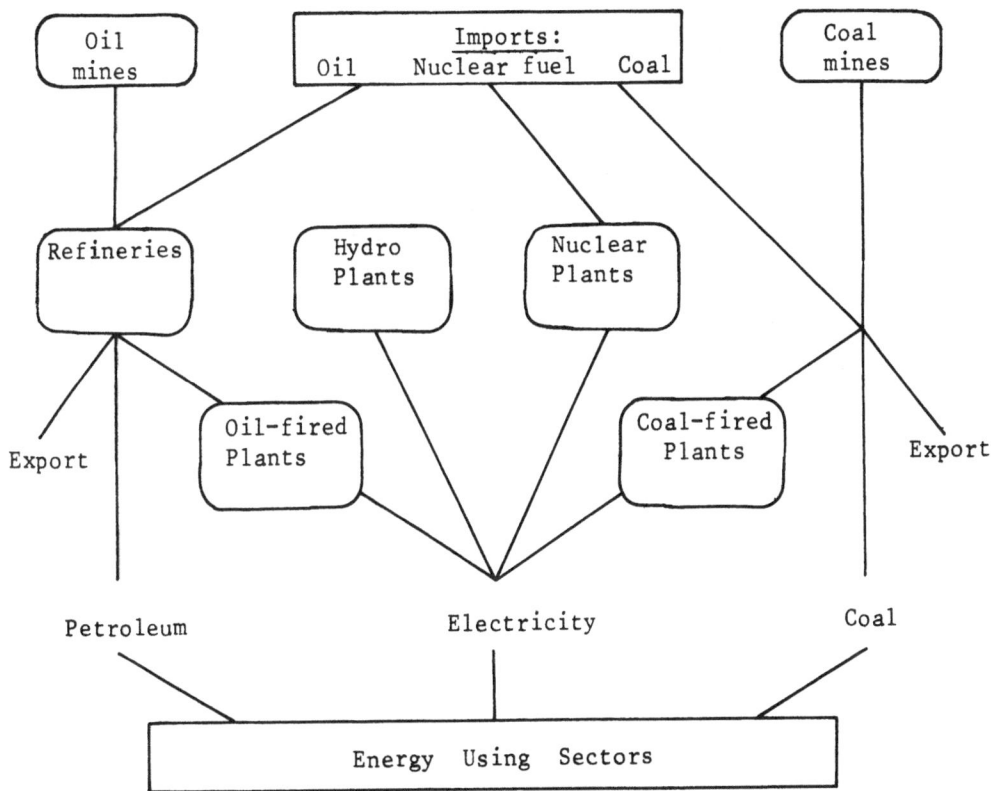

FIGURE 1 - Energy flow network

Major conversions take place in the refining of petroleum and in the generation of electricity. Transportation of petroleum or coal and the transmission of electricity have associated energy losses. The final form of energy reaches the consuming sectors as solid fuels, liquid fuels, and electricity.

Investment activities are the ones shown in boxes in Fig. 1, namely: coal and oil mining; refining facilities; power plants utilizing coal, petroleum (fuel oil and diesel oil), nuclear and hydro sources; and transmission of electricity.

The time horizon is 15 years, with another six years for end effects. There are five planning periods of three years duration.

The decision variables of the model for each period are: amount
of oil, coal and nuclear fuels imported; amount of locally extracted oil and coal; new capacity added in oil and coal mining; petroleum refining, each type of power station, and transmission network; production levels of refineries and power plants.

Fuel demands are treated exogenously, as are other quantities that are
decided on outside of the energy system. The import prices of
fuels and energy technology equipment; unit local costs of energy
investments; operating and maintenance costs; conversion efficiencies; time lags between payments and investments, and between investments and actual production; and calorific values of energy
sources and carriers constitute the "technology coefficients" of
the dynamic linear programming model. The objective of the model
is to minimize the weighted sum of: fuel import costs; energy
technology import costs; local investment and operating costs;
idle capacity costs; and costs of unsatisfied demand. The details
of the model are given in the following sections.

3.   THE MATHEMATICAL MODEL

3.1 Fuel Demand

The sum of fuels (F) of type ℓ supplied to consuming sector s at
time t and the unsatisfied demand (UD) must equal the demand (D)
of those sectors:

$$F_{\ell t}^s + UD_{\ell t}^s = D_{\ell t}^s \qquad s \in S; \qquad \forall \ell t \quad (1)$$

3.2 Energy Balance

Fuels are obtained from primary energy (P) of type j obtained from
source i and converted at facility k with thermodynamic efficiency
E:

$$\sum_k \sum_j \sum_i P_{\ell t}^{ijk} E_{\ell t}^{jk} = E_{\ell t} \qquad \forall \ell t \quad (2)$$

$$i \in I; \qquad j \in J; \qquad k \in K$$

Here, transportation of energy carriers also imply a conversion
efficiency which is inversely proportional to the energy loss per
unit energy transported. As a result, only non-energy transportation demand is specified.

3.3 Energy Reserves

Indigenous resources of energy are limited by the reserves of those resources. For fossil resources,

$$\sum_{p=1}^{t} P_{jp} \leq R_{jt} \qquad\qquad j \in J_{fossil} \quad \forall t \qquad (3)$$

So that the total amount depleted by time t does not exceed the reserves discovered up to that time. For hydraulic potential

$$P_j \leq R_j \qquad\qquad j: \text{ hydro} \qquad (4)$$

The same constraint would hold for solar, aeolian, tidal, etc. forms of energy, although in this model these are not included.

3.4 Production Levels

The quantities of fuels produced at any time are limited by the production capacity that exists at that time:

$$F_{k\ell t} \leq K\dot{\phi}_{k\ell} + \sum_{p=1}^{t-t_k} KA_{k\ell p} \qquad\qquad \forall k\ell t \qquad (5)$$

Here, $K\phi_{k\ell}$ denotes the capacity that existed at the beginning of the planning period for the production of fuel ℓ at plant k; KA is the capacity added, and t_k is the delay that occurs between the capacity increase and actual increase in the production level. This delay depends on the nature of the plant and is usually proportional to the size of the construction.

The total amount of fuel produced is the sum of the production levels at time t,

$$F_{\ell t} = \sum_{k \in KL} F_{\ell t}^{k} \qquad\qquad \forall t \qquad (6)$$

where KL is the set of facilities used for the production of individual fuels.

In the case of electricity production, certain reliability considerations apply for the balance between hydro and thermal power plants. Seasonal and yearly variations in water accumulation in the reservoirs require a detailed reliability analysis based on marginal costs of each type of power plant, and each type of commissioning; i.e. base, peak, or medium load. In this rather aggregate model, the details of the electricity subsector cannot possibly be included. However, certain special relations have been used to generate realistic results. Two of these relations concern the maximum energy and the power that can be obtained from hydro power plants.

$$F^{k_1}_{jt} \leq RE_t \sum_{k \in KE} F^k_{jt} \qquad \forall t \qquad (7)$$

> j: electricity
> KE: all electricity generating plants
> k_1: hydro power plants

where RE is the maximum ratio of electrical energy generation from hydro power plants. Similarly,

$$(1/LF^{k_1}_t)F^{k_1}_{jt} \leq RP_t \sum_{k \in KE} (1/LK^k_t)F^k_{jt} \qquad \forall t \qquad (8)$$

where RP is the maximum ratio of power generation from hydro power plants, and LF^k is the maximum load factor for each type of power plant. The RE and RP values are calculated from reliability analyses.

Although the design of a given power plant is usually based on the type of commissioning required of that plant, there is still considerable flexibility in actual operation. In the case of hydroelectricity, the same plant may be used for base as well as peak load. For the objective of global optimization of the energy system, it is preferable that the type of commissioning of each plant be included among the decision variables. To reach this objective, the electricity demand is expressed as:

a. an energy demand (DE); and
b. a power demand (DP)

$$\sum_{k \in KE} F^k_t + UDE_t = DE_t \qquad \forall t \qquad (9)$$

$$\sum_{k \in KE} (1/LF^k_t)F^k_t + UDP_t = DP_t \qquad \forall t \qquad (10)$$

where the U prefix denotes unsatisfied quantities. Since the energy demand is the time-integral of the power (load) duration curve, the model results specify the relative positions of particular types of plants within the load duration curve, rather than prespecifying their positions.

3.5 Capacity Increase

The new capacity that can be added (KA in Eq.5) in any time period
is limited. Usually, the limitation is technological in nature.
The personnel, accumulation of experience and know-how, and accumulation of capital equipment needs increase proportionally to the
required capacity additions. New capacity added in a given period
is restricted by a certain factor (KAF) of the capacity addition
in the preceding period,

$$KA_{kt} \leq KAF_k \, KA_{kt-1} \qquad \forall kt \qquad (11)$$

3.6 Foreign Currency Shortage

In the majority of developing countries a fast rate of growth calls
for large payments for the imports of capital goods and materials.
These requirements usually result in chronic trade deficit problems and foreign currency may constitute a bottleneck in certain
time periods. In such periods, a sort of "rationing" of the available foreign currency between competing sectors becomes inevitable.
In order to represent this possible restriction, the total foreign
spending for the energy system is restricted by the amount (Y)

$$\sum_{k \epsilon KI} CKA_t^k \, KA_t^k + \sum_{j \epsilon JI} CP_t^j \, P_t^j \leq Y_t \qquad \forall t \qquad (12)$$

KI: Set of plants requiring imported equipment
JI: Set of imported primary resources

where the C prefix denotes unit costs of imports.

3.7 Objective Function

The objective function of the model is the discounted sum of the
costs of: imported primary energy, imported energy technology
equipment, local investments and operations; idle capacity of
existing plants, and unsatisfied fuel demand. These costs are
weighted with different coefficients W_i, where $i = 1,\ldots,5$ in the
order given above, so that:

Objective function: Min.(Z);

$$Z = \sum_t \frac{1}{(1+r)^t} \left[W_1 \sum_{j \epsilon JI} CP_t^j \, P_t^j + W_2 \sum_{k \epsilon KI} CKA_t^k \, KA_t^k \right.$$

$$+ W_3 \left(\sum_{k \epsilon K} KKA_t^k \, KA_t^k + \sum_{\ell \epsilon L} F_{kt}^\ell \right) + W_4 \sum_{k \epsilon K} (K\phi^k + \sum_{p=1}^t KA_p^k - F_t^k)$$

$$\left. + W_5 (\sum_{\ell \epsilon L} UD_t^\ell) \right] \qquad (13)$$

The model consists of 226 rows and 186 columns. Solution time is less than one minute, using the FMPS package on a UNIVAC 1106.

4. MODEL VALIDATION

In order to analyse past decisions in the energy sector in Turkey, as well as validate the model, the parameters and coefficients of the model had to be evaluated appropriately. Furthermore, the values of some of these were altered to generate different "scenarios".

- The demands (D) for different fuels were taken to be equal to the actual consumptions of those fuels during the 1960-75 planning period (Eq. 1)

- Conversion efficiencies (E) were calculated from the average efficiencies of plants operating during 1960-75 (Eq. 2)

- Reserves of indigenous energy sources (R) were taken as: Coal: 10^9 tons; Oil: 25×10^6 tons; Hydro: 75 Twh (Eqs. 3 and 4)

- Maximum hydro energy generation ratio (RE) was assumed to be 40%, and maximum power ratio (RP) was assumed to be 50% (Eqs. 7 and 8)

- Capacity expansion factors (KAF) for any 3-year period were given values in the range 1.30-1.5 (Eq. 11)

- The prices of fuels as well as other financial data were given values equal to the actual data for the planning period (Eq. 12)

- The values given to W_1 were 2.5, 1.5, 1.0, 1.0, and 10.0 respectively.

The first solution was obtained by using the above data. This solution (Scenario No. 1) is shown with solid lines in Figs. 2 through 8, together with other scenarios and the actual data for those years, which are shown by broken lines. An analysis of this solution reveals the following: the model proposes greater use of coal and hydro, and much less use of petroleum in electricity generation. Since the model is "smart", i.e., knows beforehand that oil prices will rise, takes early measures, such as developing and building required capacity in alternative sources.

Scenario No. 2 is developed to see the influence of a gradual rise in oil prices, rather than the sharp increase that took place in 1973. The import prices of oil for the planning period were chosen

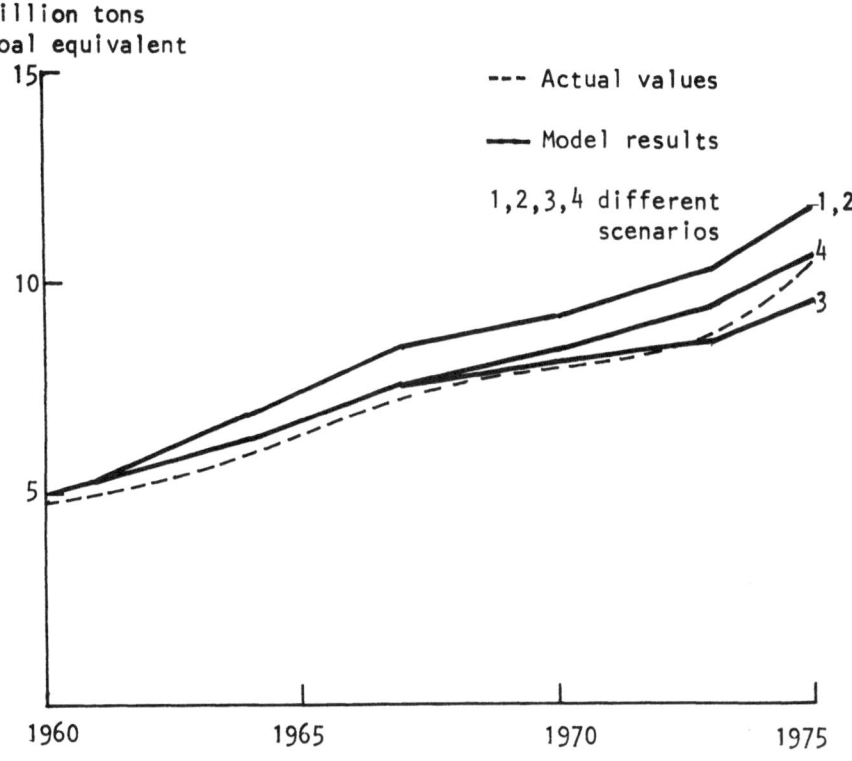

FIGURE 2 - Indigenous supply of solid fuels

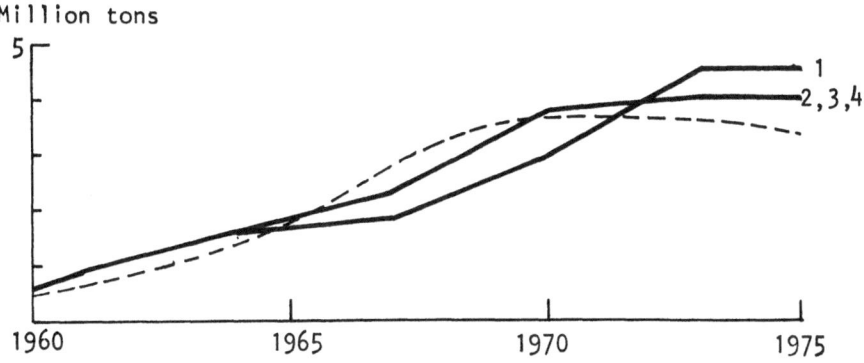

FIGURE 3 - Indigenous supply of petroleum

Million tons

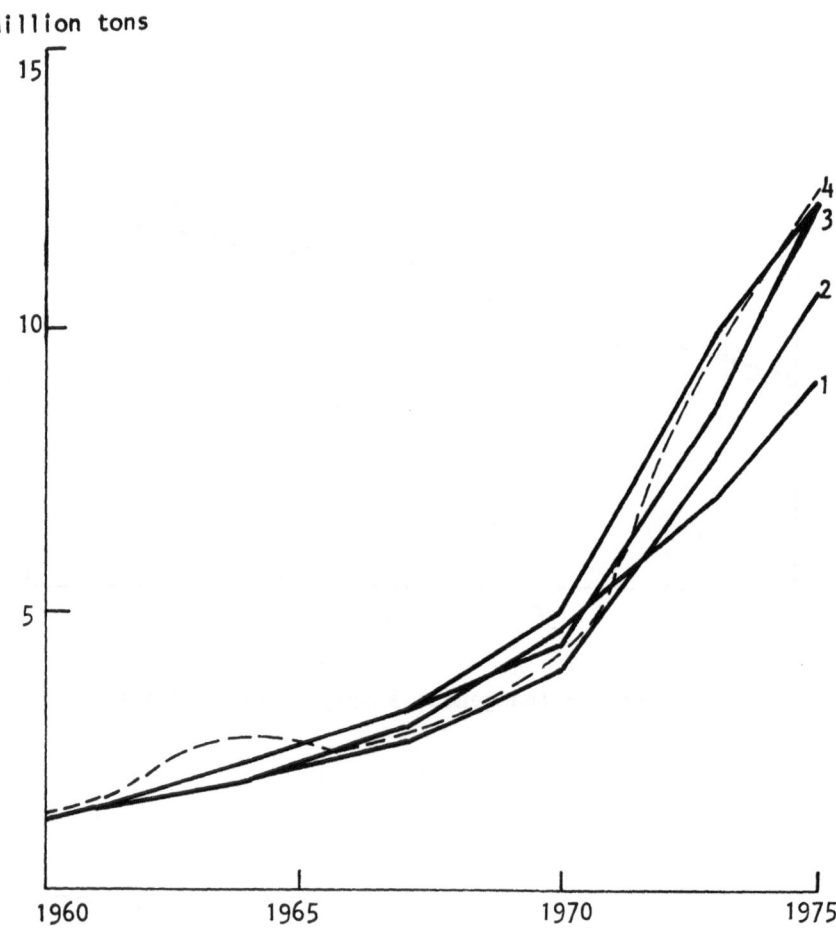

FIGURE 4 - Petroleum Imports

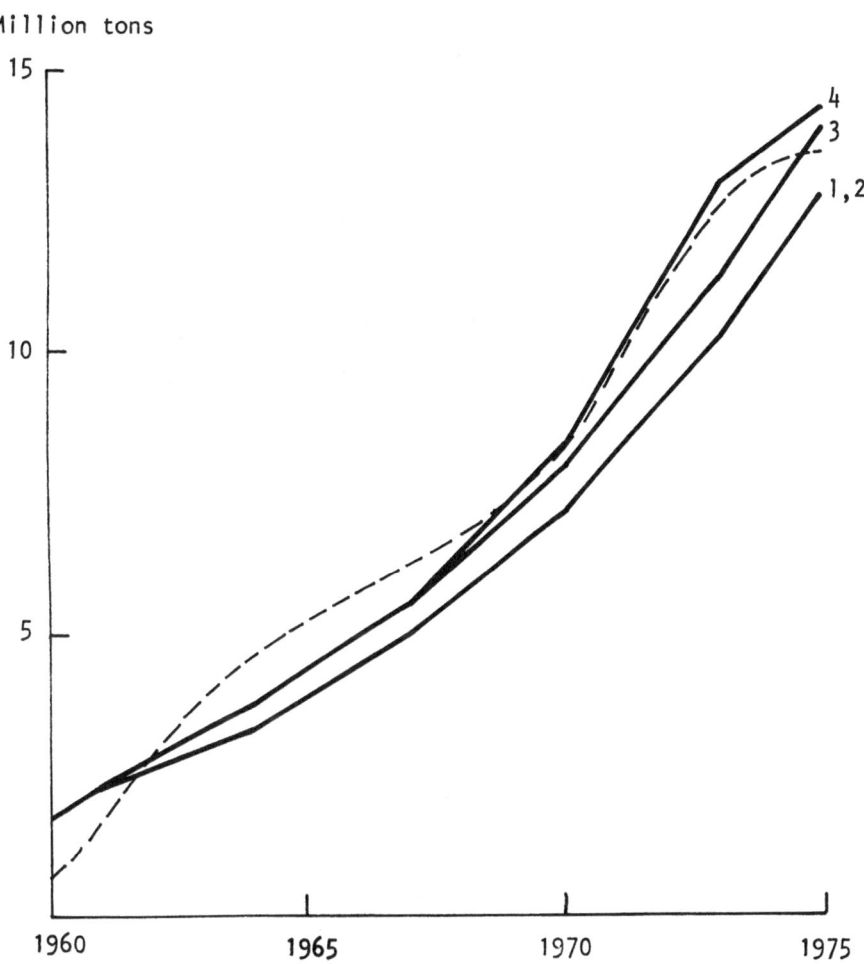

FIGURE 5 - Petroleum refining capacity

142

Twh (= 10^9 kwh)

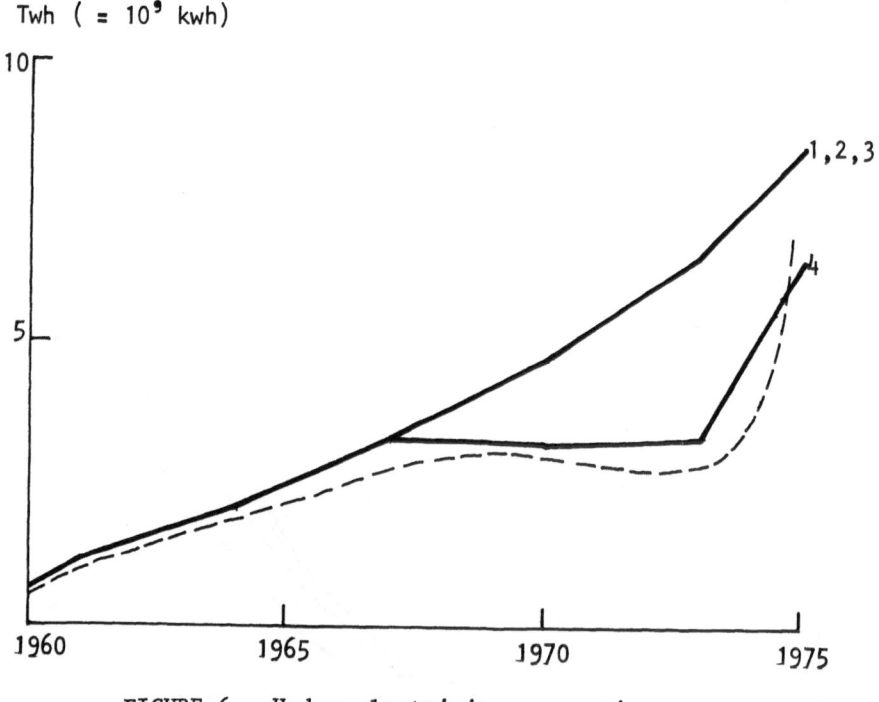

FIGURE 6 - Hydro electricity generation

Twh

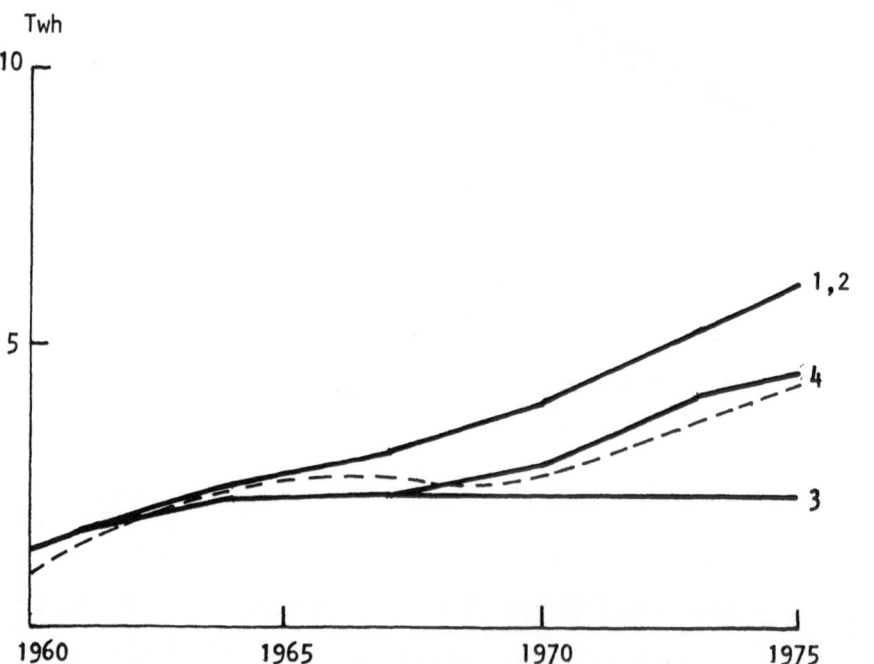

FIGURE 7 - Electricity generation from solid fuels

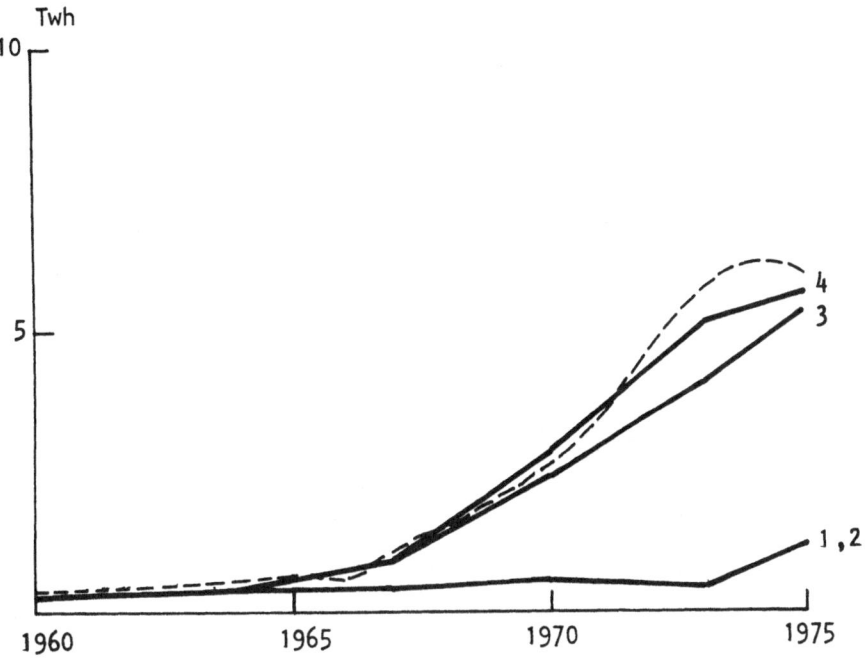

FIGURE 8 - Electricity generation from petroleum

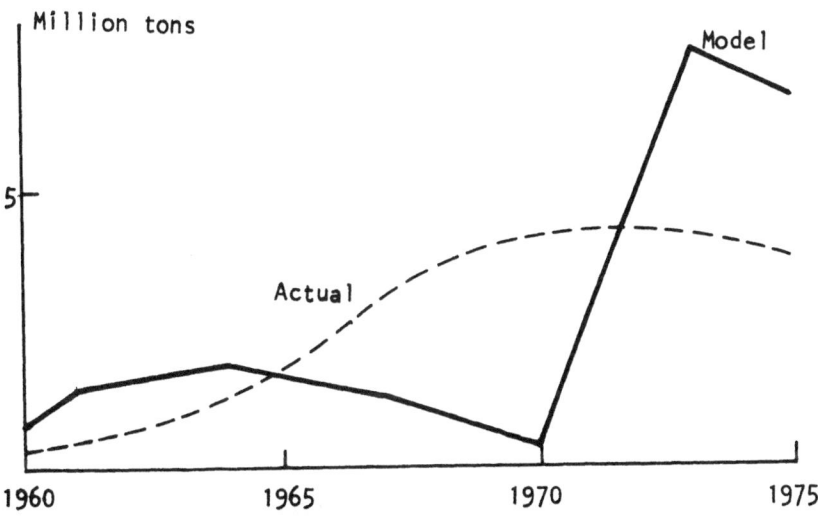

FIGURE 9 - Effect of not penalizing idle capacity in oil
extraction

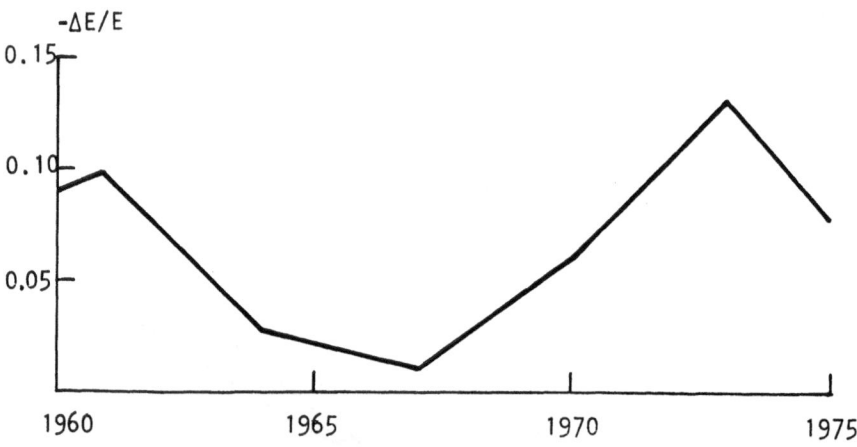

FIGURE 10 - Unsatisfied electricity demand - case of
          rapid growth

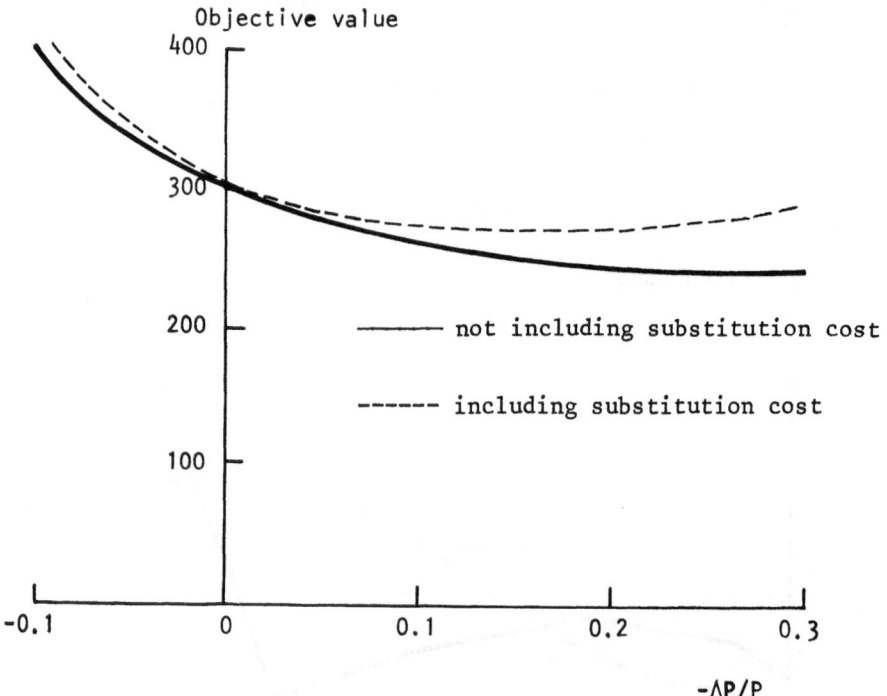

FIGURE 11 - Influence of substitution of petroleum
          by coal

so that it rises exponentially from about $ 3/barrel in 1960 to
$ 13/barrel in 1975. The results of this scenario are shown with
solid lines and the number 2. The results are almost indentical
to those in No. 1, except for oil imports and mining (Figs. 3 and
4). The relatively higher import prices in the 1965-72 period
cause greater local mining for those years at the expense of greater imports in later years, when the oil import price is even higher. The explanation of what may at first seem like a contradictory result is that the local petroleum reserves are limited, and
the discount rate (10%/annum) forces earlier use of the local
source. Had the idle capacity not been penalized the model would
have proposed rapid development of oil mining capacity and delayed
production until the 1970's when the oil import price soared. The
effect of not penalizing idle capacity is seen in Fig. 9 where oil
production declines for the early years of the planning period,
to increase rapidly after 1970.

Scenario No. 3 is based on the supposition of a constant price for
imported oil. The results are also given in the same figures, denoted by "3". The values for the local mining of coal and oil,
the imports of oil, and refining capacity are quite close to their
actual values. The overall development of the electricity sector
is also represented more accurately. The largest deviation occurs
in hydroelectricity generation where the model proposes greater
production. In fact, plans made prior to 1970 had foreseen a rapid
development of that source, but the 4-5 year delay in a major
hydro project (the leakage problem in the Keban Dam) stagnated
hydroelectricity production for that many years.

Results for Scenario No. 4 were obtained by means of an interactive procedure. For the initial nine-year period (1960-68),
the same parameters and coefficients as in Scenario No. 3 were
assumed, and therefore the results were identical to that scenario.
For the second run, the delay in Keban Dam was modelled, by fixing
the hydroelectricity level in 1969-74 to its value in 1968. All
other decision variables for 1960-68 were constrained to their
values given in the solution. After the results of run No. 2 were
obtained, the decision variables for 1969-74 were constrained to
their values in the solution, and the price of oil was increased
to the level it reached after the OPEC price rises, and the model
was solved once again. As can be seen, the results thus obtained
are very close to the actual data for those years.

The development of the Turkish energy sector for the 1960-75 period
may be explained by summarizing the results of the model:

*For the initial three periods (1960-68), the planning is
essentially based on a constant price for imported oil.
The delay in the Keban Dam forces even greater increases
in oil-based electricity generation as well as modest*

increases in coal-based electricity. The sudden rise in oil import prices comes as a total surprise to the planning authorities, and no alternatives exist to shift away from oil, as not meeting the demand would have caused even greater economic and probably social costs.

Varying the import price of oil in the final period (1973-75) has very little effect on the energy sector so long as earlier planning is based on a constant, low oil price. The basic reason is that the lead times needed for the development of alternative sources are rather significant.

In order to assess the influence of the shadow price of foreign currency, the weights W_1 and W_2 were given lower values, 1.5 and 1.17, respectively. Solutions were almost identical to those obtained with the higher values.

The model was also used to determine the consequences of faster economic growth. First, the electricity demand for each period was increased by 10% and later all fuel demands were increased by 10%. The model solutions resulted in faster deployment of indigenous coal as well as greater imports of oil. Although solid and liquid fuel demands were met except for the initial period, electricity supply was deficient for most of the planning period, as seen in Fig. 10. The basic reason for the deficiency is the bottlenecks that occur in building-up the needed capacity. Since electricity supply increased by over 11%/annum for those years, any further increases strain the resources.

The influence of substitution of fuels was also tested. Firstly, 10% of petroleum was substituted by coal, and secondly, the reverse substitution was effected. These results are shown in Fig. 11, where the total cost (the value of the objective function) is plotted against the substitution ratio based on petroleum demand. The extent of substitution is taken as what could be termed "heavy substitution" of petroleum 30% of which is met by coal. This probably represents the upper limit since roughly half of the petroleum was used in transportation, and a certain amount had to be used in densely populated urban centres where coal usage would probably have been physically impossible.

5. APPLICATIONS IN ELECTRICITY PLANNING

The model is currently being used by the Turkish Electricity Authority for the analysis of possible nuclear strategies for the near and medium term. The objective of the study is to determine the possible and likely ranges of the nuclear share in electricity generation mix to the year 2000. The latter phase of the project will involve generating "robust" strategies for investment planning.

In order to observe the consequences of probable developments in electricity generation as part of the national energy system, a rather large number of runs were completed by using the model. The parameters and certain coefficients of the model were systematically changed in order to determine the influence of:

- energy demand growth rates
- import prices of fuels and energy technology
- available reserves of domestic oil and coal
- discount rates.

One of the scenarios (for rapid economic growth) is presented below. Petroleum, coal and electricity demands are allowed to increase by 7%, 7%, and 11% per annum, respectively.

The results for primary energy mix and electricity generation mix are shown in Figs. 12 and 13. In this scenario, the technology coefficients were identical to their historical values, petroleum import prices were allowed to rise by 3% per annum in real terms, coal import prices were allowed to rise by 2% per annum, and the objective function remained the same as in the validation runs.

Although certain important results may be obtained by analysing each run separately, it is much more interesting to view the results simultaneously. For example, superimposed results pertaining to percentage shares in electricity generation mix are shown in Fig. 14. By increasing the number of scenarios, the likelihood of different outcomes can be examined in a broad context. Such a presentation of possible choices is usually of greater significance to the decision maker. Although there is considerable variance in the solutions, certain trends can easily be identified in these preliminary experiments.

After the global analysis of electrical power system development is completed, the following information will be available:

a. The likelihood of the need for nuclear energy;

b. The conditions under which nuclear energy is not called for;

c. The fraction of energy and power demand that would almost certainly have to be met by nuclear sources.

These results, however, would only be valid for average investment and operating costs of nuclear reactors. In order to account for differences in costs and efficiencies, the "nuclear power plant" node of the model will be extended to cover different reactors

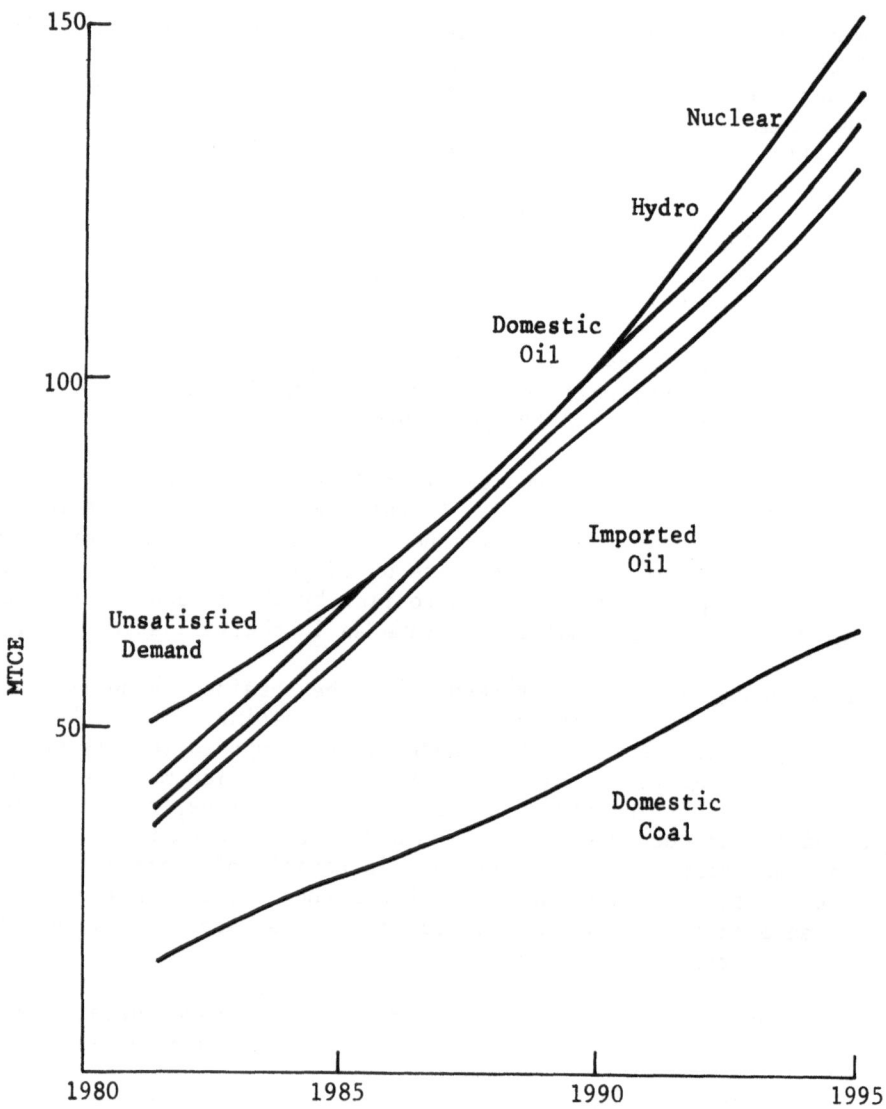

FIGURE 12 - Primary energy mix - Scenario A

such as: BWR (boiling water reactor), PWR (pressurized water
reactor), HTGR (high temperature gas cooled reactor), and FBR
(fast breeder reactor). In other words, the model shall focus on
the point of interest , analogous to a zoom lens.

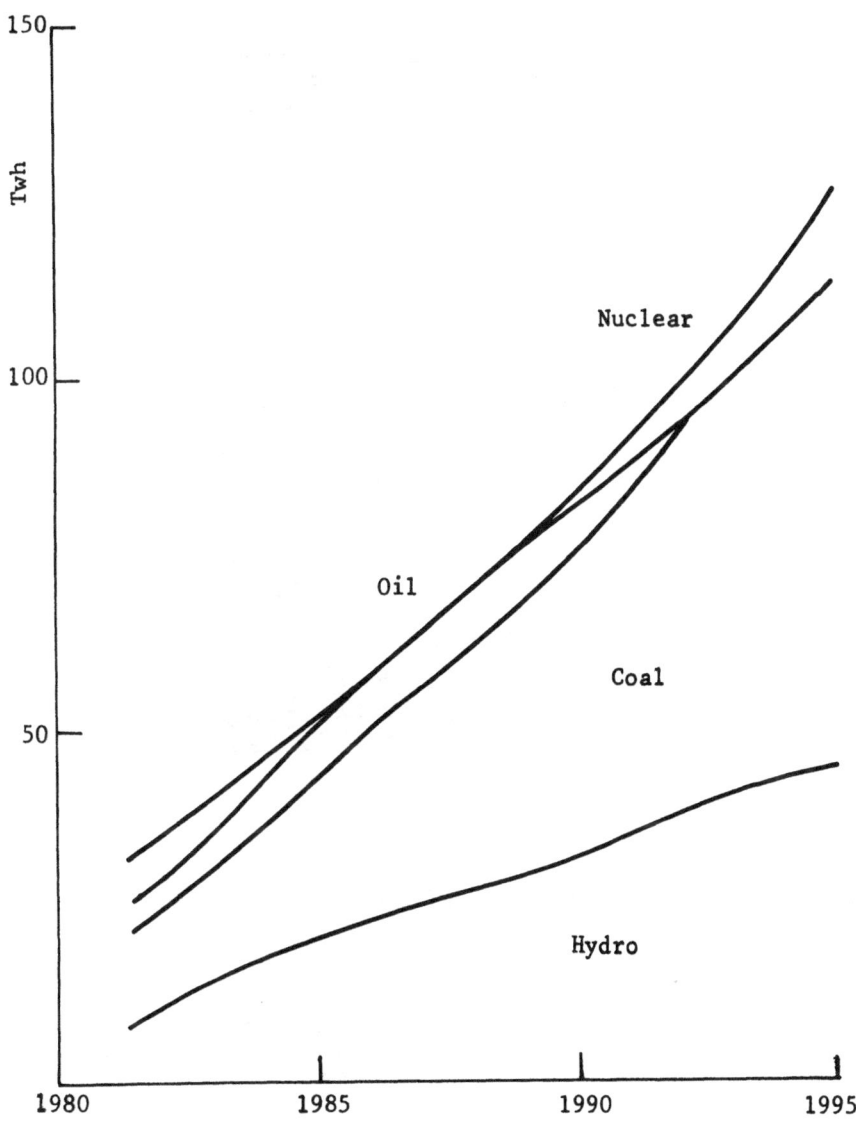

FIGURE 13 - Electricity generation mix - Scenario A

As a matter of fact, the same type of approach is undertaken in
another study where the model is used to analyse the possible
savings in cogeneration; i.e., the simultaneous generation of
electricity and heat. Possible cogeneration facilities are denoted
by expanding the definitions of the thermal power plants so that
they include the investment and operating activities of heat
generation and transmission. More detailed studies will be initiated

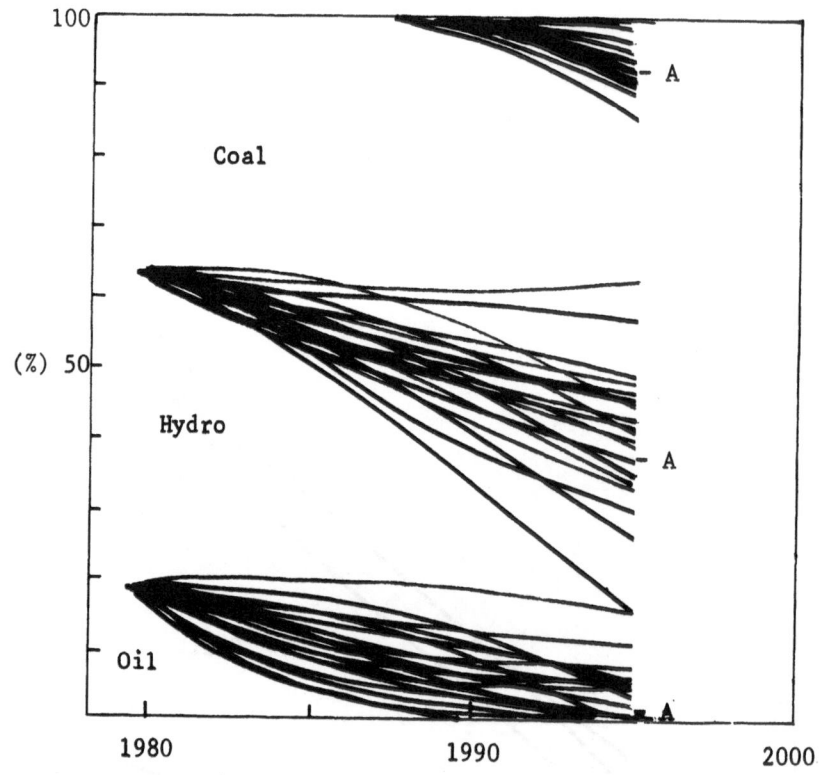

FIGURE 14 - Electricity generation mix in twenty different
                scenarios

after analysing the overall merits of such generation. This
study is undertaken by the Turkish Scientific and Technical Council
at the request of the Ministry of Energy.

6.   CONCLUSIONS

There were two basic objectives in the study presented:

   i.   To validate an optimization model

   ii. To demonstrate possible uses of the model in actual
        decision making.

As far as operational models such as this one are concerned, it
is quite important to test model behaviour against the real world,

especially if the model is to be used in advising decision makers.
The validation experience is not only useful for creating confidence
but also essential in understanding the mechanics of the system
being modelled.

Representing real systems is a demanding task and one may be
tempted to include as much detail as possible in the hope that the
results will be more "realistic". While it is more laborious to
design a large model, it is probably more difficult to develop
a small model that reproduces observed behaviour modes. On the
other hand, there are good reasons for keeping a model as small as
possible. For one, interactions with decision makers indicate
that the smaller and simpler the model, the easier it is to interpret and communicate the results of the model. Another advantage
of the small model is the ease with which modifications can be
made in obtaining alternative solutions. This is particularly
important where the uncertainties in model parameters are great,
since the decision maker would seldom be interested in a single
"optimal" solution, or strategy. Within the present computational
or theoretical limitations, a large number of solutions has to be
analysed to come up with "robust" strategies which are of greater
significance in working out actual decisions.

In analysing the consequences of a decision in any subsector of
the energy system, it is essential to avoid "modelling myopia" by
an appropriate representation of the closely related activities.
It can be argued that such myopia can be eliminated even further
by linking the energy sector with all major activities in the rest
of the economy. That way it would be possible to render energy
demand endogenous while at the same time observe the impact of
the energy sector on the economy. Such an approach, however, also
has its own disadvantages. For example, in the case of an "open"
economy; i.e. where imports and exports are great, the future
international prices of all traded commodities have to be predicted,
not just that of energy.

As far as the mathematical specifics of the model are concerned,
some observations would be in order. In modelling the electricity
subsector, for example, only peak power demand and total electrical energy demand are defined exogenously. This is a departure
from common practice in electricity planning where the load duration
curve is broken into segments.* The present approach is less
restrictive since it leaves the model to work out the average load
factors for the optimal commissioning of each type of power plant,
and requires fewer constraints and parameters.

---

*See for example, Electricity Economics by R. Turvey and D. Ander-
son, Johns Hopkins Univ. Press, Baltimore, 1977.

In recent applications of the model, certain constraints (in electrical power demand) have been modified. Equations (7) and (8), which had been included at the request of the Turkish Electricity Authority, are now changed in such a way that a reserve margin is added to each type of power plant. The reserve margin is typically 15% for thermal plants and 30% for hydro plants, the latter being larger because of yearly fluctuations in the water supply. Thus, the share of hydro in the total is no longer restricted to a priori ceilings, but is left to be determined by the model. The results obtained in this way have been found to give a sufficient reliability margin as well as reduce total costs, and the Electricity Authority has adopted this practice in their planning criteria and investment programmes.

The greatest restriction of the dynamic linear programming approach - as with most optimization models - is that it has "perfect foresight"; i.e. decisions for any given time are based on the knowledge of the future values of certain parameters. The other extreme is "perfect shortsightedness", as is the case with static models. This can be overcome by either embedding the model in a simulation framework or by running the model interactively, as was done in obtaining scenario No. 4 in the validation experiments. The procedure is somewhat tedious since it involves writing additional constraints for each run.

In closing, a few words must be said on the ultimate objective for developing and using formal models. Since actual decisions are eventually based on mental synthesis, perhaps the most constructive function of a formal model is its use as an educational tool to improve the capabilities of the mental model.

DISCUSSION

Kavrakoğlu was asked whether there was any unsatisfied demand in his model. He replied that historically, there had been only 1-2% and that this increased to 5-10% in the future projections. He was also asked whether the inclusion of natural gas, cogeneration and solar systems would make a significant difference to the results of his model. He agreed that natural gas might be included - Turkey was at the moment negotiating a pipeline from Iran. Solar could certainly cover part of domestic hot water requirements, but this only accounted for 1-2% of total demand. He also added that there was little investment in domestic oil production because of the constraint on recoverable reserves included in the model.

It was remarked that national projections were often unnerving. Were one to add up the import implications of the individual national studies, the number would certainly exceed the expectations on exports from other regions.

Several speakers expressed scepticism at the demand projections
from the Turkish central planning agency which produced the demand
figures. There was no implicit or explicit demand elasticity in
the model either. The point was made that population growth was
probably the factor driving everything else and that it would be
interesting to try other scenarios which assumed a lower rate of
population increase.

Kavrakoğlu agreed on the need for more scenarios; more were in fact
in the process of being run. He had learned from the study that
there was a high degree of inertia in the energy supply system
- historic trends did not change very quickly. He felt that this
was true even more in the demographic sector - a point on which
he was supported by the IIASA study. There was another important
demographic factor which was driving energy demand on, and that
was migration towards the urban areas. 55% of the Turkish population was at present rural and used mostly non-commercial energy.
On migration to the towns, they became commercial energy users.

It was pointed out that in 1975 the average per capita energy consumption in Turkey was 1 kW/capita. The average in Europe at that
time was 4 kW. The projections which Kavrakoğlu presented would
bring Turkey up to this level by 1990. His results underlined
the conclusions of IIASA, that flexibility in energy use was indeed
limited in developing countries.

Several participants returned to the question of demand projections
which they felt were too high. Often Turkish planners based their
projections on attaining the same standard or per capita energy
use as, say, Germany by 1990. It was important to see how energy
was used, though. The case of the electricity authority was mentioned. Electricity demand had been projected at 12% annually
(the rate of increase in the past 15 years). This rate was higher
than in any other country and one could well ask what was so exceptional about Turkey. A reduction in the assumed growth rate from
12% to 8% reduced the projected electricity demand in Italy 2000
from 200 Twh/year to 125 Twh/year. The speaker felt that some of
the explanation was to promote the electricity authority's claim
for more capital from the government.

ABOUT THE USE OF AN ENERGY OPTIMIZATION MODEL

D. Finon

Institut Economique et Juridique de l'Energie
Université des Sciences Sociales de Grenoble
France

ABSTRACT. An energy model can aid the planning activity by providing help at three stages of the energy systems [1].

" - System behaviour: impact analysis of different endogenous
and exogenous events ('what happens if?')

- System controlability: limits of the manoeuvrability of
the system considering the actual set of decision variables available to a decision maker ('what can we do?')

- Policy alternatives: to generate different policy alternatives with respect to generally formulated goals concerning the future state of the energy system ('what shall we
do?')"

An optimization model of the global energy system could theoretically provide answers to these three questions, but my experience
of building and using such a model for seven years lets me think
that its use must be limited to a precise role that we shall try
to clarify.

1. BRIEF DESCRIPTION OF THE EFOM MODEL

Since 1970, the Institut Economique et Juridique of Grenoble (IEJE)
has prepared several versions of an optimization model of the national energy system by regrouping the whole set of activities of
production and consumption of energy [2]. A model of this kind is
based on identifying the set of energy chains and economic activities associated with the satisfaction of the energy needs of a

given country, starting with the simple idea that each activity can be characterized firstly by its links with the other activities and, secondly, by a system of attributes (cost, efficiency, capacity, environmental impacts, etc.). It is therefore possible to represent such a system *by an oriented graph* and to carry out an optimization programme to find the least expensive supply structure to satisfy the energy needs of the community. A number of models of this type have been developed in various parts of the world, the best known of these being the BESOM model of the Brookhaveb National Laboratory.*

The optimization model of the energy system EFOM (Energy Flow Optimization Model) is a linear programming model aimed at defining the structures of energy activities that permit the energy needs of the community to be satisfied at minimum cost. The link between supply and demand is forged by taking explicitly into account the different energy consumption activities, that is, the various technical possibilities of consumption. The evolution of the demand for useful energy by the various consumer sectors and for the period under consideration is defined by external factors, the energy system being optimized to meet this demand.

As far as economics is concerned, the model covers only production and consumption activities located on national territory. The inputs to the energy system correspond to activities of importation or local extraction of energy resources and the outputs of the system to activities of export or internal consumption. Inclusion of the siting of activities is simplified by assuming that the production activities of a given energy chain are concentrated at one and at the same point of the territory and separated from the corresponding consumption activities by a given distance which determines the associated cost of the transmission activities.

We describe as energy chains the different combinations of economic activities (exploitation of natural resources, conversion, treatment, transport, stock piling, distribution, etc.) that succeed one another between the resources and the different uses of the energy and which may involve different types of vector. For reasons of clarification, the overall graph is broken down into subgraphs representing a branch or a clearly defined group of activities and called a subsystem. The energy flows that follow the

---

*

The Directorate-General for Research, Science and Education of the Commission of the European Communities requested the IEJE in 1976 to continue its efforts towards formalization by constructing, on the basis of its previous work, an exhaustive model taking into account as many processes as possible of production and consumption of energy within the scope of the "Energy Systems and Modelling Project".

arcs between two operations indicated by the nodes represent the variables of exploitation of the optimization programme. The other variables correspond to the plant capacities of various processes to be created in the future.

The energy requirements to be satisfied are defined by the type of use, extending as far as the product whose manufacture requires energy, as is the case with industries that are heavy energy consumers (iron and steel, aluminium, cement, glass, paper, chemicals). Substitutions between forms of energy are permitted in this last case through the possibilities of passing from one technological process to another, the interaction between final energy and capital then being very strong.

The optimization period is 45 years (1976-2020). The allowance for each year and its associated expenses would require each variable and each constraint to be defined 45 times; but, in order to reduce the size of the programme, the period 1975-2020 is divided up into six subperiods: 1976-1978; 1979-1981; 1982-1985; 1986-1990; 1991-2000; 2001-2020.[*]

The different types of constraints are:[**]

  a. the constraints of plant capacity;
  b. the constraints of limitation of flows of primary energy extracted locally, established on the basis of reserves and official extraction plans, or limitation of imported energy flows for political reasons (limitation of dependence) or technical reasons (long-term gas import contracts);
  c. the constraints of satisfaction of needs expressed in useful energy;
  d. the constraints of limitation of development associated with the inflexibilities of all kinds retarding the speed of penetration of new technologies (nuclear, hydrogen, solar, geothermal, district heating, combined power stations, etc.);

---

[*] The constraints and the flow variables are defined for the last year of each subperiod; the equipment variables correspond to the capacities of the plant created during each subperiod. The equipment created each year of a subperiod are deduced from the capacity implemented during the subperiod divided by the duration of the latter.

[**] This brief presentation does not allow us to examine certain refinements of the scheme, in particular at the level of formalization of the management of the electricity load curve or of cogeneration.

e.   the facultative political constraints resulting from
objectives complementary to the goal of minimizing
costs expressed by the objective function; these could
mean: the limitation of energy dependence, the obligation to develop a new noncompetitive technology, a
nuclear moratorium or the pollution limitations.

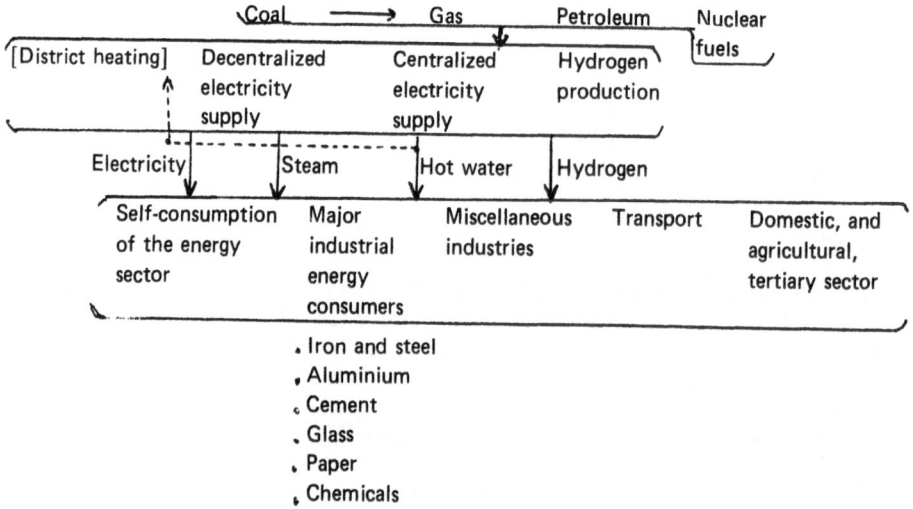

Finally, the bases of the representation are simple, obvious, even
elementary, but the model nonetheless became complex as and when
successive improvements were introduced (integration of the processes of production of the major energy consuming industries;
inclusion of new technologies, etc.). At the present time the
EFOM model is reaching maximum size, not in relation to the possibilities of solving a linear programme, but in relation to the
mastery of the interdependences between the multitude of energy
chains represented in the model. EFOM 12, which is operational
since 1976, includes 250-340 processes (it depends on the inclusion
or not of the processes of the major energy consuming industry);
the size of the linear programme is either 3200 unknowns, 3300
constraints, or 4300 unknowns, 3800 constraints.

2. THE LIMITS OF GLOBAL ENERGY OPTIMIZATION MODELS

2.1 Simplified Representation of Energy Processes (quality dimension, time dimension, space dimension...)

The representation of energetic chains in EFOM was the result of balancing the desire to represent correctly the management of energy plants against the necessity of not exceeding reasonable dimensions. Because of this, many aspects of the problem of energy supply, which are perhaps decisive in the establishment of future optimum supply, have been neglected.

However, we have tried to include energy uses in order to embrace the processes of consumption and the possibilities of substitution of energy carriers by the medium of the substitution of processes. But on the other hand, we have not considered:

- the load curve of certain uses (space heating for instance),
- the thermodynamic quality of the energy resources or needs,
- the geographical location of production and consumption activities...,

which are three key factors of the possible reorganization of national energy systems. For instance, in order to include the sources or carriers of "low power density", we have proceeded to the assimilation of each of the local problems to a typical problem determined from an empirical mean of technico-economic characteristics (mean period of sunshine, mean distance of transport, mean residential density, mean cost of distribution...). Now the space and time dimension of the exploitation of energy resources and the supply to different types of area ranging from districts with sparse population to the industrial and urban megalopolis should be the key factor of an energy optimization approach:

a. a space dimension, since the local characteristics of the sources and needs will be the deciding factors of the local supply structure;

b. a time dimension, since the demand of certain economic agents is irregular and the ability to stockpile local sources/carriers is not always ideal.

Furthermore, many new energy chains can develop only through the intermediary of rigid infrastructures (electricity, gas, hot water supply systems) the economics of which can very often be appreciated only in relation to local conditions (proximity of the source, existence of a supply network, etc.). It is certainly not at a global level that the best choice can be made between local networks; ultimately it may seem absurd to wish to superpose different

distribution networks linked by a rigid infrastructure on regional
or national transport networks, all the more so as the distribution costs are likely to benefit from substantial economies of
scale. But, in this field, we must certainly distrust all a
priori judgements, since no serious economic study has yet proved
the virtue of monopolization of a district market by a single
carrier of high thermodynamic quality.

On the other hand, lack of consideration of the specificity of the
local factors also means serious drawbacks for a meaningful integration of the environmental factors into a global energy approach,
pollution being related essentially to the problem of concentration
of economic activities in specific areas.

Faced with these limitations, different options could be defined:

    a.   to "regionalize" the model (but, at the level of the
        region, the problem could be the same as at the level
        of the national system; moreover, the problem of data
        and of energy need forecasting could be very important);

    b.   to obtain a crude global model which could be broken
        down, by a zoom effect, at the level of study of pre-
        cise processes or of a particular branch whose inter-
        actions with the rest of the energy system would then
        be considered;

    c.   to obtain a succession of very detailed submodels
        demanding more sophisticated calculation tools and
        studying chains or particular processes, it being
        possible to link the submodels in a formal or infor-
        mal manner to form a global model.

2.2 Simplified Representation of the Optimum of a National Energy System in a Situation of Dependence

-   *The decision-making hypotheses:* The optimization programme
could be interpreted as the quest for the equilibrium of the energy
market in a situation of more or less pure and perfect competition.
We prefer a dual interpretation: the EFOM model would be the
formalized expression of the problem confronting a body responsible
for drafting and implementing energy policies. And one of the
conditions of optimality in the two cases is the alignment of
selling prices with the marginal costs.

Reality is often far from this scheme; the strategy of enterprises,
whether they be public or private, is not always compatible with
what could be the public interest in the energy field. Their
capability of expertise is often greater than the public adminis-

tration's capability, and, it is not seldom that public interest
is sometimes assimilated to the interest of certain firms which
dominate the energy sector in various countries. (Development of
energy systems took place around a concentration of the major
activities in the hands of a small number of firms on either a
national or international basis).

The results of the model could demonstrate the interest of development of new energy sources of carriers; but large enterprises
could not wish to face the risk of reduction of their autonomy or
of their potential development.

The chances of transformation of the energy system of generation
and consumption should therefore be assessed in relation to the
inertia of the structures which have arisen around this precise
type of energy utilization. This inertia has to be analysed just
as much in relation to the organizational forms created around
the structures (large public or private groups, research bodies,
controlling administration, framework of regulations, etc.) as in
relation to existing plants, to future technologies permitting
the reproduction of the present energy system on the same bases,
and to the skills and knowledge developed around these structures.

It appears crucial to define certain constraints of development
applied upon certain technologies in order to keep the results
credible. But this problem gave evidence that energy policy is
not only the result of a technico-economic optimization.

-   *The oversimplification of the objective function of the ,*
    planner in a context of deep incertitude: Generally speaking,
the objective function of an optimization model is the cost of
energy supply which has to be minimized in order to determine the
sectorial optimum for the whole.

It would be naive to think that optimization of the national
energy system is perfectly separable from optimization of the
remainder of the economy. Energy programmes are tied to much
more general political decisions, such as economic and industrial
policies, national independence, respect for the quality of life,
the level of social acceptance of risks, etc. Some elements of
what might be called the function of collective utility have been
or could be more or less faithfully reintroduced into the model
either in the form of constraints or in the form of social costs
in order to complement or moderate the simplistic objective of
minimization of the discounted cost of long term supply: limitation of oil dependence, outflow of currency, possibilities of
financing of investments, environmental impacts, etc.

But it would be illusory to believe that the function of the
collective utility so represented is truly relevant. The essence

of the key factors of an energy policy rests, in fact, on a profound uncertainty. The level of future needs, the fossil and fissile fuel resources/reserves, the relations between countries importing and exporting energy, the role of large international firms, the energy policy of the United States, the feasibility and cost of new technologies (breeder reactors, gasification, solar power, hydrogen, etc.), the transfer of technologies, and the social acceptance of certain risks are - for countries that depend on imported energy - so many unknowns which should induce these countries to minimize the risks of any sudden interruptions or changes. The fact that the model rests on a hypothesis of perfect forecasting prevents it from providing a satisfactory reply to this adaptability problem, although it enables a reference energy scheme to be described for well-defined and well-forecast situations of its environment. But, it anticipates future events a long time in advance, which enables it to set up an already appropriate structure when these events supervene. Many events are for the most part impossible to effect with an objective probability and the state of the system can only be inappropriate at the date of the eventual ruptures and its response can only be very slow, in view of the enormously long periods of development, construction and life of energy plants for production and consumption. The decision trees method would not allow major improvements to face this problem.

The model is then incapable of ascribing an objective advantage to solutions which would improve adaptability or would prevent some irreversibilities. Energy systems which are too specialized - and for this reason too rigid - are condemned to be extremely vulnerable to sudden changes in their environment; the consequences of the rises in world oil prices of 1973-74 illustrate this problem very well. An energy strategy based on the reproduction of a centralized and specialized energy system with two or three dominant vectors or chains which depend on either imported resources (coal, gas, petroleum) or on future processes based on technical gambles (gasification, fast breeder reactors) presents unsuspected disadvantages to weigh against the advantages of an alternative based on a diversification of energy sources and the use of carriers adaptable to different sources. This alternative would of itself give rise to a de facto stability because it would lead to a distribution of the risks and limit the vulnerability of the whole.

Our purpose is not to initiate here a comparison of the advantages and drawbacks of the one or the other process, but simply to emphasize that economic calculation in general and optimization models in particular are blind to this type of preoccupation, since they do not know how to treat satisfactorily the problem of choice in an uncertain future not analysed as to probabilities".

Moreover, they never assess the socio-political advantages of more neutral technologies that are more easily politically controlled.

3. USE OF THE MODEL FOR THE STUDY OF SYSTEM BEHAVIOUR AND CONTROLABILITY

These various theoretical and practical limitations of an optimization model ought never to lead users to frame energy policy proposals on the basis of the model's results. It would be naive to think that a model of this type would allow one to decide technocratically and once and for all:

    a.   the investments to be made at different periods in each energy chain;

    b.   the distribution of the market between the various chains in the long term;

    c.   reference prices of each of the products (with the help of the dual programme).

It would be a very Cartesian reaction to wish to determine a single optimal solution to such a complex problem in such an uncertain environment. But is it necessary to throw the baby out with the bath water?

The model must be used as the outcome of a global and detailed study of the political, economic and technological aspects of the energy supply, a study which is made by trial and error and by successive approximations between the whole and the parts, between a global approach and partial studies at the margin of the system. To do this, several interactive stages could be followed in succession:

    1.   Establishment of the geopolitical, economic, technological, social, and institutional framework by a study of interactions of the national energy system with its environment (for example, a study of the world energy system of the world resources, the role of the USA, the role of large companies, the evolution of

---

*The approach based on decision trees offers the advantage of being able to determine a strategy which is a function of the achievements of the various uncertain parameters. But it scarcely takes into consideration the advantages of flexible and adaptable structures.

under-industrialized economies, the national industrial
system, etc.). It is at this level of study that the
possible evolutions of the national economic system and
their connections with energy requirements would be investigated.

2. Description of the whole of the energy chains and partial studies of some of them. To do this, an evaluation of the resources and uses by regions and districts
   would have to be performed at the same time as the possibilities of development of new technologies were
   analysed. We could proceed at this level to an initial
   selection of technologies to consider. This second
   stage should already enable us to outline the frontiers
   of the field of possible energy systems.

3. Placing the different energy chains in competition,
   doing so within each energy use with the help of an
   EFOM-type model within the framework of different consistent scenarios established at the first stage, and
   with reference (among other things) to geopolitical
   relations, economic development, the extent of institutional innovation and the predictable state of the
   new technologies.

The EFOM model could be fitted into such an investigation process.
The model thus has as its principal contribution the automatic and
global formalization of the competition of the various energy
chains: firstly, within an inter-time framework enabling time
constants to be taken account of; secondly, within the coherent
framework of the neo-classical theory of the partial equilibrium
at the level of the energy market. The primary purpose of the
EFOM model is to obtain a better grasp of the totality of the
problems in their interaction and their globality and to integrate
the rigidities of the system since the specifications cover plants
(capacity, age, lifetime), and eventual political or sociological
inertia. The fact of placing its use at the last stage of a process of studies which would enable the whole of the data and the
economic, technological and political parameters to be established
reliably leads to the running of "what if?" and "what can we do?"
tests rather than to the normative search of an optimal energy
structure ("what shall we do?" test).

The elaboration of sets of contrasting hypotheses on the price
of petroleum, the cost of nuclear power, energy needs, innovations
both technological and institutional, social acceptance of the
nuclear risk (necessary hypotheses for defining the constraints
and parameters of the model), depends on extensive studies preceding the use of EFOM or of any other model of this type.

Each test of a variant may involve the variations of one or more sets of parameters or constraints. It would thus be possible to test different pairs of hypotheses on the price of petroleum and the cost of nuclear plants ensuring that there is consistency between the level of future prices, that of the needs and the institutional policies adopted to help in the development of local systems, all of which means setting fairly loose limits on the development of the latter.

The model also enables us to appreciate the global cost of readaptations necessitated by taking this or that decision (limiting the dependence on petroleum or energy at a given level of consumption,* temporary nuclear moratorium, cessation of nuclear fuel reprocessing, etc.) by comparing the objective value of the basic case with that of the variant considered. Should this value prove too high, it is always possible to pass to a new variant where the demands of the policy expressed by the constraint considered would be reduced.* This trial and error procedure is closely related to a procedure of implicit revelation of collective preferences.

The "what if..." use of the model does, however, have some normative features, because the underlying foundations of the optimization remain normative. As a consequence, if the conjoint optimization of all the energy chains lead to results that diverge appreciably from the partial studies of the second stage or from the conclusions and recommendations in the matter of official energy policy,** it would be wise to ask the reasons for this divergence:

a.   Might the official programmes result from partial optimizations of the agents of the different sub-systems considered that were not compatible with the general economic optimum?

b.   Might the economic information implicit in the bases of the model (shadow prices) be too remote from the information really exchanged between the agents?

---

*

It is important to stress also that the test may apply to the study of the adaptation of the system to an objective to be attained in a given year. For example, oil imports may not exceed a certain amount: 50% of the total primary energy in 1990. Because of the rigidities of the structures, these have to be adapted in advance to the imposed reduction of imports of crude oil, the model determining the route as a function of the target so defined.

** These results would, of course, be obtained by keeping the same hypotheses concerning the economic and political parameters.

c.   Would the partial ignorance of the structural, insti-
     tutional, or industrial constraints in the model be
     sufficient to explain the divergence between the results
     and the official plan?

d.   Could not the simplification of the different energy
     chains and, in particular, of the transport and distri-
     bution activities explain in part the divergences be-
     tween the model results and the official forecasts? In
     this case might it not be necessary to improve the
     studies upstream of the model?

The results from the model would give those who decide on optimal
energy choices cause for thought, by making them assess the margin
of manoeuvre they have in different structural and institutional
contexts, by making them appreciate the significance of the rigidities of all orders and the commitment to measures aimed at reducing their impact in the event of the divergence between the optimum of the model and the official plan proving too serious.

The special study of the place of one or more new technologies
(Technology Assessment Studies) would be closely related to the
"what if" approach, but it could be carried out more empirically
in order to appreciate their speed of penetration, or to reveal
their lack of competitiveness in different contexts bearing on the
economic parameters (price of imported fuels, cost of the innovations considered, associated "learning factor", etc.) and institutional factors (regulations, incentives, etc.). It would be
possible, for example, to set up several tests on the price of
petroleum, or on the cost of the technology(ies) studied to examine at what levels this or these technology(ies) begin to make
their appearance. In this last case little importance is attached
to the date of breakthrough of the technology considered or to its
speed of penetration into the energy balance, although the value
of the parameters remains defined in the model with respect to
time. Likewise it may be profitable to study the speeds of penetration as a function of constraints reflecting the hypotheses made
concerning the mechanics of the innovations, the rigidity of habits
and behaviour, the rules in force and the effects of the domination
of certain firms on the energy market. Further, the cost of these
new technologies depends on their speed of penetration of the market as a result of effects of series, and this introduces an amp-
lifying factor that is impossible to take into account in a linear
programming model.

The model could help to assess the discounted advantage of the
future development of one or more new technologies in different
economic contexts, advantage which has to be balanced with the
R & D outlays. The study of the dual programme could also provide
useful informations on their break-even costs.

4. IN PLACE OF A CONCLUSION

The EFOM model is simple in its design but complex because of the number of energy chains considered. It can be used in a process of studies of global energy systems only following more specific studies of the environment of the system and of different sub-systems and specific chains. Partial studies of this type should permit the establishment of the data and technical parameters to be considered. Work on the EFOM has been perpetually in advance of these studies; for this reason, it can only breed mistrust and scepticism on the part of the decision makers which are aggravated by the impressive dimensions of the programme.*

Indeed, the relation of the decision maker to new methods to aid the decision making may be compared with that between the worker and a new machine that might transform the organization of work. The decision maker regards them with much mistrust as he fears that he will find his expertise or his modicum of power reduced or, simply, find himself obliged to modify his habits and his ways of reasoning. Confidence is all the more difficult to establish as, on the one hand the decision maker does not have the necessary time to participate in developing the model and, on the other hand, the model is too big or too complex to be quickly understook by a busy person.

Also, presently, we prefer the use of a simplified model in order to convince French decision makers of the advantages of such an approach.

REFERENCES

1. N. Ericsson,"Interactive structural energy models: a decision maker approach,"Lund Institute of Technology, Sweden, 1978.

2. D. Finon, Optimization Model for the French Energy Sector, in Energy Policy, Vol. 2, pp.136-151, June 1974.

    D. Finon, Un Modèle Energétique pour la France, Editions du CNRS, Paris, 1976.

---

*The French decision making process is also very peculiar.

DISCUSSION

One of the major questions which Finon's model set out to answer was the choice of future energy supply technology. An important determinant of whether tomorrow's technology would be centralized or decentralized will depend on the cost of energy distribution. It was suggested that distribution costs could be included in the model as a function of load density. Finon replied that load density was only one of several factors which had to be taken into consideration. Regional differences and social costs of large centralized energy supply should also be taken into account and this was more difficult in a linear programming model. Other factors such as environmental costs and safety risks in new technology should also be included in the decision function.

Finon was asked whether it was possible to validate such an optimization model, as was done by Kavrakoğlu and by Deam. He did not think that this was possible in his case because the energy market he was describing was not perfect while an optimization model assumed perfect competition. The oil market was a different case. There, one did have perfect (or very near perfect) competition. This had enabled Deam to obtain a high correlation between the output of his linear Programming model and observed data.

Finon's model took demand as given and would thus be a more relevant planning tool in a centrally planned economy. The assumption of inelastic energy demand was questioned. Sweeney's analysis had shown that if a model were to address the question of alternative technologies, it was critical to take elasticities into account. If the overall energy elasticity was low, one could easily underestimate the usefulness of supporting the development of alternatives.

Finon's model concentrated on <u>useful</u> energy. These were no empirical studies on the elasticity of useful energy although Danzig in the PILOT model had make some attempts. Useful energy had never really been measured consistently either so the task would be difficult.

Finon had indicated in the course of his presentation that he considered simulation models to be of more use than optimization models. This postulate was questioned by several speakers. It was generally agreed that the usefulness of a model was a function of its objective, and its objective can never be prediction. Decision makers were most interested in analyses of the "what if" type and the objective was normally to take robust decisions. Optimization models gave point solutions which was really more useful as it for example represented a least cost solution for that supply configuration. If optimization models were well structured, trade-offs could be studied in a coherent fashion. A third advantage of optimization models was that one obtained shadow prices directly.

Other people felt though that optimization models placed a limit on the number and type of "what's" and "if's" which could be answered. This, however, reduced to the problem of defining and solving multi-objective functions. Most often it was not the modelling approach which gave rise to divergence, but the scope and parameters of the model.

MODELLING ENERGY DEMAND FOR A FAST GROWING ECONOMY

J.E. Samouilidis

Chair for Industrial Management
National Technical University of Athens
Greece

ABSTRACT. The Greek economy has scored high growth rates in the last twenty-five years. The per capita gross national product has more than tripled during that period and reached the level of 3,000 US $ in 1978. The manufacturing sector, and in particular, energy intensive industries have been the leading growth sectors of the economy. Government.'s economic development plans for the next decade indicate target growth rates much higher than those considered satisfactory in many developed countries, with a primary role allocated to the continuing growth of the industrial sector. In this context, in which energy becomes acutely critical for future growth, a system of models has been developed to provide forecasts of the energy demand, under alternative scenarios of economic growth. The system combines conventional econometric techniques with an optimization model - a goal programming model - whose main role is to operate as an "information integrator". Thus, it integrates information included in the econometric models with exogenous information, such as energy policy objectives, feasibility of technical reports and judgemental analysis, to provide a set of consistent energy demand forecasts.

1. INTRODUCTION

The Greek economy has been a fast growing economy for the last twenty-five years. The per capita gross national product has more than tripled during that period and reached the level of 3,000 US $ in 1978, which corresponds to 50% of the average income per capita in the European Community. This development process was accompanied, as one could expect, by considerable structural changes in the economy, which were particularly reflected in two indices:

i.e. the product of the manufacturing sector expressed as a percentage of the GDP and the product of all energy intensive industries as a percentage of the manufacturing sector.

Government's economic development plans are based on the assumption that the economy will continue to grow with a target annual growth rate in the range of 5 to 6% in the next ten years. Manufacturing is forecast to be the leading growth sector, with particular emphasis given to some metal extracting and processing industries, which could thus contribute to a higher utilization rate of indigenous natural resources.

Though some of the above plans could be criticized as being rather optimistic, particularly in the light of events in the world oil industry since the end of last year, there seems to be a concensus among politicians, government planners, and independent economists that the country's growth rate should, and could, be higher than the expected average in the European Community. Also, that economic development, at least for some years, should be predominantly based on industrial growth.

As a result of the above, it is more than obvious that the growth of energy demand, which was high in the past and is expected to continue so in the future, should be given primary consideration. It could be said that the role of energy in economic growth is more critical in the Greek economy than is the case with the developed economies in which the composition of the GNP will remain either constant or even change in favour of the service sector.

Within the above context a system of models was developed to provide alternative scenarios of the energy demand by the Greek economy. The purpose of this paper is to present the main features of this forecasting system, and to highlight its possibilities and limitations. Before presenting this system, we start with a short overview of the Greek energy sector, and we continue with a literature survey on energy demand modelling.

2.   TRENDS IN THE GREEK ENERGY SECTOR

Energy demand in the Greek economy has been growing with high rates in the last twenty-five years. Average annual rates were in the range of 10-12% before 1973 and decreased to 7-9% thereafter. In 1978, energy consumption reached 1.5 tons of oil equivalent per capita, which is still very low when compared to the corresponding figure in the U.S. or the E.C. in the past. Income elasticity of demand has remained rather stable on a level which is characteristic of developing economies, i.e. 1.5-1.6.

Total expenditure for energy which amounted to only 5% of the GNP in 1965, reached 9% in 1973 and passed the 10% level in 1976. This overdoubling in the energy contribution to GNP is due both to an increase of energy use per unit of product and to an increase of energy prices. In relation to the expenditure for energy, it must also be pointed out that 90% of it represents foreign exchange payments. Thus, expenditure for imported energy which was only 90 million US $ in 1966, reached the level of 1.3 billion in 1978. This last figure corresponds to 20% of the value of imported goods.

The shares of various primary energy carriers have not changed drastically in the last years. Oil accounts for 70%, coal - mainly lignite - for 26%, and hydroenergy for 4% of total primary energy consumption. Significant changes, however, have taken place in the final energy side. The share of electricity has increased from 27% in 1965 to 35% in 1978. At the same period, the share of lignite in electricity generation has gone up from 44% to 58%.

Considerable changes have also occurred in total demand. In 1978, industry's share was roughly 45%, transportation accounted for 24%, the remaining 33% having been covered by domestic and miscellaneous users.

3.   A BRIEF SURVEY OF THE LITERATURE

Recent literature on energy forecasting offers a very wide range of tools. On the one extreme of the range, there are rather simple systems. Energy demand is analysed by use and forecasts are made based on assumptions about energy technology at the consumers' end. Effects of changing life styles are also considered [1,2]. On the other extreme, rather complicated analytical methods, using dynamic simulation models, forecast energy through a global analysis of energy-economy interactions [3,4].

There is one major characteristic that distinguishes the great variety of energy forecasting methods; the existence of feedback loops (Fig. 1). There exist open-loop systems: those that analyse energy through one-direction relations. This group comprises the early econometric approach of the late fifties and the sixties, or the more recent socio-technical system approach [1]. There exist also closed-loop systems, those that consider the interaction between energy demand and economic growth. This group comprises both optimization models [5] or dynamic simulation models mentioned above [4].

Models in the last group have a broader purpose than estimating energy demand. They also estimate other variables of the energy system, as energy supply, investment, etc. Some go even further to give estimates of variables of the macroeconomic system. To

174

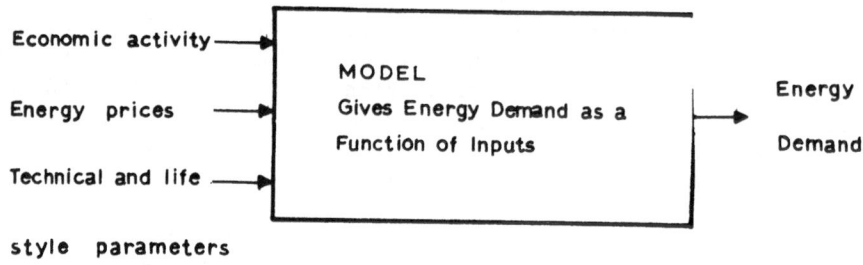

Economic activity

Energy prices

Technical and life

style parameters

MODEL
Gives Energy Demand as a
Function of Inputs

Energy

Demand

(a) Open Loop Forecasting Systems

- Conventional econometric methods

- Socio technical system approach

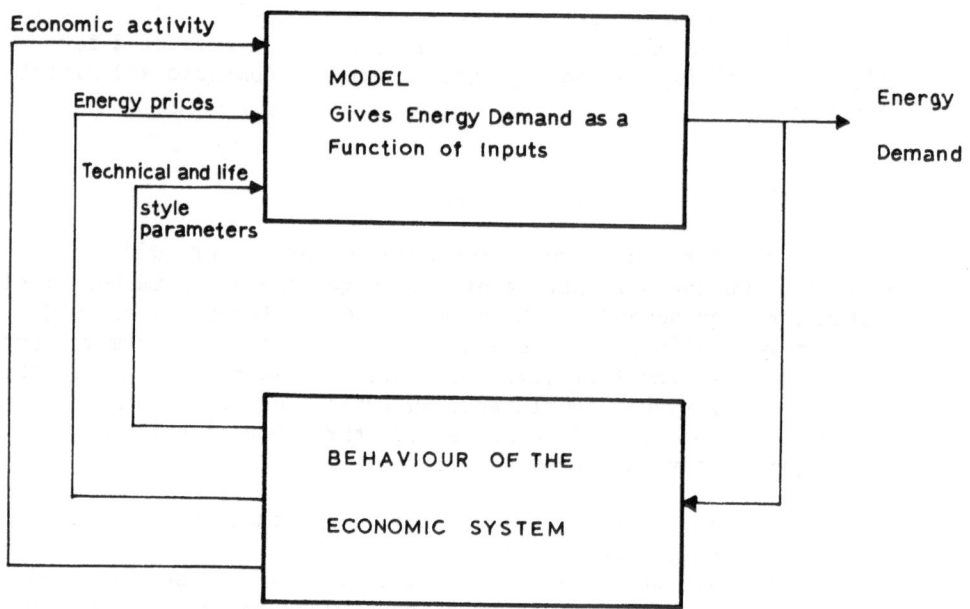

Economic activity

Energy prices

Technical and life

style
parameters

MODEL
Gives Energy Demand as a
Function of Inputs

Energy

Demand

BEHAVIOUR OF THE

ECONOMIC SYSTEM

(b) Closed Loop Energy Forecasting System

- Optimization models

- Dynamic simulation models

FIGURE 1 - A concise classification of energy forecasting systems

achieve all that, they require a very high level of sophistication
and complexity, and considerable computational effort [3].

4. OVERVIEW OF THE FORECASTING SYSTEM

In this paper we describe a system for forecasting final energy
demand, which is a refinement to the classical econometric models.
Our approach is not as complex as a dynamic simulation model; on
the other hand, it is not so broad in its scope, either. We do
not forecast the whole state of the energy sector, just energy
demand.

The forecasting system proposed in the paper consists of two major
components: a set of econometric models, and a goal programming
model. The econometric models provide "preliminary" energy forecasts, which are revised by the goal programming model (GP), as
described in the following paragraphs.

An overview of the forecasting system is given in Fig. 2. It consists of seven steps. STEP 1 refers to the analysis of energy
demand by consuming sectors (i.e. industry, transportation, etc.)
and by energy commodity or fuel, (i.e. electricity, solid fuels,
etc.). An energy requirement matrix is thus formulated. Element
a_{ij} represents demand of the i th fuel by the j th consuming sector.

Demand analysis, with the aim of identifying explanatory variables,
is carried out in STEP 2, using alternative theories of demand.
STEPS 3 and 4, which follow, consist respectively of the formulation and evaluation of econometric models. STEP 5 sets up scenarios for future paths of the explanatory variables of the econometric models. In STEP 6 the system calculates preliminary energy
forecasts, which are fed into the goal programming model, formulated in STEP 7, considering all available information exogenous
to the set of econometric models. Optimal solutions of GP give
scenarios of energy demands.

5. THE ECONOMETRIC MODELS

The econometric models express energy demand as a function of a
set of explanatory variables, exogenous to the forecasting system.
Alternative sets of variables are used, depending on the theory of
demand. A short description of the rationale behind each model
is given below.

The first group of models rely on the classical theory of demand.
It is interesting to note that energy is both an input to production, giving rise to intermediary demand, and a consumer's good,
creating a final demand. Energy as an input to production is

176

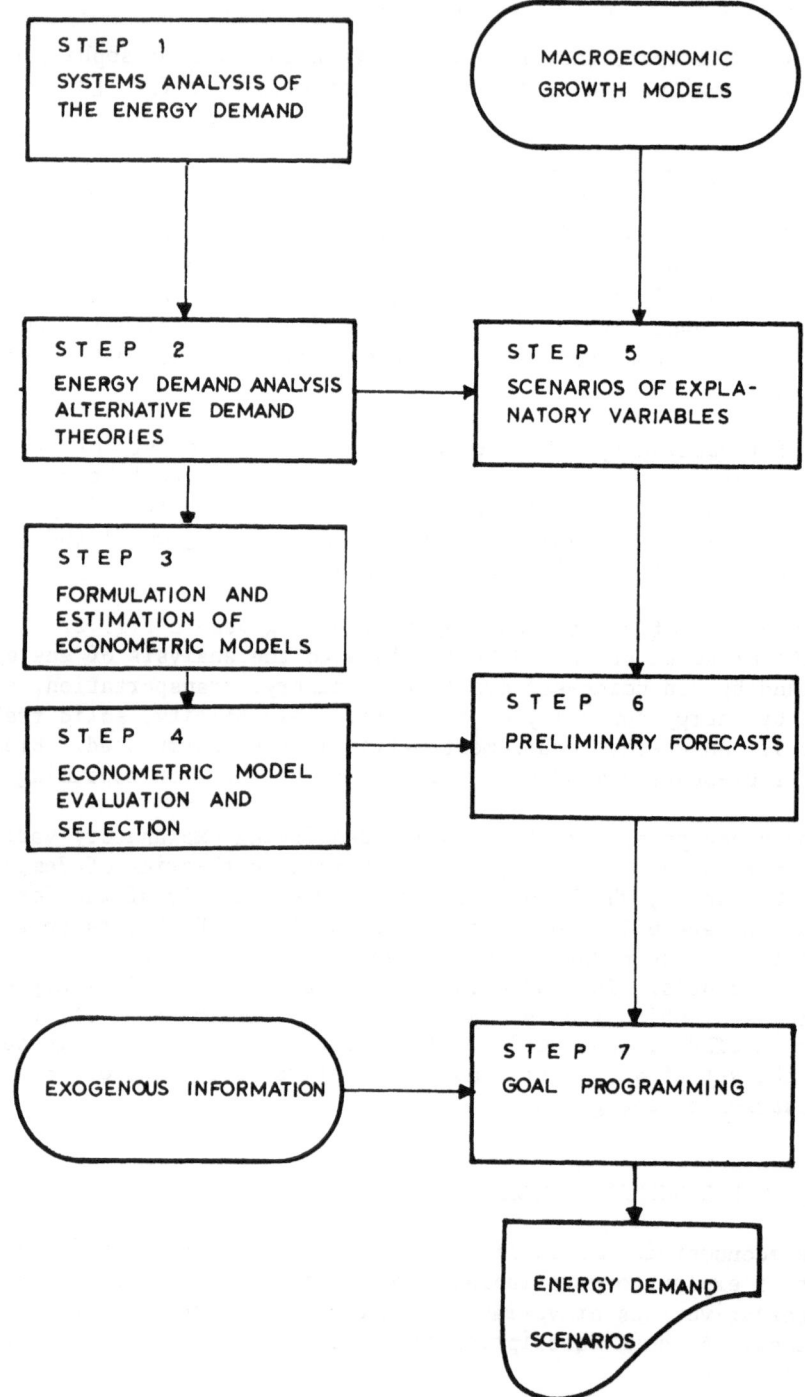

FIGURE 2 - A block diagram of the energy forecasting system
(Source [11])

represented as a function of output, of the price of energy, as well as of the prices of other basic inputs, such as capital and labour. On the other hand are models that analyse energy as a consumer's good use, as explanatory variables, personal income and the price of energy. In both cases, the dependent variable is the demand for total energy, or the demand for each fuel separately. When we model the demand for each fuel, we include prices of all fuels in the model to take account of the interfuel substitution effects.

As a special case of classical demand theory, we have also included models of "energy intensity", using the energy content per unit of output as a dependent variable. The price of energy is, of course, an important independent variable. Time is also included among the independent variables to reflect the influence of technological change.

When analysing the energy intensity of aggregates, such as the economy as a whole or the manufacturing sector, we include certain specific variables to take into account changes in the structure of gross domestic product. The "gross domestic product mix" is defined as the ratio of the product of the manufacturing sector to gross domestic product. Similarly, we introduce the "manufacturing product mix", defined as the ratio of the output of the five energy intensive industries (paper and pulp; chemicals; cement and others; metal processing including iron and steel, and aluminium; and oil refineries) to the output of the whole manufacturing sector.

The second group includes models of the "fuel shares". The share for each fuel is defined as the ratio of the demand for this fuel to the demand for total energy. The models are based on the hypothesis that the share of each fuel depends on the ratio of prices – e.g. price of the particular fuel to the price of total energy – and the existing technology as reflected either by the time variable or by lagged values of the dependent variable [5,6].

The third group of models takes the view that the growth of energy demand in a fast developing economy, moving from a stage of agricultural predominance to industrialization, cannot be explained only by substitution phenomena of the type expressed by the classical theory of demand. Capital formation in the economy as a whole and particularly in the manufacturing sector, must have played a major role and should be used as an explanatory variable. Thus we formulate models analysing demand for total energy as a function of the price of energy and changes in capital stock. We also formulate models of the energy used per unit of capital and include as underlying variables the price of energy, as well as a time trend.

Finally, models in the fourth group, considering energy as an input
to technical systems, try to establish relationships between energy
demand and the systems output as expressed in physical terms. Thus,
in transportation energy is related to passenger-miles or ton-miles
or car-miles. In residential uses, in a broader context, we use
volume of buildings as an underlying variable, while in agricultural applications variables such as irrigated land, or quantity
of fertilizer, or horsepower of agricultural equipment are included.

Model formulation and parameter estimation is followed by a stage
of model evaluation and selection. Models are evaluated to test
the reliability of the parameter estimates using standard criteria
of the conventional econometric methods, i.e. using the boundary
values of such statistics as the t-value of the regression coefficients, the correlation coefficient, the Durbin-Watson index
etc.

Only models passing these tests are kept for further consideration.
But still, all those models that are "econometrically" acceptable,
that is, explain reliably the behaviour of the energy demand system in the past, might not be appropriate for forecasting future
states of the demand system. Conversely, it should also be pointed
out that in this stage two or more models for a given demand
might pass the evaluation test.

The limits of econometric methods in the current energy context
have been extensively developed in recent literature [2]. Some
of the main arguments are retained for a short discussion here.

Conventional models are designed to provide forecasts under conditions of a steady environment. However, steady economic growth
and relatively stable energy prices which have dominated the
pre-1973 era, cannot be extended to the future. Thus, values of
parameters estimated on historical data are not a reliable indication of future performance. This is particularly true for the
two key parameters of energy demand, viz the income elasticity
coefficient and the price elasticity coefficient. The former indicates the effect of personal income - in the case of final demand - on energy demand. The latter is a measurement of energy
price impact on demand. Both of these parameters will tend to
have in the future values which are different from past values [1,3].

It should also be added that traditional econometric forecasting
was usually applied to give separate forecasts for each energy
commodity. This was made possible because of a strongly partitioned
energy market. It is expected that in the future substitution
between energy commodities will be much more important than in
the past.

In conclusion, the break with past energy pattern, brought on by the events of 1973 and thereafter, has greatly curtailed the usefulness of conventional econometric models for energy forecasting. To take care of this fact, we have formulated the goal programming model described below.

6. THE GOAL PROGRAMMING MODEL

Goal programming (GP) tends to become a really popular technique among OR practitioners. It has been and continues to be used for a wide variety of purposes, as literature surveys indicate [7,8,9]. Among others, GP models for forecasting purposes have also been cited [10].

The GP model used in this paper provides revised energy forecasts based, on the one hand, on the preliminary forecasts given by the set of econometric models and on the other, on exogenous information. Here goal programming plays the role of an "information integrator".

The model we formulate revises the forecasts provided by the set of econometric models taking into consideration factors that could not be included in the specification of the models (cf. Fig. 3). All these factors and their impact on energy demand constitute, what might be called, "information exogenous to the econometric models". This information contains various types of elements:

- objectives of energy policy,
- constraints of technical nature,
- educated guesses about future values of key parameters, such as income or price elasticities of energy,
- direct estimates of certain energy demands.

The GP model contains five types of constraints,

   i. simple energy demand goals;
  ii. goals for the share of each fuel in the energy aggregate;
 iii. energy demand bounds;
  iv. bounds for the share of each fuel; and
   v. identity equations.

The objective function minimizes the total sum of the deviational variables, thus expressing that energy demands and fuel shares should be as close as possible to their goals.

The simple energy demand goals formulate a goal for each energy demand variable. These goals might be either projections provided by the econometric models or they might be results of technical

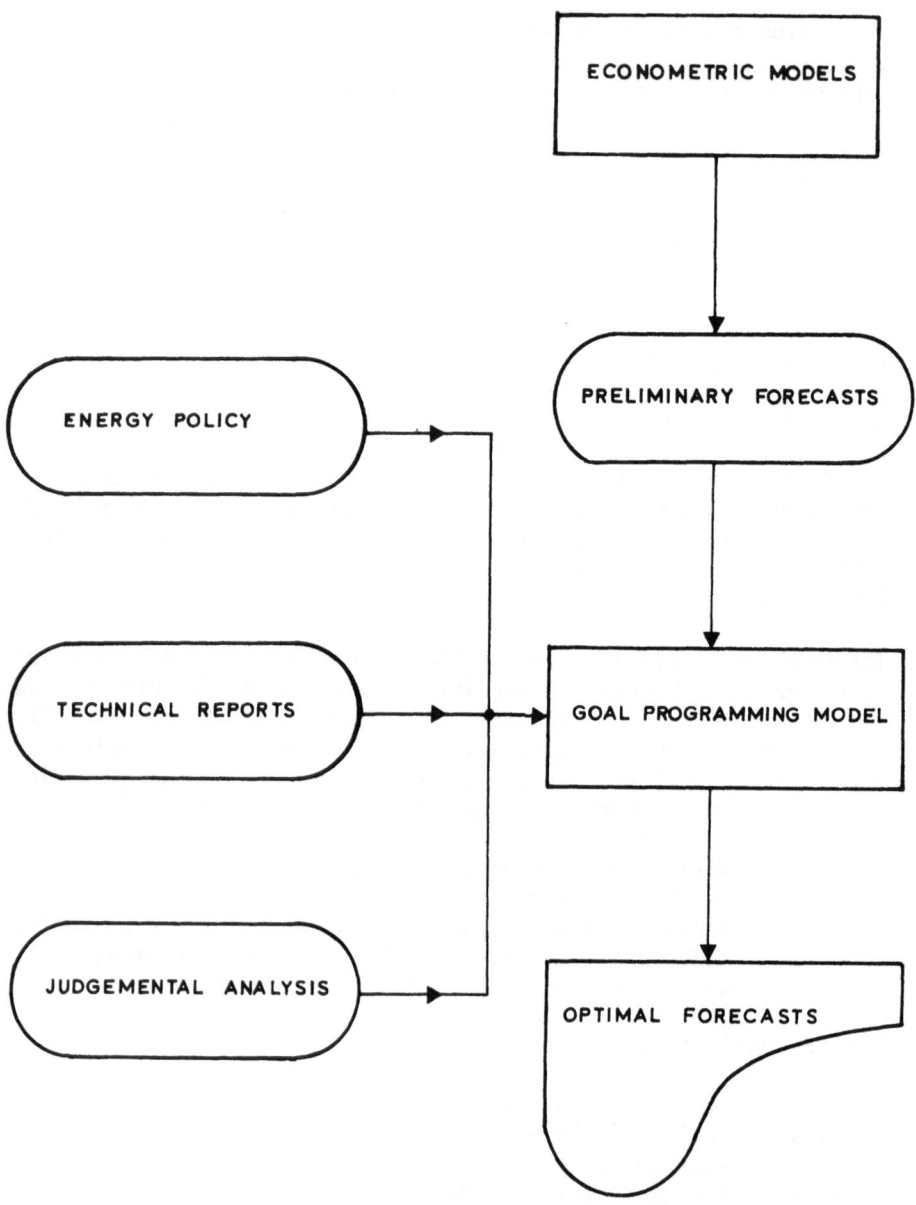

FIGURE 3 - Goal programming as an "Information Integrator"

reports. For example, the Greek energy system does not include natural gas in its present state and no econometric forecasting is therefore feasible for the demand of this fuel. In this case a goal of natural gas demand will be based on a feasibility report, resulting from some type of cost-benefit analysis.

Constraints of the second group formulate goals for the share of each type of fuel in the energy mix. These goals will be calculated either from the econometric projections or through some policy objectives. Let us refer to a particular example; it is a policy objective for almost all oil importing countries that the share of oil in total energy should decline in the future. The exact value of this goal can be arbitrarily estimated taking into consideration past performance of the system, and assumptions about policy measures. Goals of this group are crucial in our forecasting system and can be the subject of post-optimal parametrization procedures.

The third and fourth groups of constraints formulate upper and lower bounds for energy demands and for fuel shares in the energy aggregate, respectively. These bounds might also be either projections of the econometric models or results of exogenous calculations based on technical considerations. Let us take two examples. Electricity demand by the transportation sector is actually very low in Greece. It is only used in a limited scale for city transport in the Athens area. There are plans, however, for extensive use of electricity both for metropolitan and intercity transport. Assumptions about the progress of implementation of these plans, based on the relevant technical reports, can give us bounds for electricity demand in the transportation sector. The share of diesel oil in energy demand for space heating in the residential sector could be used as the second example. At present, this is the dominant fuel for space heating. It is an obvious policy objective to have it replaced by other energy commodities such as solar energy, natural gas, or even electricity. Thus the present share of diesel oil is taken as an upper bound for the future. The lower bound can be calculated on certain assumptions about investment in new buildings and retrofit installations in existing buildings.

The identity equations of the goal programming model provide for the internal consistency of the forecasts. There are two types of identity equations; identity equations across the lines of the energy requirement matrix, and identity equations across the columns of the matrix. The first type states that, for each sector, demands for the various fuels should sum up to the demand for total energy. The second type requires that the sum of energy demands - either for total energy or for each fuel separately - by each subsector should equal energy demand by the consuming sector.

It could thus be said that goal programming is used as an integrating device. It integrates information provided by a wide variety of sources in order to produce a consistent set of energy forecasts. Among the sources of information used one could distinguish:

   a.   the econometric models, giving information about the
        past behaviour of the energy demand system,

   b.   governmental plans for future energy policy,

   c.   feasibility reports on introducing new technology,
        or switching from an old technology to a new one, and

   d.   subjective estimates of parameters.

It should also be pointed out that the level of sophistication of the GP model totally depends on the wealth of information provided by the integrated sources.

And a last remark on what the GP additionally offers to the forecaster. It gives him the possibility to make concurrent use of the information provided by two or more econometric models explaining the behaviour of the same energy demand. Thus a model of classical demand theory and a model of capital formation can be both used to provide preliminary forecasts for total energy demand by the manufacturing sector. Bits of information included in these models are fed into the GP model through preliminary forecasts and are integrated to give a set of consistent results. This is important, because it helps to relax the assumption of stable econometric structure, underlying the use of conventional econometric forecasting.

7.   IMPLEMENTATION AND RESULTS

The energy forecasting system described above was applied to provide long range forecasts of the energy requirements of the Greek economy.

A Greek energy requirement matrix was developed consisting of four columns, representing four energy commodities, and twenty lines, representing twenty consuming sectors in a hierarchical structure. A data bank was then constructed containing all matrix elements for the period 1958-75. Only annual energy consumption was considered.

Demand functions, for both final and intermediary demand, contained lagged values of the dependent variable, which is quite common in econometric work, especially when time series are being used [5]. All demand functions were of the double-logarithm form, i.e. of the so-called Cobb-Douglas type. These functional forms, being the most frequently used, are linear in logarithms and thus render the estimation process by means of ordinary least squares relatively simple.

A great number of econometric models were formulated and their parameters estimated to fill the requirements of the goal programming model and at the same time represent all demand theories. Our attempt was to be left, after this evaluation and selection stage, with at least three models for each element of the energy requirement matrix.

The forecasting period, extending up to 1990, was divided into two subperiods by considering the year 1982 as an intermediate planning horizon. The scenario approach was used to give us preliminary energy forecasts for the two above mentioned years. Economic growth and energy prices were designated as the key scenario variables.

Future assumed paths of the scenario variables were combined to give four scenarios. Scenario 1 assumes a rapid industrialization process oriented towards energy-intensive sectors under a regime of stable energy prices. It is, more or less, an extension of past trends. Scenario 2 also takes into account a rapid growth of the manufacturing sector, but with only a moderate development of energy-intensive industries and under conditions of considerable increases in the price of oil. Scenario 3 is similar to 2 in terms of economic growth, but assumes only a moderate rise of the price of oil. Finally, Scenario 4 assumes a high rate of oil price increase, as does 2, but sets forth a balanced growth of all sectors of the economy; thus, under this scenario the process of industrialization of the Greek economy will stop, and the manufacturing sector will grow at the same rate as will agriculture and the service sector.

It can be observed that the four scenarios were formulated in a manner so as to illustrate the main answers that would be given to the big dilemma faced by economic policy makers in Greece, i.e. industry and in particular heavy industry, industry but light and heavy industry in a balanced mix; or an equal growth rate of all branches of economic activity. The last alternative is obviously the best from the point of view of energy consumption but it has many other considerable drawbacks, above all of which is its questionable feasibility.

Two other points should also be remarked regarding the scenario selection. Scenarios 2 and 3 differ only with respect to future energy price increases, so that comparisons of their respective results could help in drawing conclusions about the effect of price on energy demand.

Scenarios combine price increases and economic growth in patterns that look most probable. One is inclined to think, at the one extreme, that the very energy thirsty economic growth of the first scenario is most probable in a context of stable energy prices. At the other extreme, a pattern of equal growth in all sectors could be opted for only under steeply rising energy prices.

Having formulated the future scenarios and having projected accordingly the econometric models to give us the preliminary energy forecasts, our next task was to formulate the GP model which would integrate all available information. As it is now clear, we had to formulate and solve eight different models each corresponding to each of the four scenarios and each of the two planning horizons.

In a concise version of the forecasting system already completed, each GP model included 58 variables and 38 constraints (LP rows). To be more specific, each model included:

- 16 structural variables (i.e. energy demands, and

- 42 deviational variables.

The set of 39 constraints is analysed as following,

- 16 simple energy demand goals;

- 5 goals for the share of each fuel;

- 9 fuel share bounds;

- 8 identity (or consistency) equations.

Eighteen bounds on energy demand variables were also included.

For an analysis of the results given by the optimal solutions of the GP models the interested reader is referred to [11], or even to [12] for further details. In this paper we only include a diagram, which shows total energy demand in the period 1965-1990 (see Fig. 4) and give some short comments.

To have an idea of the proximity of the optimal demands, given by the GP model, to the preliminary demands, given by the econometric models, we compared the optimal value of the objective function with the sum of demand goals. The ratio of the former to the latter, expressed as a percentage, is approximately 5% in the 1982 planning horizon and 10% in the 1990 horizon. Thus, it can be seen

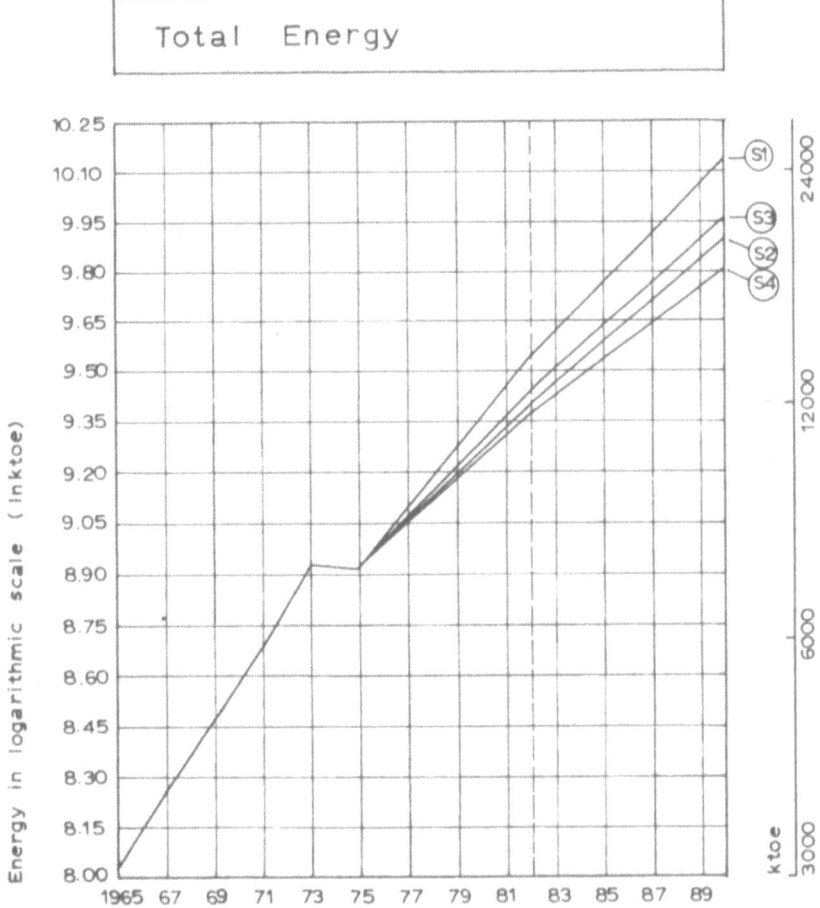

FIGURE 4 - Demand for final energy by the Greek economy 1965-90

1965-75 : Observed   S1 : Scenario 1   S3 : Scenario 3

1976-90 : Forecast   S2 : Scenario 2   S4 : Scenario 4

(Source: [11])

that in both planning horizons the GP solutions are not very distant from the preliminary forecasts. As one expects, the planning horizon affects the proximity of the two sets of forecasts. Horizons that are closer to the starting year do not create great deviations between preliminary and optimal forecasts. It is also obvious that the number of constraints which derives from the quantity of exogenous information, affects the sum of optimal deviations in the GP solution. The larger the quantity of information, the higher tends to be the optimal value of the objective function.

Post optimal analysis was also carried out to study the effect of the planning horizons on the forecast demands. It was demonstrated that the choice of the planning horizon did not alter the forecasts.

8. CONCLUDING REMARKS

Goal programming, in addition to the great variety of its uses as a decision model, proves to be of considerable help in forecasting, too. In this paper it was used as a main component of an energy demand forecasting system. In this context, GP performed the function of an "integrator" of information concerning the future growth of energy demand. Information was retrieved from many sources. The most important was the set of econometric models, which explained past behaviour of the energy demand system based on different approaches. Other sources of information used were government energy policy objectives, feasibility reports about new uses of energy in the Greek economy, and even the forecaster's guess. This last source was required in particular to fill the gaps left by the shortage of technical reports.

The use of goal programming as an energy forecasting system can be further extended, and we believe that it can become a component in some of the approaches to energy forecasting. In the system described in this paper, GP was combined with conventional econometric models. It is evident that it could also fit into the socioeconomic approach, which is the other type of the open-loop methods.

Another very challenging task is to investigate the possibility of integrating a GP model into a closed-loop system. It would seem that GP could be a most appropriate tool for an energy policy model designed to give the impact of the key policy variables on the economy. Its suitability is due to the fact that GP renders the formulation of a multi-objective process - like the design of energy policy - rather easy. But this is still a subject for further research.

We would like to conclude this paper with a final remark on the terms of this Advanced Study Institute. Energy modelling is a rather new field of application of applied sciences, which in our days has become a booming activity. It is viewed as a field of application to which four applied Sciences, macro-economics, micro-economics, management science, and engineering, have major contributions.

REFERENCES

1. B. Chateau and B. Lapillone, Long-term Energy Demand Forecasting: A New Approach , Energy Policy, pp. 140-157, June 1978.

2. W. Haefele and P. Basile, Modelling of Long-Range Energy Strategies with Global Perspectives , in: K.B. Haley, (ed.), OR'78, North-Holland, Amsterdam, pp.493-529, 1979.

3. Energy Modelling Forum,"Energy and the Economy", Vol. 1, Institute for Energy Studies, Stanford University, Stanford CA, 1977.

4. E.A. Hudson and D.W. Jorgenson, U.S. Energy Policy and Economic Growth, 1975-2000 , The Bell J. of Economics, 5(2), pp. 461-514, 1974.

5. Federal Energy Administration,"National Energy Outlook", U.S. Government Printing Office, 1976.

6. J.A. Hausman, Project Independence Report: An Appraisal of U.S. Energy Needs up to 1982 , The Bell J. of Economics, 6(2), pp. 517-551, 1975.

7. A. Charnes and W.W. Cooper, Goal Programming and Constraining Regression - A Comment , Omega, The Int. J. of Management Sci., 3(4), 1975.

8. J.P. Ignizio, A Review of Goal Programming: A Tool for Multiobjective Analysis , J. of Opl. Res. Society, 29(11), pp.1109-1119,

9. J.S.H. Kornbluth, A Survey of Goal Programming , Omega, The Int. J. of Management Sci., 1(2), pp.193-205, 1973.

10. J.E. Samouilidis, An Approach to a Flow of Funds Forecasting Model , M.Sc. Thesis, Imperial College of Science and Technology, London, 1970.

11. J.E. Samouilidis and I.A. Pappas, "A Goal Programming Approach to Energy Forecasting", a paper presented in the Third European Conference on Operations Research, Amsterdam, Apr.1979.

12. J.E. Samouilidis, Mathematical Models for Energy Demand and Applications , Unpublished Doctorate Thesis, National Technical University, Athens, 1979.

DISCUSSION

In the ensuing discussions, the following points emerged:

- All prices used in this study were exogenous and not derived from supply-demand equilibrium.

- Cobb-Douglas type of functions gave good estimates basically because of the nature of available data. however, care should be exercised when using this type of function.

- The weights chosen in goal programming affect the "index of proximity" to the preliminary forecasts. So, the index of proximity defined in this way will not give any information other than the case where all objective weights are one.

- The goal programming model in this study did not actually have any "decision" variables. It was used as a computational tool for the consistency of demand forecasts using:

  a. preliminary forecasts, and
  b. exogenous information.

  In other words, goal programming was used as an "information integrator". The lower and upper bounds for the demands were obtained from the preliminary forecasts and goals were derived from judgemental analyses, preliminary forecasts, energy policy scenarios, and technical reports. Equal priorities were given to each goal for simple analysis and in order for the index of proximity defined to be meaningful.

- Demand function estimations were based on ordinary least squares and not on simultaneous equations derived from utility functions.

INTEGRATION OF OPTIMIZATION AND SIMULATION MODELS

G. Egberts, R. Heckler, H.P. Schwefel, A. Voss

Programmgruppe Systemforschung und Technologische
Entwicklung (STE), Kernforschungsanlage Jülich
Federal Republic of Germany

ABSTRACT. Following a more general discussion of the methodological peculiarities and differences between simulation and optimization methods in energy planning, an outline of the principal
possibilities which allow coupling of simulation and optimization
methods is given. The Jülich Energy Model System then is used to
demonstrate practical possibilities for interfacing descriptive
and normative modelling approaches.

1. INTRODUCTION

A great number of mathematical models dealing with energy demand,
supply and utilization have been developed. The incomplete IIASA
review of energy models [1] shows a great variance with respect to
objectives, aggregation, time horizon, area of definition and methodology. As an aid for energy planning all models, despite their
differences, principally try to answer one of the following questions:

- What might our energy future look like?
- Which actions should be taken to achieve a certain
  goal?

Models which belong to the first category may be considered as
prognostic or descriptive, and display a kind of autonomous system
behaviour which is modified by boundary conditions. Whereas the
first approach belongs to a concept of "planning as reaction",
the latter category involves goal attainment, and thus assumes
autonomous planning, "planning as an action". The first category
comprises all kinds of econometric analyses like correlation or

regression analysis, input-output techniques, but it also includes the various approaches of simulation. The second question, however, lies within the domain of optimization methods.

But not only the general objective of model applications determines whether a descriptive or normative approach should be taken. Very often the application of optimization methods, though otherwise desirable, must be rejected due to a lack of well defined objective functionals or because of the nonexistence of control over important subsystems.

Thus, for energy planning a typical application area for optimization tools could focus on the technical system of energy supply and be directed towards the evaluation of guidelines for a goal-oriented investment or development policy. The environmental system, however, dealing with emission patterns, diffusion functions, enrichment chains etc., can conceptually only be treated by means of a purely descriptive modelling tool. As a consequence, a holistic approach to energy modelling cannot restrict itself to a single most appropriate methodology. Instead, a modelling project dealing with the technical supply system, with the environment, the national economy, or even the public acceptance should use a variety of different methodologies appropriate to the nature of the subsystem dealt with. Here a question arises about how to reflect the interactions which exist between the energy system and its related sectors by the use of a system of interconnection on the modelling level between modules of different goals and methodology. This is both a conceptual and e methodological problem.

We want to illustrate some examples of possible linkages between optimization and simulation models for energy systems, including both data and control flows. A software concept for the integration of data and models developed at Jülich will be shown and guidelines for further work will be displayed.

2. THE DIFFERENT NATURE OF SIMULATION AND OPTIMIZATION

2.1 Simulation

In many uses one cannot do experiments to investigate the behaviour of a real-life system. This might be true for reasons of cost, the risk of a large scale accident or one caused by the irreversibility of time. However, if a consistent and strict theory for predicting the behaviour of a system is also lacking, one has to study a model to obtain any kind of detailed information about the system. Simulation as a modelling technique is, generally speaking, experimenting on a stand-in or substitute for the object under investigation. Although this broad definition of simulation also includes, for example, aerodynamic investigations on automobiles or aeroplanes using wooden models, we want to concentrate on mathematical simulation, wherein the system is represented by a model consisting of mathematical relations, which are suitable to be run on a computer. Normally in operations research a mathematical simulation model is understood as a computer test facility where the real world environment of the system to be tested is represented by random events. Although this description is important for many practical applications, we will understand simulation of energy and economic systems here as deterministic simulation. That means that our system - energy and economy - is represented by a set of differential and ordinary equations appropriate for determination of the system behaviour over time. Thus, it is believed that the general laws of the historical development of our system can be expressed in an analytic way, such that the lack of predictability is only caused by incomplete knowledge about boundary conditions or structural coefficients. Hence, dynamic simulation as a tool is directed towards better prediction of the system development by a reduction in its complexity. This means that knowledge about internal relations of the systems is used to reduce the amount of necessary exogeneous input to a minimum.

2.2   Optimization

Although optimization as a tool for goal-oriented decision making is a normative approach, it has some descriptive elements, too. This is especially true when we are dealing with national economic sectors of which the energy problem is one example. An optimization model of a whole economic branch can be interpreted as a simulation of market processes. If one remembers the classical doctrine of liberalism, that the market place converts the sum of "private vices" into "public virtue", the conceptual similarity with an algorithmic optimization process to obtain a globally extremal system configuration is evident. Thus, dual activities resulting from a linear programming model can be interpreted as simulated market prices. In addition, an optimization model can yield other valuable information. In the field of energy planning it may be that the most important yield could be the order of preference for single or groups of new technologies. A comparison of scenarios, assuming availability and nonavailability of certain technologies, can define their weight in terms of goal criteria like overall costs, oil imports, or environmental damage. This information can be valuable for the assessment of economic suitability of development expenses. Perhaps the most valuable information is obtained by parametric studies. At first, the stability of a certain solution with respect to individual or collective parameters must be investigated. This is especially true for long term energy planning with its considerable data uncertainties. Another important application of parametric programming is the determination of tradeoff curves between possibly conflicting objectives. For example, the marginal costs of oil saving as a function of the savings already gained is certainly one of the most important pieces of information needed for economic considerations with respect to energy.

Thus, in contrast to a pure simulation approach, which is only able to display consequences of well defined action, optimization can yield selection rules for the decisions themselves. It gives in a quantified manner recommendations of actions to be taken once a goal is defined. This may be the reason why optimization methods, especially linear programming, have gained such a wide area of application, especially in corporate planning.

3. JES-JÜLICH ENERGY MODEL SYSTEM

For many years computer assisted decision aids for energy economics and energy politics have been developed in the Programme Group of Systems Analysis and Technological Development (STE) [2-8]. The investigations began with the preparation of worldwide energy models and environment models for the analysis of world energy demand and the possibilities available to meet it. After that we also started research for energy models for the Federal Republic of Germany. At present, the accent is on permanent development and practical application. In most cases the use of a model allows one to analyse and answer only a certain category of questions. In the last few years the wide spectrum of problems concerning energy economics and energy politics has not only led to the continuous enlargement of the existing models and thus leading to an ever broadening range of applications, but it has also led to the development of new models for actual problems. We think that it is self-evident that different methods like simulation, optimization, input output techniques - to refer only to some of them - have been used; just as there does not exist a model which answers all the questions, there does not exist a best method. Each of them has specific advantages and disadvantages, and the choice must always be orientated towards the question.

With the growing number and the increasing complexity of the models, the bulk of data to work with became gradually larger, and data processing became more and more important. The problem of the coupling of separate models arose and also problems grew from the increasing requirements for accessibility to the user, and the need to provide plot-and-report software.

This led to the development of an integrated system of data, methods and model base, which will be explained in the following. Because it is of importance, the coupling of different models especially of simulation and optimization models will be discussed here.

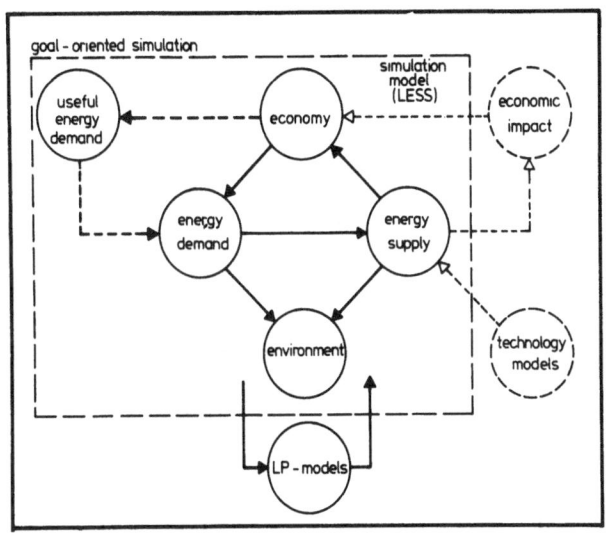

FIGURE 1 - Energy model system

In Fig. 1 the different models and modules of our energy model
system and their most important interfaces are represented sche-
matically. First, there is the long term energy simulation model
which consists of four modules. The modules of energy consumption
and energy supply represent the center of considerations. They
are the two essential aggregates, because they determine the relations of supply and demand for each energy carrier and the results of technical changes in the production, transformation, distribution, and utilization of energy carriers. The energy sector
is referred to in the model by means of the presentation of many
separate processes. To elucidate the fact that the energy sector
is embedded in the rest of the economy, and to be capable of comp-
rehending the ecological consequences of alternative energy strategies, the module of energy consumption and energy supply are
coupled with an environmental module and a macroeconomic module.
The structure of the long term simulation module is described more
precisely in [9] and will not be explained here.

At the moment the useful energy demand model shown in Fig. 1 is
being worked on. The Economic Impact Model which had been developed by Y. Kononov [10] is used to determine the direct and indirect requirements of alternative strategies of energy supply on
the other economic sectors. The technology models, for example
for different heating systems, form a further addition to the model
system. If the simulation model (LESS) enables one to use a method
which is largely oriented on events seeking to answer questions

such as "What happens, if ...?", then in the goal-seeking simulation, which will be referred to more closely later on, the question "What must be done to ...?" is brought into focus.

The linear optimization models, which are part of the model base
system, show similar normative characteristics. From this figure
the coupling of simulation and optimization models becomes evident.
Later on, the coupling of linear optimization models with dynamic
simulation models, and the enlargement of dynamic simulation models to so-called goal-seeking models by superposition of heuristic optimization processes will be discussed.

Fig. 2 gives the organizational structure of the integrated system
of data, methods, and model base. In addition to the energy bal-
ances, the data base consists of a large number of statistical
time series from both the energy and the economic sector. The
Interface for Regression and Correlation Analysis (IRECA) is an
interactive method base for nine different methods of linear and
nonlinear regression and correlation analyses. DAIMOS (Data
Interface for Modular Simulation) represents a control system under which dynamic simulation models can be run. The most important
characteristics of DAIMOS are:

- automatic delivery of all time-dependent and time-
  independent input data combined with diagnostics
  about the completeness,

- automatic storage of the output data,

- automatic handling of unsorted and coupled equations,
  and an

- interactive output mode.

The third important interface of the whole system is called OASIS
(Optimization-And-Simulation Integrated System). OASIS contains
all important, derivative-free parameter optimization techniques
which can be superimposed onto the DAIMOS-guided simulation model.
No changes are necessary within the simulation model itself, and
during a search for a maximum or minimum the optimization method
used can be changed interactively.

The last component of the model system are the two time-dependent
linear optimization models MARKAL (Market Allocation Model) and
MESSAGE [11,12]. Both models are similar in structure, although
MARKAL is much more detailed. Both models represent the energy
supply system, comprising of all important conversion, transportation, and distribution steps from primary to final or useful
energy demand in various end use sectors. Both models use the
standard MPSX software.

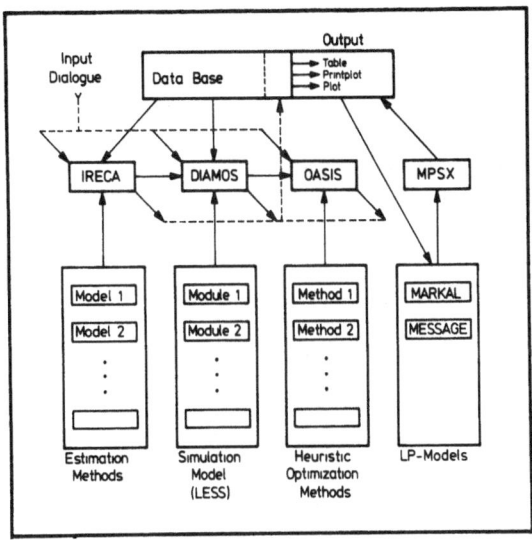

FIGURE 2 - Organizational structure of the model system

The whole model system JES is programmed in standard FORTRAN and as far as possible, all interfaces operate with identical input and output formats.

In the following, we will explain in some detail the coupling of the long term dynamic simulation model (LESS) with the linear programming model and the extension of the simulation model to a goal-oriented model, by superimposing direct search methods.

4. COUPLING SIMULATION MODELS WITH LINEAR OPTIMIZATION MODELS

Before dealing with methodological aspects of interfacing different kinds of models, let us give a short description of the linear energy optimization model MARKAL. MARKAL is a dynamized linear programming model of energy production, transformation, and utilization developed jointly at BNL and KFA. It covers at present a time horizon of 40 years, from 1980 up to 2020. Thus, with a time spacing of five years, it consists of 9 static submodels, interconnected by a set of interperiod constraints. Three different objective functions are used at present:

- Total discounted system costs
- Cumulative oil imports
- Cumulative thermal heat releases.

These functionals should represent the three most important goals of energy policy, namely economic suitability, security, and environmental neutrality of the energy supply. The model covers for every energy carrier considered: extraction, transportation, transformation or refining; distribution, and utilization. Including the sectoral useful energy demand and the nuclear fuel system, 30 energy carriers are incorporated into the model. The load duration structure of the electricity demand is modelled by a disaggregation of electricity supply and consumption into six energy categories and six power categories according to the seasonal demand variation. District heating is modelled in a similar fashion.

The main sectors of the model can be summarized as follows:

- Residential and commercial: space heating
- Residential and commercial: other applications
- Transportation
- Industrial applications
- Electricity and heat production
- Petroleum refining
- Coal gasification and liquefaction
- Nuclear fuel fabrication and reprocessing
- Indigeneous production
- Imports and exports.

In total, about 70 distinct technologies are included in the model at its present stage. This led to an overall matrix size of about 2500 rows and 3000 columns.

Coupling simulation and linear programming optimization by control transfer, there are essentially two distinct types of interfacing:

- Optimization algorithm calling simulation procedures,
- Simulation algorithm calling optimization procedures.

The first approach, applying simulation as a procedure, is a tool to extend the scope of existing optimization procedures. Using a linear programming model, it might be used to integrate a sector by simulation which would be too complex to be formulated within a linear programming model. Thus, the simulation procedure could serve as a linearization of a nonlinear subsystem. Another application could be the simulation of an endogeneous setting of bounds to the linear programming based on linear programming results obtained up to that point. The area of environmental planning in connection with energy supply could be envisaged as a possible application of this feedback. Another example of calling a simulation programme would form a kind of decomposition approach.

The simulation procedure then generates solutions to subproblems; whereas the linear programming model is restricted to the reduced master problems. This procedure seems meaningful whenever sub-systems with relatively few degrees of freedom can be identified. Thus, if one can pretty well distinguish mainly-driving sectors from mainly-driven ones, a considerable reduction of problem size should be obtainable.

The second approach, calling optimization procedures by a simu-lation algorithm, can be used to simulate a sequence of short term optimal decisions. Within the energy system such an optimi-zation procedure is able to effect a sequential updating of struc-tural coefficients in the simulation as a result of sequential investment decisions of utility companies. Such an optimization routine could yield important feedback to the overall system, in-cluding, in addition to the technical development of energy supply, such data as price levels of energy carriers, investment require-ments, or pollution levels.

Coupling by means of a simple data transfer between simulation and optimization systems is easier to establish. Thus, let us discuss possible data flows between simulation and optimization in energy planning using the example of the Jülich Energy Model System (Fig. 3).

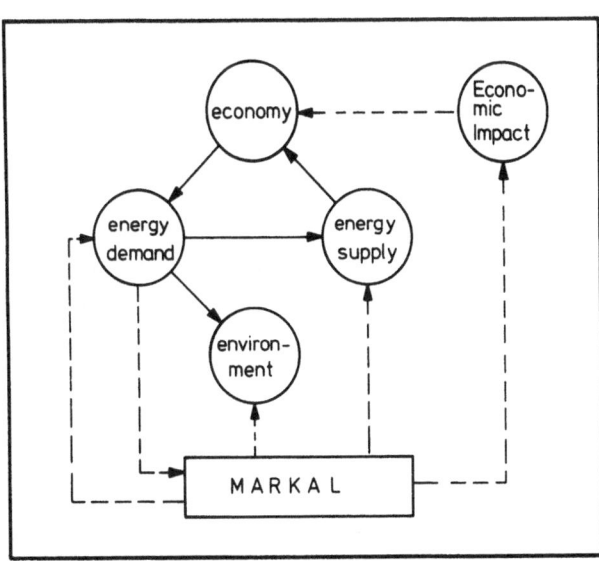

FIGURE 3 - Data transfer between the simulation and
optimization model

The set of couplings provided by data transfer, which we consider
to be most important, is shown with the following diagram that
includes different simulation and optimization models dealing
with energy supply, macroeconomic interaction, investment planning
and environmental effects.

Thus, the possible and rational feedbacks by data transfer within
our modelling concept can be summarized as follows:

- Determine the useful energy demand as an input to the
  energy supply optimization model provided by the simulation model.

- Use the fuel allocations and changes in the technology
  mix as evaluated by the optimization model to set
  structural coefficients of the simulation model.

- Let the shadow prices of energy carriers given as an
  output to the optimization model drive cost structures,
  behaviour relations or substitutional effects within
  the macroeconomic part.

- Take the activity levels in the technical energy system as output from the optimization model to determine
  environmental quality scenarios by the ecological
  module.

- Necessary investment calculated by the optimization
  model may serve as inputs for the Economic Impact model.

5. SUPERIMPOSING DIRECT SEARCH METHODS ONTO DYNAMIC SIMULATION MODELS

In addition to the integration of the optimization and simulation
models which have been discussed in the previous chapter, one can
choose a different method. This is the extension of the descriptive simulation model to a goal-oriented normative model by way
of the direct methodical coupling of simulation and optimization.
Extensive simulation models of the energy system, for instance
LESS, are normally very comprehensive; that is to say, they contain a great number of variables and relationships, which are
partly nonlinear and which frequently contain feedback relations
between important variables. Furthermore, in the case of simulation models, they deal preferably with a nonanalytical mathematical description of a system, which does not allow in general an
analytic determination of the partial derivatives.

If one does not intend to adapt the simulation model to the specific requirements of special optimization methods, one may in

general consider for the optimization of simulation models only
those methods which require a sequence of values of objective
function as internal information because the partial derivatives
cannot be generated and because nothing can be said about the
topology of the system of equations. Those methods whose (operational) characteristics are only based on a comparison of values
of objective function are called "direct search strategies". They
are partly of heuristic nature and there is no theoretically found-
ed guarantee for the convergence to the absolute optimum, as in
the case of linear programming. But they have proven to yield
practical solutions even when other methods fail.

Present known direct search methods include quite a number of different strategy concepts. Fig. 4 lists the derivative free procedures included in OASIS. It is not possible to go into details
here about the different direct search methods, but let us say a
few words about a rather new one, the so-called"evolution strategy".
A detailed description and comparison of the different strategies
may be found elsewhere [13]. It is based upon a simple imitation
of the basic rules of biological evolution: mutation, selection,
and recombination.

List of Strategies Compared

Code	Description of strategy or variant
FIBO	Univariate strategy with Fibonacci search
GOLD	Univariate strategy with golden-section search
LAGR	Univariate strategy with Lagrange interpolation
HOJE	Strategy of Hooke and Jeeves (pattern search)
DSCG	Strategy of Davies, Swann, and Campey with Gram-Schmidt orthonormalization
DSCP	Strategy of Davies, Swann, and Campey with Palmer orthonormalization
POWE	Strategy of Powell (conjugate directions)
DFPS	Strategy of Davidon, Fletcher, and Powell (variable metric) modified by Stewart
SIMP	Simplex strategy of Nelder and Mead
ROSE	Strategy of Rosenbrock (rotating coordinates)
COMP	Complex strategy of M.J. Box
EVOL	(1+1) evolution strategy
GRUP	(10,100) evolution strategy without recombination
REKO	(10,100) evolution strategy with recombination

FIGURE 4 - Direct search strategies included in OASIS

The simplest concept of an imitation of biological evolution is the binary evolution strategy. Mutation and selection are regarded as regulations for the variation of the parameters and for the recursion of the sequence of iterations. This simple strategy can be described in the following way: A population consisting of two individuals, the parents and one descendant shall be given. The descendant differs insignificantly from his father.

The variations are random and independent from each other. Both the individuals have a different fitness because of their variations. Therefore only one of them is able to get further descendants and this is the one who represents the greater value of vitality.

Extended strategies which represent higher levels of limitation of the evolution events start from the idea of a larger population and realize as well the recombination of characteristics which are possible with sexual propagation.

The evolution strategy is not a Monte Carlo method, although it contains some stochastic elements. Mutations are not pure random settings of the parameters but changes of the variables from one iteration (generation) to the other belong to a Gaussian distribution. The parameters of that distribution, variances and covariances are attributes of each individual, just like the object parameters of the function to be extrematized. And they are changed from one generation to the other, too. By the selection of the fittest, the population does not only creep towards the optimum, but also adapts the parameters of the random mutability and thus accelerates the convergence, for example on ridges or in narrow valleys. Moreover, if the population is large enough, this method gives a rather good chance of finding a global out of several local optima, and there are nearly no restrictions to the type of objective functions. The evolution strategy has proven to be the most reliable one out of all known direct search methods, especially when the number of variables is large.

Combining simulation and direct search optimization may be done in two principally different ways, optimization within simulation, and simulation within optimization. For preassigned time steps within a simulation run the optimization algorithm may be called. In this case the optimization algorithm can be incorporated as a subroutine of the simulation model.

For a dynamic model, it is not always sufficient to optimize the system for one single moment, and even a sequence of optimizations for consecutive time points will usually not lead to an overall optimal solution. The path of a dynamic process within a definite system will be determined by system parameters, i.e. initial values and coefficients of differential equations. To achieve overall optimization, it is necessary to run the model over the whole period for each parameter setting.

In principle, the optimum seeking technique handles the simulation program as a "black box". It generates consecutive parameter settings $p = \{p_i; i = 1(1)n\}$ as input and receives output values $F(p)$ depending on the objective chosen. Instead of a series of optimizations within one model run, a series of model runs within one optimization task is performed.

OASIS is constructed in such a way that there is a minimum of linkage between the simulation model and the optimum seeking programme. The specifications necessary are:

-   The optimization strategy chosen.

-   A time limit for execution as termination criterion in addition to the normal convergence criterion.

-   Accuracy parameters for the direct search method chosen.

-   List of names of parameters to be varied.

-   The name of the objective function, including information whether a minimum or a maximum is searched for.

-   Names of items to be used for evaluating constraints.

In the following, we will now demonstrate the integration of simulation and direct search methods by means of an example. This example should only be taken as a demonstration of the methodological procedure; and not from an energy policy point of view.

What will be shown is, that direct search techniques enable the user to find those parameters or time series within a dynamic simulation model which maximize or minimize an integral criterion under restrictions given to other resulting variables or derivatives of them. In principle, a solution by hand is possible, too, but would cost even more simulation runs and give no certainty of having arrived at the desired solution.

Using the dynamic simulation model LESS, the following objective function was chosen.

$$\int_{t_i}^{t_f} (MPO(t) + MPM(t))(t - t_i)dt \rightarrow min$$

$$t_i = 1985; \quad t_f = 2000$$

This is the integral over the mineral oil (crude MPO and refined MPM) imports weighted with time. As free parameters two times series were chosen:

FCTX(t) : the quota of methanol added to motor spirit,

CATNL(t) : the capacity of high temperature reactors used
for production of process heat to gasify lignite,

each of which was given as base points the years 1990, 1995, and 2000. The values for 1985 were set to zero.

Constraints were given to

$MPN(t) \leq RMPN(t)$: the imports of natural gas

$MPC(t) \leq RMPC(t)$: the imports of hard coal

$MGB(t) \leq MCB(t)$: the indigenous mining of lignite

according to exogeneous time series.

Methanol production as a new conversion technology uses gas which could be imported as natural gas or produced as synthetic natural gas by nuclear lignite gasification. Other possible options were not used in this case. Lignite now mainly is used for producing electricity. The indigeneous mining being limited (imports are negligible) lignite gasification reduces lignite electrification which has to be compensated by other fuels. In this case, hard coal had to fill the gap, but mining and imports of hard coal were restricted, too. On the other hand, lignite gasification by means of nuclear process heat produces electricity and coke (to be used in blast furnaces e.g.) as byproducts, thus changing the balance for other energy carriers. An additional constraint had to be added in order to ensure that the remaining amounts of lignite for production of electricity would always be positive. There is not enough space here to explain all other relations within the energy supply module being affected by a combined methanol production and lignite gasification strategy.

Fig. 5 shows the development of the gas input for methanol production and of the amount of gas produced by lignite gasification. The latter being higher, especially towards the end of the time period is due to the restriction of natural gas imports.

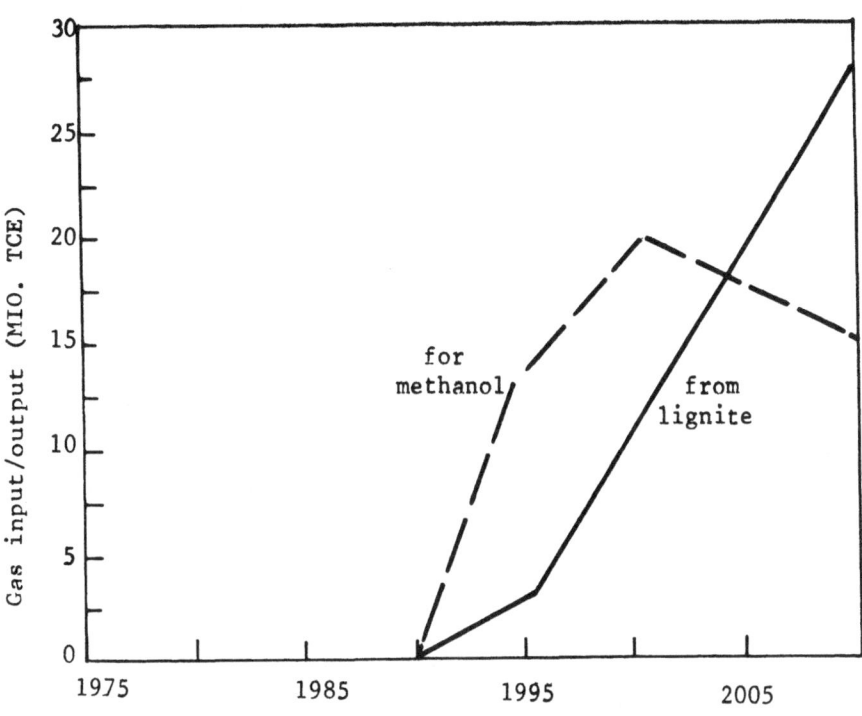

FIGURE 5 - Gas input for methanol production and gas
output by lignite gasification

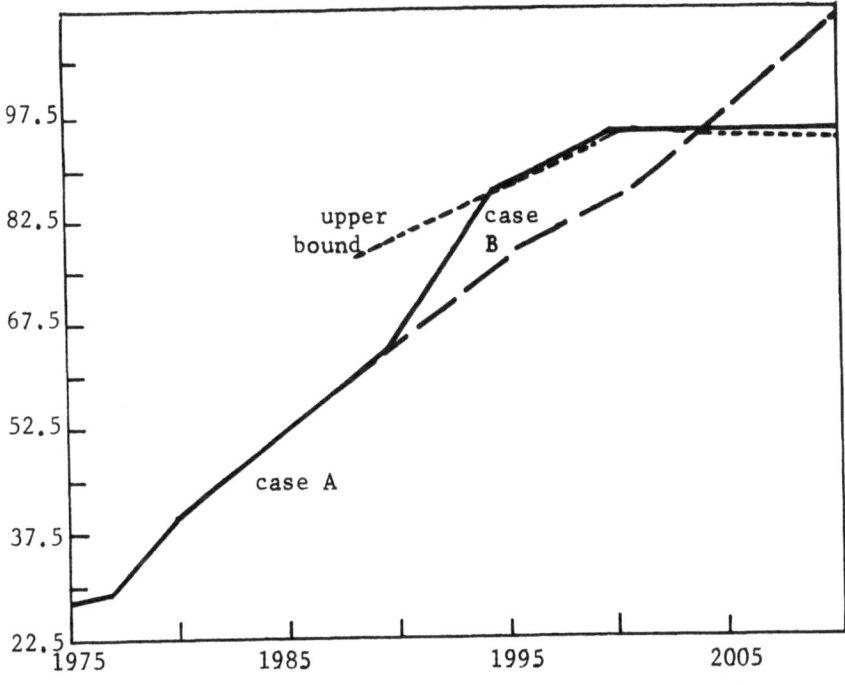

FIGURE 6 - Imports of natural gas

204

From Fig. 6 one can see that this constraint is violated in case A (no methanol and no gasification, which is the initial state). That means that the optimization had to start from a nonfeasible point.

Fig. 7 demonstrates how the missing lignite for electricity production has to be replaced by a corresponding amount of hard coal.

Finally the imports of crude oil and petroleum products which were minimized are shown in Fig. 8 for the initial (case A) and the final state (case B) of the optimization task.

Let us finish with a résumé. The field of energy planning comprises such a variety of different aspects and objectives that one cannot restrict oneself to a certain methodology of mathematical modelling. Using different methodologies such as simulation and optimization in parallel, the problem of interfacing immediately arises. As model development and application progress this problem area becomes more and more important. This includes both methodological and organizational aspects. As an example of models used at STE, some possible and rational interconnections have been discussed. The task of making a formal integration of our models into an integrated operating system for data and control transfer is rather new. Thus most work still remains to be completed. However, we do not aim for a kind of integrated "supermodel", perhaps with doubtful results. Our intention is more to establish easy transfer of data or control without excluding explicit control functions by the user himself.

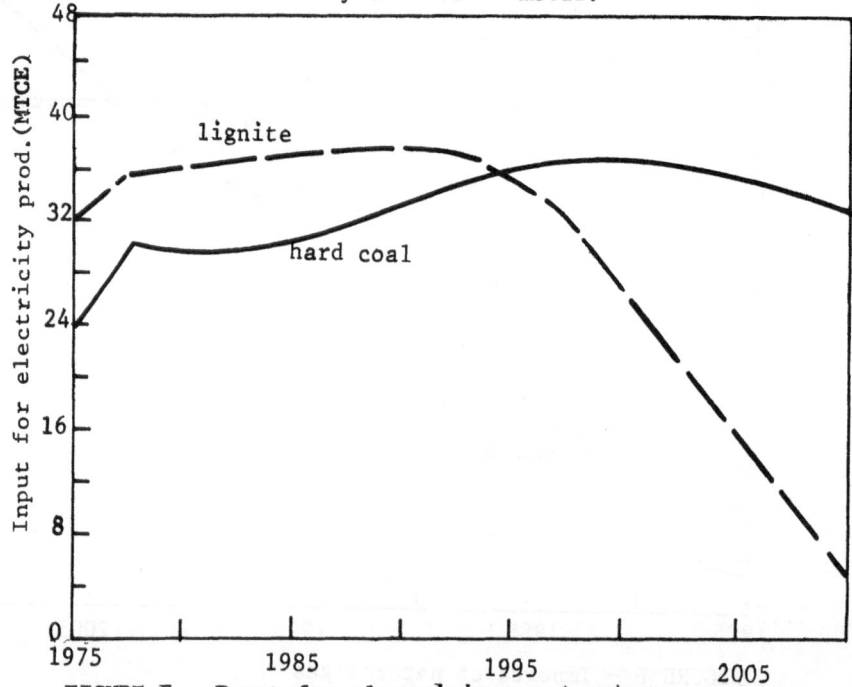

FIGURE 7 - Input for electricity production

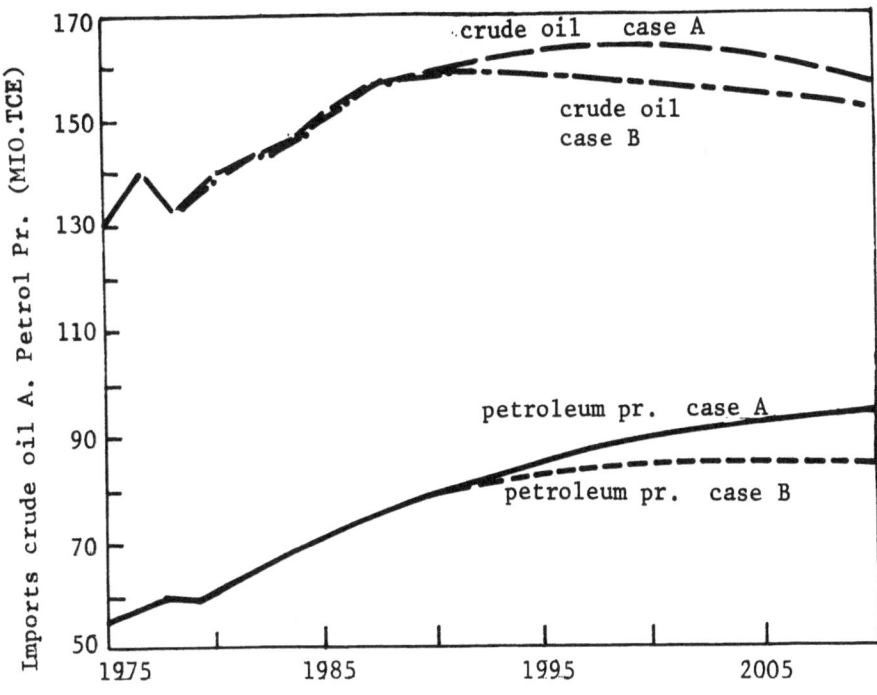

FIGURE 8 - Import of crude oil and petroleum products

REFERENCES

1. J.M. Beaujean and J.P. Charpentier,"A Review of Energy
    Models No. 4,"July 1978, RR-78-12, International Institute
    for Applied Systems Analysis, Laxenburg, Austria, 1976.

2. St. Rath-Nagel,"Alternative Entwicklungsmöglichkiten der
    Energiewirtschaft in der Bundesrepublik Deutschland," ISR
    28, Birkhauser Verlag, Basel und Stuttgart, 1977.

3. A. Voss,"Ansatze zur Gesamtanalyse des Systems Mensch-
    Energie-Umwelt", ISR 30, Birkhauser Verlag, Basel und
    Stuttgart, 1977.

4. A. Voss, et.al.,"Dynamische Energiemodelle als Planungs-
    und Entscheidungshilfe dargestellt an einem Energiemodell
    für die Bundesrepublik Deutschland," in: Energiemodelle für
    die BRD, ed. by Ch. König, ISR 42, Birkhauser Verlag,Basel
    und Stuttgart, 1977.

5. K. Schmitz,"Langfristplanung in der Energiewirtschaft," ISR
    65, Birkhauser Verlag, Basel und Stuttgart, 1979.

6. K. Schmitz, H.P. Schwefel, Finding Reasonable Energy Poli-
    cies by Means of a Dynamic Simulation Model, in Proceedings
    of the International Symposium "Simulation '77", ACTA Press,
    Zürich, 1977.

7. U. Schöler, et.al., A Dynamic Energy Model for the Countries
    of the European Communities, EUR 5953, Commission of the
    European Communities, Brussels, Luxembourgh, 1978.

8. G. Egberts,"Kostenoptimale Entwicklungsperspektiven des
    Raumheizungssektors im Energieversorgungssystem der Bundes-
    republik Deutschland - Ein Optimierungsmodell", to be pub-
    lished.

9. K. Schmitz, W. Terhorst, A. Voss,"Simulation Techniques in
    Energy Analysis", Advanced Study Institute on Mathematical
    Modelling of Energy Systems, İstanbul, Turkey 10-12 June,
    1979.

10. Y. Kononov,"Modelling of the Influence of Energy Development
    on Different Branches of the National Economy,"RR-76-11,
    International Institute for Applied Systems Analysis,
    Laxenburg, Austria, 1976.

11. G. Egberts, et.al., MARKAL,"A Dynamized Linear Optimization
    Model of the Energy Supply for the Purpose of the Interna-
    tional Energy Agency", to be published.

12. M. Agnew, L. Schrattenholzer, A. Voss,"MESSAGE, A Model for
    Energy Supply Systems Alternatives and Their General Envi-
    ronmental Impact", WP-79-6, International Institute for App-
    lied Systems Analysis, Laxenburg, 1979.

13. H.P. Schwefel,"Numerische Optimierung von Computer-Modellen
    mittels der Evolutionsstrategie", ISR 26, Birkhauser Verlag,
    Basel und Stuttgart, 1977.

DISCUSSION

Several participants were interested in hearing more on how the
technological models were coupled with the energy supply model.
Voss replied that this was mainly a question of data transfer.
Different cases (e.g. better insulation standards in housing)
were run on the technological models and the results used to
form inputs into the larger model.

Voss was asked whether he had considered using multi-objective
functions in the model. He replied that they had enough problems
at the moment with the two objective cases: cost effectiveness and

future oil imports - but this was certainly something that warrented closer study.

Voss' solution algorithm was essentially a hill-climbing algorithm. He said that their group had carried out an extensive comparative study of different solution methods before arriving at the EVOL technique. In response to a question, Voss said that the EVOL technique did not guarantee that one arrived at the global minimum. There was a danger of landing at a local minimum although the function that they had in their model was well behaved and did not contain local minima. One participant pointed out a general drawback of using a gradient approach and that was the case when the objective function was piecewise linear. The discontinuities in the function could produce wild results.

PART III

THE WORLD OIL MARKET

Understanding Energy

Long Range Pricing of Crude Oil

Choice of Modelling Technique: An Example
of a Simulation Model for World Oil Supply
and Demand

UNDERSTANDING ENERGY

R.J. Deam

Energy Research Unit
Queen Mary College

ABSTRACT. The world petroleum market is modelled, using a linear programming approach. The fundamental decision variables of the model are: extraction of different crude oil, transportation of crude oil and petroleum derivatives by different ship categories, refining activities, and prices. The model has been tested against real data for the years 1967-71.

1. INTRODUCTION

It is reasonable to assume that the customer wishes to minimise the amount of money he pays for energy and therefore looks for the lowest price. The price of oil products is determined by the price of the crude oil, the price of freight and the price of refining. Price dictates consumer behaviour; cost - of the crude, the freight and the refining - is only important insofar as it sets a lower limit on price.

Starting with this unexceptional view of consumer behaviour a picture - or model - of the oil and gas markets can be built up, and this can be further extended to encompass all other energy sources.

2. THE ENERGY MARKETS AND GIBBS' PHASE RULE

Elementary science teached us that - at equilibrium - if water and steam exist together then we cannot alter the temperature and pressure independently. If we set a particular temperature, then a corresponding pressure results. If we change the pressure, the temperature must alter to maintain equilibrium.

This limitation of freedom is most familiar to chemists in the form of J. Willard Gibbs' famous Phase Rule:

$$F = C - P + 2$$

where   F : no. of degrees of freedom (how many parameters we can vary)

            C : no. of components (one, water, in our example)

            P : no. of phases (two, water and steam, in our example).

For the water/steam equilibrium described

$$F = 1 - 2 + 2 \ .$$
$$= 1$$

If we fix the temperature the pressure is fixed and vice-versa. This system - at equilibrium - has one degree of freedom. It is termed <u>unidimensional.</u>

Even though oil and water do not mix, the oil market can be understood in closely analogous terms which form the basis for the analytical methods developed by World Energy Models (WEML) Ltd. The price of energy - like the physical parameters, temperature and pressure - is unidimensional.

The implications of this simple observation are profound. If there is only one energy price that dictates the price of all forms of energy in international trade, then given this price all other prices throughout the world can be calculated if we have the necessary data and computing expertise.

3. THE MARGINAL REFINERY

The logical basis for this model (picture, theoretical representation) of the international energy market can be demonstrated by involving the simple refinery shown in Fig. 1. Crude oil is split into three fractions. The top fraction, naphtha, boils at less than 200°C. Middle distillate is taken from the center of the distillation column, boiling in the range 200-350°C. And heavy fuel oil, the residue with a boiling point in excess of 350°C, is available from the base of the column.

From geological information and laboratory research we will know that a particular crude oil will yield α barrels of naphtha β of middle distillate and (assuming no loss), therefore $1 - \alpha - \beta$ barrels of heavy fuel oil.

We are interested in prices and there are four products in our simple refinery system with four prices which we will call P_C (for crude), P_N (for naphtha), P_M (for middle distillate) and P_H (for heavy fuel oil). But these prices are not independent of one another. A number of equations can be written relating them.

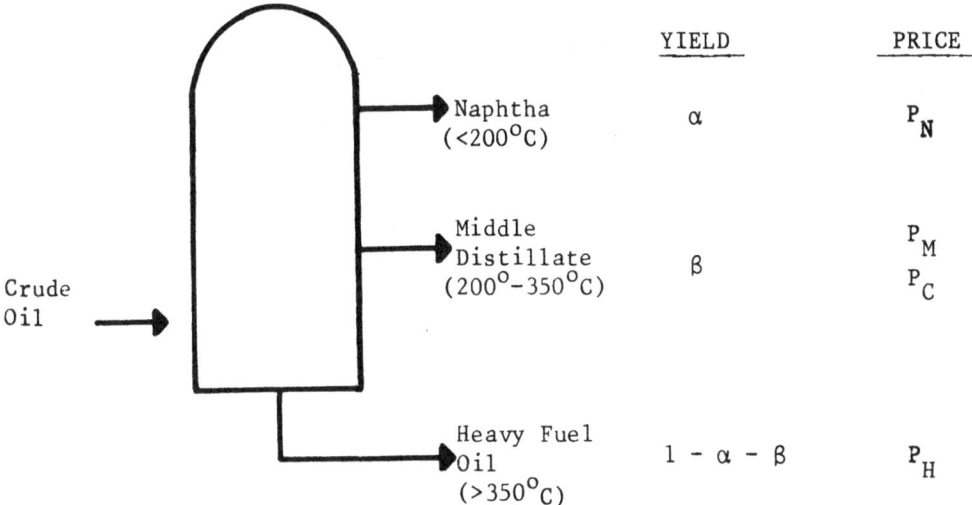

FIGURE 1 - The marginal refinery

Assume that the refinery produces:

W_N ton naphtha
W_M ton middle distillate
W_H ton heavy fuel oil

and processes 100 tons of crude oil.

If the refinery is just breaking even

$$100 P_C = W_N P_N + W_M P_M + W_H P_H.$$

Naphtha can be made from middle distillate by hydrocracking - at a cost which we will call C_H. (This cost can be accurately determined from engineering cost data and technical information.)

$$P_N = P_M + C_H \tag{b}$$

Similarly, we can make middle distillate from heavy fuel oil by thermal cracking - at a quantifiable cost C_T.

$$P_M = P_H + C_T. \hspace{4cm} \text{(c)}$$

In our very simple refinery there are three equations and four unknown prices; all other parameters can be fixed and determined.
We know from elementary mathematics that there is no solution to
these three simultaneous equations unless one of the four prices
is defined.

Although this model refinery is very simple, each additional complexity we seek to introduce - such as adding to the number of products - will produce a new equation and also a new, unknown price.
The number of unknown prices (degrees of freedom) will always exceed the number of equations by one. This energy system has only
one degree of freedom - it is unidimensional. For example, if we
had considered the price of regular motor gasoline, then we could
have equated this with the price of naphtha plus the cost of up-
grading through catalytic reforming - another product, another
equation.

The price of the crude entering our refinery is obviously its cost
plus a "profit" which cannot be determined mathematically. The
profit element can only be determined by those who own the oil or
have control over it. In the Middle East it used to be the oil
companies that set the price; now it is OPEC.

But there are many crude oils. (In WEM's* computer model 75 are
represented). Only one crude is independently significant; this
is called the "marker" crude. Prices of all other crudes depend on
the price of the marker crude.

The marker crude must always meet the following criteria:

1. It must be obtainable at short notice in large quantities.

2. It must be continuously available over a long period.

3. Its owner must be indifferent to wealth, i.e. the pro-
    ducing country must have no financial problems.

4. It must be cheaper, at least slightly, than other com-
    peting energy sources.

In the 1950's the marker crude was Iran's Agha Jari. In the 1960's
it was Kuwait's. Now it is Saudi Arabian light crude.

To summarise WEM's basic premise: the world energy system is unidimensional in price. All prices depend on one price - the price
of the marker crude - which cannot be deduced but has to be dictated. Once dictated, all other prices - at equilibrium - are fixed.

---

*"World Energy Model"

Shortly we will compare theory with practice and see if the real world is in equilibrium.

Reverting to our very simple refinery, we can test the effect of trying to set two prices simultaneously. If the system is uni-dimensional it should react against this.

For example, we could say that the price of middle distillate should be less than the price of naphtha, or that the price of middle distillate should be lower than the price given by equation (c). In either case the refiner would find it uneconomic to put the bottom end of the naphtha fraction into the middle distillate fraction and a shortage of middle distillate would result. The system would react by creating a shortage.

There are many examples of abortive attempts at setting more than one price in the real energy market. The price of natural gas in the USA has long been set by government decree ignoring market forces. Shortages of gas have resulted. Under President Nixon in 1973, the US Government imposed a maximum price limit on middle distillate (called Number 2 Furnace Fuel Oil in the US). Crisis conditions followed with acute shortages of this product. More recently the British Government decided that all the increase of crude should be borne by gasoline. The whole increase of about 8p/gallon was whittled away in about three months through price wars at the pumps.

What is the lesson? That in the free world the competitive na-ture of the industry makes it impossible to dictate higher or lower prices than equilibrium conditions allow. Sooner or later the system will react through shortages or surpluses to compensate for non-equilibrium prices.

Equations such as the ones written above for the very simple re-finery specify the equilibrium prices but these equations cannot be considered immutable. They have to change in the light of events. If we have equations which are an accurate reflection of the situation in the real world, then knowing the price of the marker crude, all other prices can be calculated. The problem is to determine these equations. Changed circumstances - such as when naphtha was surplus and was used as heavy fuel oil in Europe - necessitate new equations, which of course, generate new prices.

To solve this problem WEM has used linear programming (LP), an operations research tool used by the oil industry for many years to solve logistical problems of transportation and refining. WEM's analysis of the world energy system involves some 18000 prices with 18000 - 1 equations. LP permits the determination and solu-tion of these equations.

4. THE LINEAR PROGRAMMING APPROACH

A simple picture of the world oil system is shown in Fig. 2. There are two sources of crude oil, A and B, two oil refineries, 1 and 2, three installations or three markets and a number of ships carrying crude oil and oil products plying the seas between these locations. This microcosm of the world system can be described mathematically using a series of equations.

If, over three months, we use 100 000 tons of motor gasoline at installation 1, then our first equation is that the sum of the amount of gasoline from refineries 1 and 2 equals 100 000 tons. As shown in Fig. 3, this can be developed into a series of equations where x_1 is the amount of crude oil being transported to refinery 1 and x_2 to refinery 2 from crude oil source A and x_3 is the amount of crude oil from crude source B to refinery 1 and x_4 from B to refinery 2. x_5 is then the gasoline transported from refinery 1 to installation 1, x_6 from refinery 1 to installation 2 and so on.

Equation (1) is then:

$$x_5 + x_6 = 100\ 000. \tag{1}$$

Similarly, if the refinery capacity of refinery 1 is 1 000 000 tons, then the amount of crude from A and B coming to refinery 1 cannot be more than this:

Equation (2) is:

$$x_1 + x_2 \leq 1\ 000\ 000 \tag{2}$$

which can be rewritten:

$$x_1 + x_2 + z_1 = 1\ 000\ 000$$

where z_1 must be a positive quantity representing the unused capacity in refinery 1. (None of the x's can be negative, of course, since they are all physical quantities.)

In a similar way, if crude A can provide 1 500 000 tons then the amounts of crude being transported from A to refineries 1 and 2 cannot be more than this.

Equation (3) is:

$$x_1 + x_3 \leq 1\ 500\ 000 \tag{3}$$

or

$$x_1 + x_3 + z_2 = 1\ 500\ 000.$$

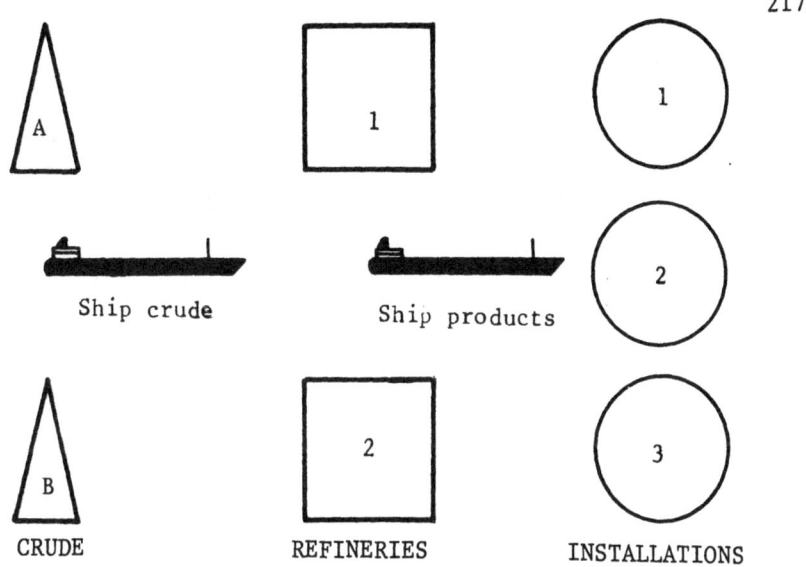

FIGURE 2 - An oil system

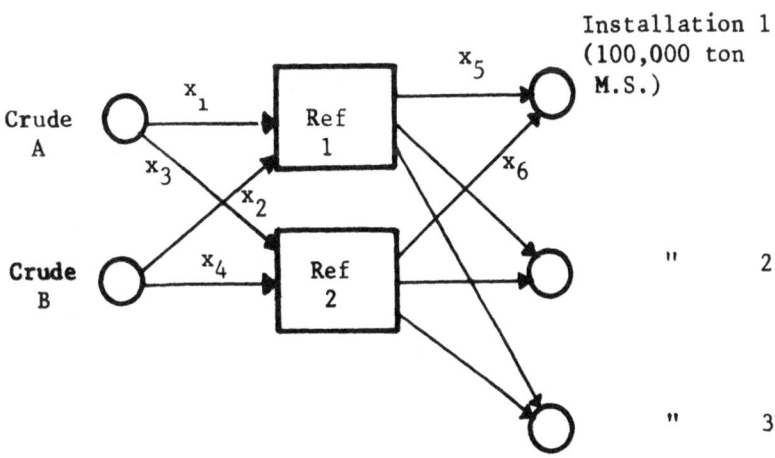

Equations

1 .... $x_5 + x_6$ $= 100,000$ (instln. 1 M.S.)

2 .... $x_1 + x_2$ $\leq 1,000,000$ (Ref. 1's capacity)
$x_1 + x_2 + z_1 = 1,000,000$

3 .... $x_1 + x_3$ $\leq 1,500,000$ (crude A's availability)
$x_1 + x_3 + z_2 = 1,500,000$

FIGURE 3 - The model - its mathematics

In this way a whole series of equations can be written. Several thousand are needed to specify the world system and within this system there are many more unknowns than there are equations. To find the equilibrium equations we introduce the criterion that the market (i.e. installations 1, 2 and 3 in the simple picture) would minimise the energy bill throughout the system. We then select values for x_1, x_2, x_3, etc. which will minimise the cost to the consumer having taken the price of Saudi Arabian light crude (the marker) as specified by OPEC. The computer can then solve the equations and print out equilibrium prices for a very wide range of crude oils and products. Under conditions of a perfectly competitive market, these prices will pertain in the market place.

How well does the model perform against the real world? This has been tested historically.

5. TESTING THE MODEL AGAINST THE REAL WORLD

Prices of oil products in Rotterdam are published daily and historical data can be used to test the mathematics outlined above. The question is: will the prices determined by solving the equations be the same as the ones actually found in the market place?

Prices for heavy fuel oil, middle distillate (gas oil) and gasoline (petrol) were determined in Rotterdam for the period 1966-1972 from the known prices of the marker crude oil (Kuwait at that time), freight and refining. There is a lag ("friction" in the market) before actual prices correspond to these determined prices but the trends are clearly comparable.

If the actual and calculated prices are plotted with the time lag eliminated as in Fig. 4, then there is a close fit for heavy fuel oil. For middle distillate there was a lack of competition in the market from 1966 to mid-1968. But finally attempts by the oil companies to keep the price above the equilibrium price broke, the price fell and the fit became satisfactory from 1969. A lack of competition is even more pronounced in the case of motor gasoline during 1967. At this time there were the beginnings of a price war at the pumps in Western Germany and the industry was endeavouring to preserve its distribution network. There followed a fall in the price with an over-reaction which makes the fit here less good than in the previous two cases. Not surprisingly the gasoline market proves to be the furthest from unfettered competition.

These comparisons show that the WEM model is a good mathematical representation of the real world, determining equilibrium prices - the prices towards which the market tends to move. It also shows that the published prices of oil products in Rotterdam are a valid determinant of prices generally.

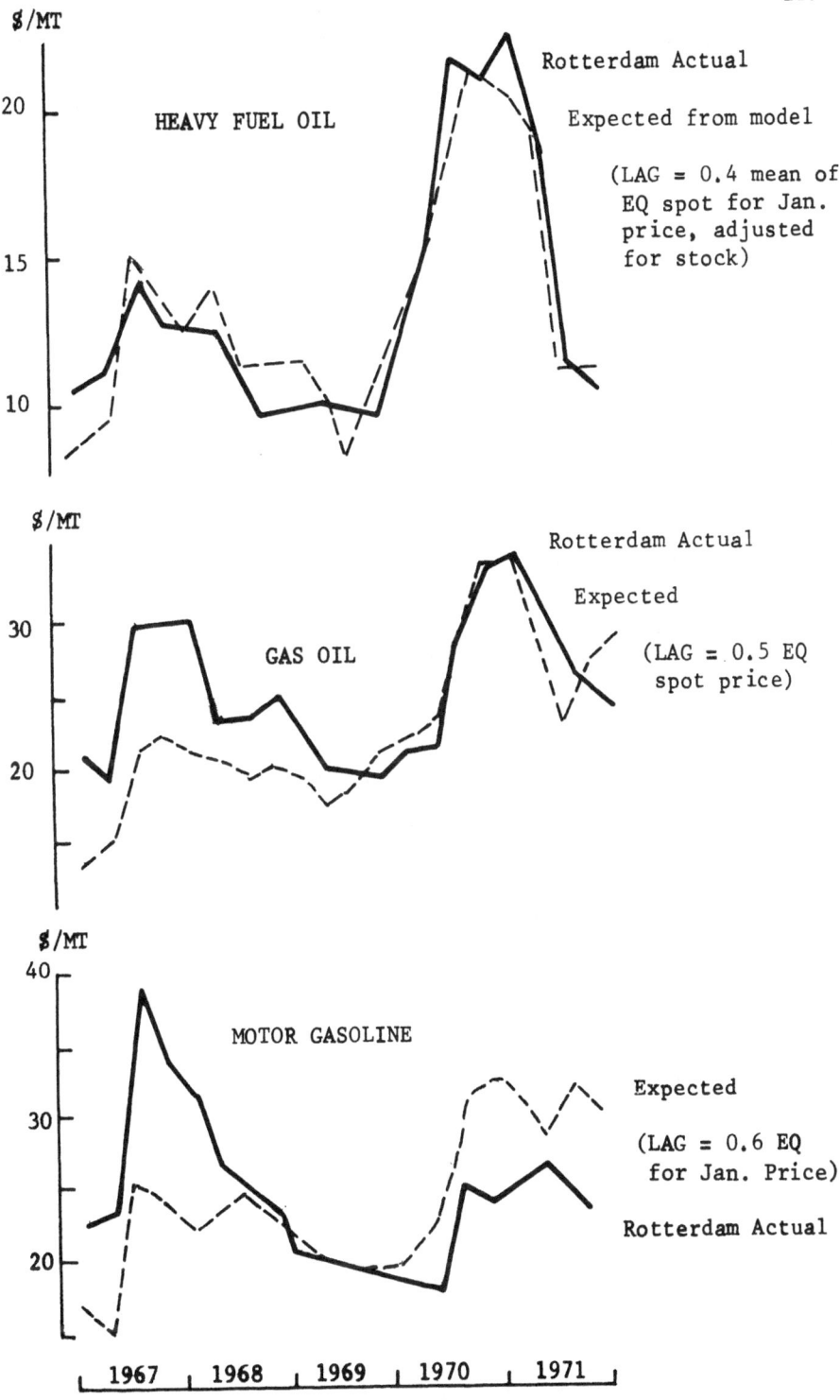

FIGURE 4 - Results from dynamic quarterly model

Given the present price of the marker crude and a mass of well-documented freight, refinery and engineering costs data, the model can be used to forecast prices three months ahead. As another test of the model's real-world value, its forecasting powers were used in June 1975 to look forward to the last quarter of the year, the actual prices for that quarter now being known.

At this time the price of Saudi Arabian light crude (the marker crude) was US $ 10.46. Putting this in the model, a production figure of Saudi Arabian light crude of 294.61 million tons/year is reached. The model specifies 75 crudes throughout the world, some listed separately (as for all OPEC crudes) some aggregated (Texas crudes for example). For this test WEM concentrated on OPEC nations' crude prices.

For example, the Nigerians wished to "lift" 69.46 m tons/year of their Nigerian Light and the model determined a price of US $ 11.75/bl. Similarly, a lifting of 40.210 m tons/year of Nigerian Medium determined a price of US $ 11.50. But these were not the real prices at the end of the last quarter of 1975. Plotting the percentage variances above and below actual prices against percentage variances above and below on lifting for each OPEC crude gave a straight line. This shows that when equilibrium prices are not used, surpluses and shortages result. The tonnage of oil produced and the asking price are related. Raise the price above the equilibrium value and the market will - sooner or later - not take up the anticipated production. Set the price below the equilibrium value and competition for this oil will - sooner or later - force up the price to the equilibrium value.

All crude prices are interrelated. The price of the marker crude is externally determined and the production of this crude is then fixed. For all other crudes selection of particular production levels will fix the prices - at equilibrium. Over and under liftings are related to the differences between market prices and equilibrium prices.

Another test is to forecast product prices three months ahead.

The model specifies 22 areas throughout the world and 12 different products - premium and regular gasoline, gas, middle distillate oil, bitumen, high sulphur fuel oil, low sulphur fuel oil, etc. Therefore 264 (22 x 12) different product prices can be calculated. WEM's test centered on four products in Rotterdam. Prices at Rotterdam calculated by the model are shown in Table 1 for the four products in the third quarter of 1975, along with the actual fluctuations over this period. Premium gasoline prices ranged from $ 147 to $ 149/ton, spot on the calculated $ 148. At the beginning of the quarter the actual prices of middle distillate was $ 130/ton which dropped to $ 100 by the end, nearer to the calculated figure of $ 79.73. High sulphur fuel oil started at $ 50 and rose to $ 68 equalling the calculated figure. These comparisons show clearly the predictive capability of the model and the tendency in the market for prices to move towards the equilibrium prices at differing rates.

TABLE 1 - Product equilibrium prices ($/Tonne)

	LDF	PG	GO	HSFO
Australia				
|				
Benelux	130.80	148.10	79.73	67.77
|				
Se Asia				

22 areas x 12 products

Rotterdam Actual		147/9	130/100	50/68

These are,of course,prices of products crossing national borders. Nothing can be predicted about the retail prices of premium gasoline within particular countries because taxes and excise duties cannot be calculated. The prices that can be calculated, those on products entering countries, are the prices that determine the balance of payments of a country.

The model can generate product prices, as we have seen. Similarly it generates refiners' rents and spot freight rates. Information can be aggregated to give an economic summary for a country or group of countries in the free world.

6. ECONOMIC SUMMARIES

In the third quarter of 1975 the international petroleum bill for energy was a staggering US $ 184 000 million per annum. With the price of Saudi Arabian light crude set at $ 10.42, the quantity lifted is determined and simple arithmetic reveals that $ 22 768 million was paid to the Saudis for their light crude. We also know liftings of all other crudes and their prices: $ 142 938 million went to suppliers of all these crudes together. The quantities of such products as steam-cracked spirits and natural gas liquids being fed into the energy system are similarly known and so therefore are their prices. Customers paid $ 5730 m. for these.

The model also provides rents on individual refining plants throughout the world, a total figure for refiners' rent of $ 9 859 m being
obtained. There is no way of determining how much of this rent
is profit because fixed costs and depreciation interest charges are
unknown but a rise in the refinery rent will increase profitability
of the refiners - the oil companies. If it drops, profits will
drop. This correlation has been checked historically for some
major oil companies and found to be accurate.

We calculate bare boat charter rates of ships and since we know
the total tonnage by classes from Lloyd's Register we can calculate the current shipowner's rent. It turns out to be -US $
1722 m. This negative figure is reflected in the problems of some
oil companies and the collapse of some shippers. About $ 67 m
is paid to port owners and $ 3 m to outside people such as seamen
on vessels.

7. SHIPPING

Lloyd's Register of Shipping provides us with oil tanker capacity
by size of vessel. For example, in the third quarter of 1976
there were 19 m deadweight tons of tankers up to 25 000 tons,
35.29 m dwt in the 25-50 000 dwt category and so on to a total
of 264.62 m dwt. When this information was fed into the model
some classes of vessels were found to be totally used, notably
the 50-80 000 tonners primarily because this is the largest vessel
that can enter ports on the east coast of the USA. But in other
cases only 70% of capacity was in use.

Customers' bill		184 435.23
Divided between marginal crude	22 768.28	
Non-marginal crude	142 938.40	
Non-crude suppliers	5 730.28	
Refiners' rent	9 858.92	
Ship owners' rent	-1 722.32	
Port owners' rent	66.97	
Resource cost	3 003.08	
		184 435.23

FIGURE 5 - Economic summary ($ million per annum)

The model identifies six classes of ships in tonnage and constraints are built in on the size of vessel that can enter ports. Various routes are allowed for and freight rates can be calculated both by voyage and class of ship. An example is the Ras Tanura to Rotterdam run. On this route the generated spot freight rate (third quarter 1976) for a 25 000 ton vessel was $ 11.15 per ton of delivered oil.

Table 2 shows the crude allocation by ship class for Nigerian Light. The equilibrium price from the model is $ 11.75/bl and the model permits us to generate the optimal allocation and distribution of this crude as shown in the table. Some 2.24 m tons should be delivered to the Caribbean in 125 000 ton vessels and 18.67 m to US central in 80 000 ton vessels and so on. It is interesting to note that of the total availability of Nigerian Light, which is given as 69.46 m tons/annum, 66.6 m tons ought to go to the US east coast/US central, just the sort of pattern of supply being found in the real world.

Within the model we could find how much the Nigerians would lose by sending this oil to Japan in 200 000 ton vessels for example - a non-optimal pattern. Not only does the model determine the equilibrium price - the price that is best for both the buyer and the seller in the long term - but it shows where marketing efforts should be directed in this competitive world.

8. CRISES IN THE SYSTEM

What would happen to the world oil market if OAPEC denied oil to the USA or if the Japanese seamen went on strike? Perhaps unexpectedly, the latter event would have a much greater impact on the USA than the former. The model can plot the reverberations.

The effect of the Japanese seamen going on strike would be to deprive the world of 10% of its shipping capacity at a stroke. Prior to the crisis, the spot freight rate Gulf to Rotterdam for a 50-80 000 ton vessel was about $ 10-$ 13 per ton of crude delivered. After the crisis this would rise to $ 33. This and other rises are summarised in Table 3. Freight rates can be calculated from these charter rates.

For the Ras Tanura to Rotterdam route the differentials between large and small vessels are found to have narrowed enormously. In general these rates rise by about $ 22/ton. This, incidentally, is not a very different result from the so-called energy crisis conditions when the demand for energy went up by about 10% when shipping was about 10% less than demand. At that time (1973/74) freight rates between Ras Tanura and Rotterdam were as high as

TABLE 2 – Crude allocation by ship class MTA

	-25,000	25-50,000	50-80,000	80-125,000	125-200,000	+200,000	TOTAL
Caribbean	–	–	–	2.24	–	–	2.24
USA East	–	–	47.93	–	–	–	47.93
W. Africa	–	0.62	–	–	–	–	0.62
USA Central	–	–	18.67	–	–	–	18.67
TOTAL	–	0.62	66.60	2.24	–	–	69.46

Nigerian Light

$ 42/ton. In the example we are dealing with we see that freight rates rose to about $ 33/ton.

TABLE 3 - Example Solution

Bare Boat Class	Equilibrium Base $/DWT	Reduced Variant (10% less) $/DWT	Increase $/DWT
- 25,000	(Unused 13.7 DWT)	76.0	76.0
25 - 50,000	(Unused 10.0 DWT)	93.1	93.1
50 - 80,000	13.0	108.8	95.8
80 -125,000	19.2	120.2	101.0
125 -200,000	23.9	125.7	104.8
+200,000	24.0	118.1	94.1

Spot	Freight Rates ($/Tonne)	Ras Tanura - Rotterdam	
- 25,000	18.05	35.11	19.06
25 - 50,000	12.03	34.05	22.05
50 - 80,000	11.87	33.57	21.90
80 -125,000	10.80	33.33	22.53
125 -200,000	10.33	30.24	22.91
+200,000	9.58	30.11	20.53

The impact on crude prices is dramatic. With a base rate for Arabian Light of $ 74.77/ton, Nigerian Light rose from $ 86.5 a ton to $ 102 a ton, this being a $ 16 rise. For the Nigerians this represents about a 20% increase in income. By and large all the short haul crudes rose in price: Ekofisk oil rose $ 21; Sarir rose $ 17.8. All the North African crudes rose greatly.

As a result of this tidal wave disturbance in the system, the consumers end up paying $ 345 billion* a rise of 40% or $ 86 billion on the previous sum of $ 258 billion.

The non-marginal suppliers, the Algerians and the Nigerians and the short haul crude owners increase their revenue by $ 23 billion but the "crisis" makes no difference to the revenues received by Saudi Arabia - the suppliers of the marker crude. The revenue to suppliers of steam-cracked spirit, natural gas liquids etc. rises by $ 2.4 billion. But very surprisingly we find that the refinery

---

* billion = 10^9

rents, the index of refinery profits, rise from $ 11 to $ 42 billion. This rise of $ 31 billion is effectively a rise in the payment by customers to the refiners, whose profits should thereby be increased accordingly. This is exactly analogous to the situation in 1973 during the energy crisis which was really a shipping crisis, when the profitability of the oil industry rose enormously as the freight rates rose.

TABLE 4 - 10% Less ships effect on crude FOB prices
Effect on crude FOB prices ($/Tonne)

Crude Oil	Base	10% less ships	Difference
Ekofisk	88.7	110.5	+21.8
Sarir	86.4	104.2	+17.8
Nigerian Light	86.5	102.5	+16.0
Arabian Light (Marker crude)	74.7	74.7	-
Kuwait	73.5	72.6	- 0.9
Gippsland	85.4	98.6	+13.2

As events take place in the world, operators need to develop quickly new supply logistics for crudes and products if they are not to lose large sums of money. Refiners and the oil companies clearly benefit when shipping is short. This can be understood from classical economics and the oil companies, so-called windfall profits of 1973/74 which prompted congressional hearings in the US came about under absolutely competitive conditions.

The oil market is highly sensitive to political events, the timing of which cannot be predicted with any accuracy however much they may be anticipated. However, if we assume an event such as the shutting of the Suez Canal, the denial of CAPEC crude to the USA or the bombing of an important pipeline, then the model can be readily used to recalculate a new set of equilibrium figures towards which the world will seek to move.

9. THE LONGER TERM - INVESTMENT DECISIONS

For investment purposes the energy industry needs to look 5, 7, 10 years ahead. Typical decisions taxing the oil/gas industrialists might be whether to convert natural gas to LNG, whether to transport it across the Mediterranean Sea by pipeline, whether to convert it to methyl fuel. Similarly, a producing country wants to know if it is going to get more revenue for itself by selling crude oil by refining it and selling the products. The model can provide optimal solutions under competitive conditions to these sorts of problems.

Also, and very significantly, the model can throw light on the
long term price of the marker crude, Arabian Light - significant
because this is the one indeterminate parameter in our model.

It is possible by introducing into the model the possible crude
oil substitutes and their costs and availabilities etc., to determine the equilibrium price of Saudi Arabian Light crude. This
information is of prime importance for Saudi Arabia, obviously,
and also for all other producers and for consumers worldwide.

DISCUSSION

Deam was asked on the implications of his results for the shipping
industry - should the ship owners sink their ships? In reply it
was said that it would be better to lease them to the U.S. Government for a peppercorn rent as strategic storage. If 70 million
deadweight tonnes were removed from the list, this would stimulate
freight rates, the shipping industry, and would also lend to higher refinery profits. The losers would be the U.S. consumers who
would have to pay a little more for their oil. 95% of refinery
capacity was owned by the U.S. however, so there would be secondary benefits for the U.S. economy.

The question of the importance of accurate estimates for demand
elasticities on the simulation results was raised.* Deam did not
consider accurate estimates to be necessary. The general effect
was in the region of a 1-3% reduction in demand. However, if
"accurate" estimates were available they could easily be entered
into the model parameters.

In theory, Deam's model assumed perfect knowledge and perfect competition in the refinery industry. In the real world neither
of these conditions were fulfilled and the model did in fact contain a small empirical correction for crude burnt off as fuel oil.

---

*The model, as presented in the lecture, contained demand elas-
ticities rather than fixed demands. Editor.

LONG RANGE PRICING OF CRUDE OIL

R.J. Deam

Energy Research Unit
Queen Mary College

ABSTRACT. In the world oil market the price of one crude oil is related to the prices of all others. There is a complex logic which defines the relationship between the prices of different crudes and between the overall price of crude oil and other energy sources. The fundamental nature of the world energy system is determined by its having, at equilibrium, only one degree of freedom - it is undimensional in price.* With one price fixed (the price of the "marker" fuel) all others are determined through the various complex mechanisms that link different energy sources. At present this marker fuel is Saudi Arabian Light crude oil. Far into the future it will not be crude but an alternative energy source.

1. INTRODUCTION

The amount of any energy source being produced depends upon its price. A rise in the price of the marker crude would stimulate demand for alternatives and liftings of crude oil would be reduced. At an extremely high price, production of alternative energy, say synthetic crude oil from coal, would be stimulated to such an extent as to replace completely imports of crude oil into a particular country. On the other hand, if the price of the marker crude is set very low, say near cost, there will be no economic incentive to produce any alternative form of energy. In this case liftings of oil will be maximised, but revenue will be minimal.

---

*"Understanding Energy", Mathematical Modelling of Energy Systems, this volume.

Between these two extremes, there is an optimum price which will result in maximum revenue (net of cost) accruing to the producer. WEM's* long range World Energy Model will determine this optimum price. The logical reasoning behind the model can be understood by considering the mechanics of interfuel substitution in a highly simplified case.

2. INTERFUEL SUBSTITUTION

Why do consumers choose, for example, oil rather than coal; or substitute natural gas instead of oil? What factors control the choice and what would cause them to change?

Let us assume a closed, isolated energy system (i.e. one with no imports or exports) with only two energy sources:

- crude oil, and
- coal.

Reserves of both are adequate to meet demand, but the crude is the cheaper to produce. In this closed system there is a demand for secondary energy:

- 20 tons of motor gasoline
- 30 tons of kerosene
- 50 tons of heavy fuel oil.

All the heavy fuel oil is burnt to generate electricity. Only three processing routes are available (Fig. 1) to achieve this mix: one using oil only; one part oil and part coal; and the third all coal.

Within this system the government encourages the energy industries to choose the processing route which minimises the total energy bill. The producer of crude can determine which route the energy industries will select by his pricing policy, as we shall see.

Considering Alternatives I and II, the consumer will choose which-ever is the lower bill of

110 x price of crude + $ 110

and   70 x price of crude + 70 x price of coal + $ 900

If both bills were the same, the consumer would be indifferent as to the choice of alternative routes. In this case:

110 x price of crude + $ 110 = 70 x price of crude +
                                70 x price of coal + $ 900

---

*WEM = World Energy Model

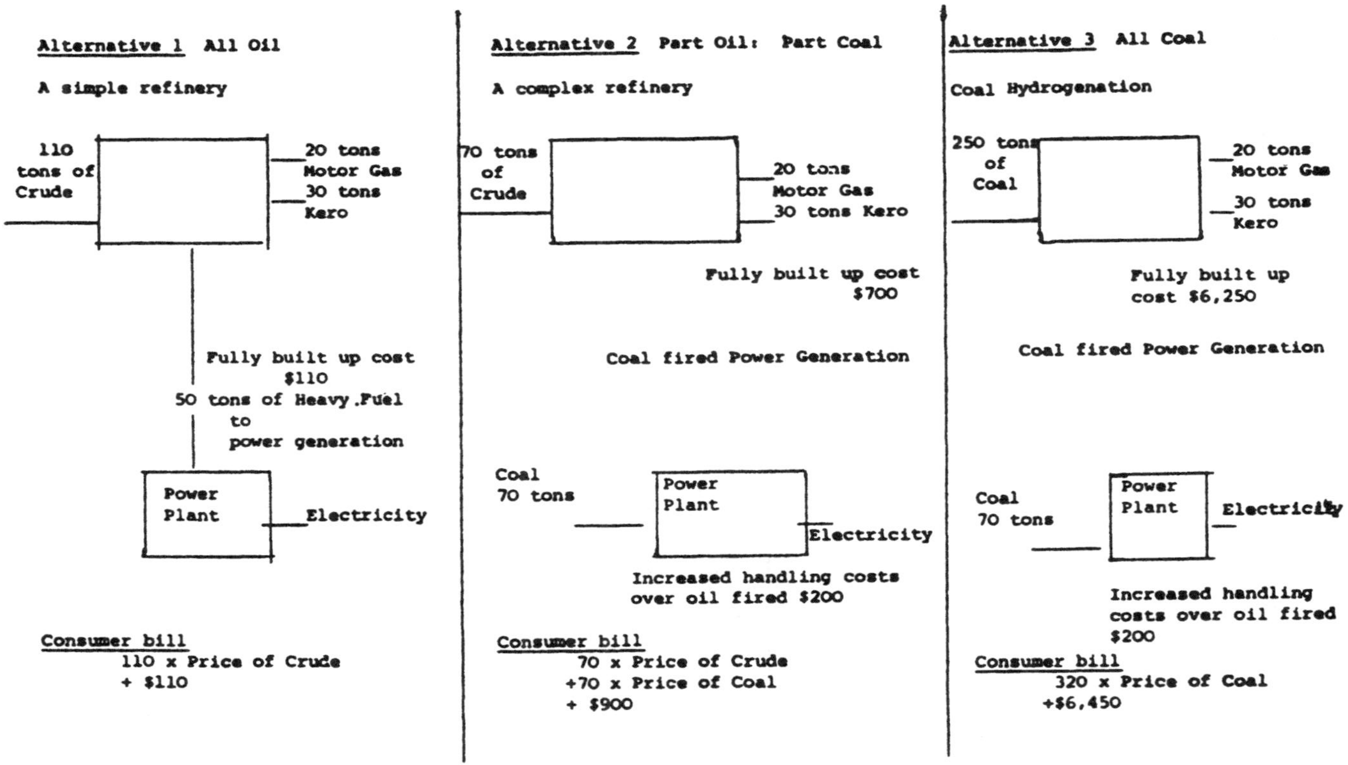

Alternative 1 All Oil

A simple refinery

110 tons of Crude →

20 tons Motor Gas
30 tons Kero

Fully built up cost $110

50 tons of Heavy Fuel to power generation

Power Plant → Electricity

Consumer bill
110 x Price of Crude
+ $110

Alternative 2 Part Oil: Part Coal

A complex refinery

70 tons of Crude →

20 tons Motor Gas
30 tons Kero

Fully built up cost $700

Coal fired Power Generation

Coal 70 tons → Power Plant → Electricity

Increased handling costs over oil fired $200

Consumer bill
70 x Price of Crude
+70 x Price of Coal
+ $900

Alternative 3 All Coal

Coal Hydrogenation

250 tons of Coal →

20 tons Motor Gas
30 tons Kero

Fully built up cost $6,250

Coal fired Power Generation

Coal 70 tons → Power Plant → Electricity

Increased handling costs over oil fired $200

Consumer bill
320 x Price of Coal
+$6,450

FIGURE 1 – Possible processing routes

Price of crude = 1.75 x price of coal + $ 19.75.

The coal producer cannot lower his price below cost, so to ensure that the consumers follow Alternative I, the crude producer would set his price just below Price A, where

Price A = 1.75 x Cost of coal + $ 19.75.

If the crude producer wished Alternative II to be chosen, he would put his price above Price A. The question is: how much above? If he sets the price too high, Alternative III will be chosen and he will have no crude production. Alternative II will be preferred to Alternative III, when the consumer's bill for II is just below III:

70 x price of crude   70 x price of coal + $ 900 < 320
                       x price of coal + $ 6450

or   Price of crude < 3.5 price of coal + $ 79.30.

Since the crude producer wishes to ensure that Alternative II is followed, he would set his price just below the cost of coal, stifling competition from coal:

Price of crude must be slightly less than

Price B = 3.5 x Cost of coal + $ 79.30.

The crude producer therefore has two alternative pricing policies:

Price A: to price crude at 1.75 x cost of coal + $ 19.75, in which case he sells 110 tons of crude

Price B: to price crude at 3.5 x cost of coal + $ 79.30, in which case he sells 70 tons of crude.

Being a rational person, he will select that pricing policy which maximises his net revenue, i.e. the larger of:

110 (1.75 Cost of coal + $ 19.75 - cost of crude)
or   70 (3.5 Cost of coal + $ 79.30 - cost of crude).

This situation can be shown graphically as in Fig. 2, where the price of crude is plotted (horizontally) against the volume of crude lifted (vertically).

For any price up to P_A, 110 tons of crude are produced. Obviously, the maximum revenue occurs when the price is at its limit (1.75 x cost of coal + $ 19.75). At any price between P_A and P_B, 70 tons of crude are produced. The maximum revenue occurs when the price

is set at the upper limit. The production of an alternative
energy source is triggered at its substitution price.

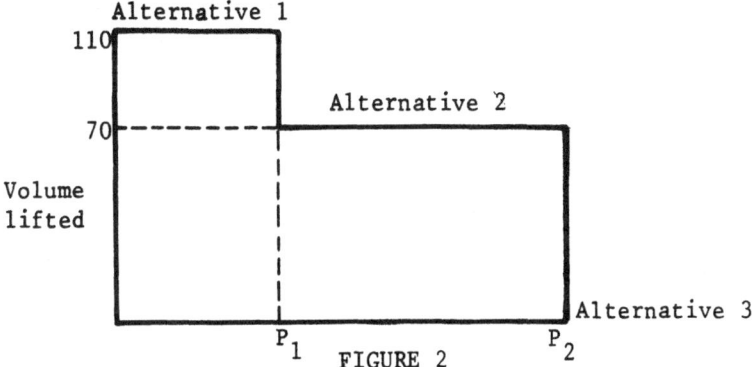

FIGURE 2

This simple closed system illustrates the principles and the logic
of the argument. The real world is, of course, much more complex
because:

- there are numerous alternative energy sources
- extraction costs vary from place to place
- relative geographical locations of the source
  and the consumer have to be considered
- there are many proven conversion processes
- there are many crude oil sources
- environmental laws apply
- political considerations overlay the techno-economic
  factors.

All these parameters have to be considered in analysing the real
world. And our highly simplified illustration has shown the
importance of refinery costs in the determination of P_A and P_B.

The real world system is so highly interactive that only by using
the World Energy Model can meaningful answers be obtained.

The results from the model can be represented graphically as
shown in Fig. 3, plotting price against volume of crude lifted.

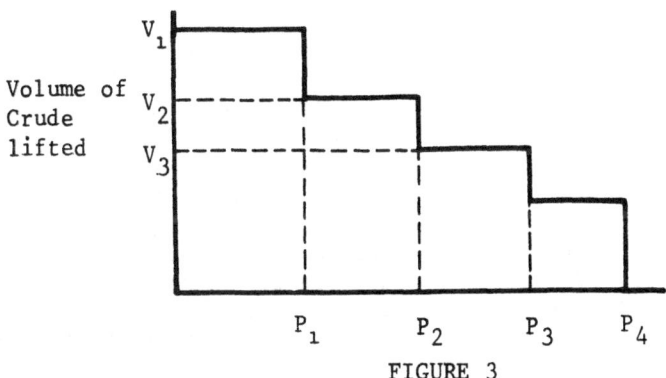

FIGURE 3

234

At a price P_1, it is just profitable for the cheapest alternative
to be produced in volume equivalent to $(V_1 - V_2)$. At P_2, the
second cheapest alternative is produced in volume equivalent to
$(V_2 - V_3)$ and so on. The Model's solutions give full details of
these alternatives at each discontinuity.

Having the price elasticity "curve" for the marker crude, we know
the revenue at each discontinuity, e.g. P_1V_1, P_2V_2, P_3V_3 and so on.
These can be plotted against the price of crude (P_1, P_2, P_3, etc.),
the highest peak occurring when the price maximises revenue to the
producer as shown in Fig. 4. (In practice, net revenue is plotted.)

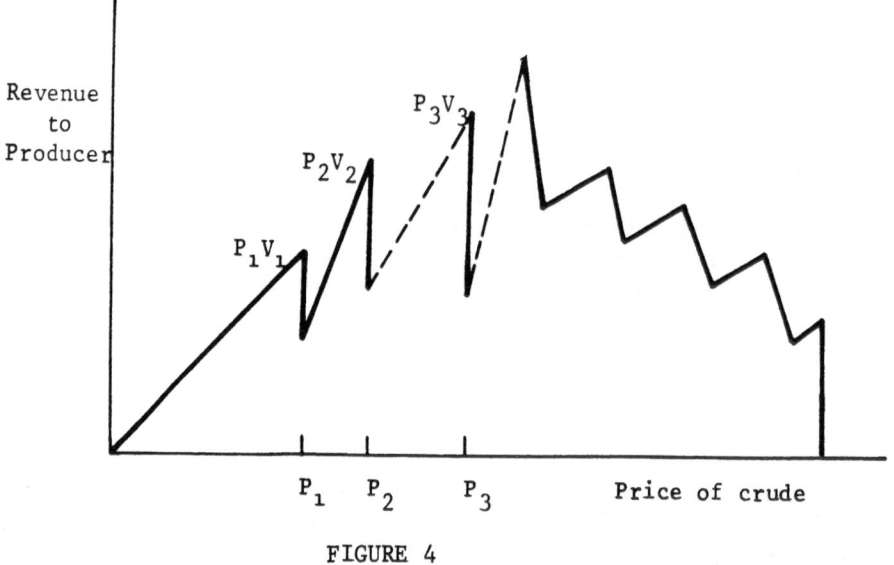

FIGURE 4

It is fortunate that this price maximises the net revenue to the
producer and minimises the total fuel bill to the consumer in the
long run. If crude oil is in short supply it should get its
substitution value (see later).

To summarise, the marginal energy source at any time (currently
the crude oil, Saudi Arabian Light) has an upper and a lower price
limit. The lower price corresponds to its cost (currently estimated at about 10.15 ¢ per barrel for Saudi Arabian Light).
Should the price fall below its cost, no oil would be
lifted. Should the price rise above that of an alternative marginal source (syncrude, tar sands or whatever) then in due course,
when sufficient of the alternative is being made, the production
of the original marginal source would cease. Thus the realistic
long term price of light Arabian crude lies between these two very
wide limits. The energy analyst's problem is to decide what will
be the next marginal source of energy. One price in the energy
system - that of the marginal source - determines all other prices.

3. NATURAL GAS

Natural gas is a strong candidate to be the next marginal energy
source in the medium term, bridging the gap until other forms of
energy can be developed. Its attractions are two-fold:

1. It is available in substantial quantities in many
    geographical regions (see Table 1).

2. It can be transported through pipelines, or as LNG -
    Liquified Natural Gas, or it can be converted into
    "methyl fuel" through catalytic reaction with steam.
    Methyl fuel is an approximately 2/3 to 1/3 mixture of
    methanol (methyl alcohol, CH_3OH) and higher alcohols,
    particularly iso-butanol.

Reserves of natural gas have not been so rigorously proved as
reserves of Middle East crude oil. However, present proven
exportable reserves of gas are equivalent to two-thirds of Middle
East crude oil reserves and they are widespread.

The relationship between the price of crude oil and natural gas
is quite complex. One possible relationship in the international
market is via substitution of No. 2 furnace fuel oil by "methyl
fuel" as a burning fuel provided they are at the same BTU equivalent price.

This relationship can be shown simply in the calculation in
Table 2, which is based on the assumptions:

1. 1 ton of furnace fuel oil made by cracking requires
    1.2 tons of crude.

2. The cost of catalytic cracking (conversion in the
    refinery) of $ 15.00 per ton, used in the calculation,
    is a historical European average for 1967-71. The
    actual figure would depend on the extent of cracking
    required.

3. "Methyl fuel" has 60% of the calorific value of
    furnace fuel.

4. Natural gas is generally closer to the market than
    crude, and one third the distance has been assumed.

5. The capital cost of "methyl spirit" plant depends on
    location. The figure in 1973 dollars of $ 67 per ton
    per annum, obtained privately from several contractors,

TABLE 1 - World ownership potential gas export sources

Export "favourable" areas	Proven resources Jan. 1, 1975 10^{12} cu.ft.
USSR	812
Canada	52 (including Arctic)
Algeria	229
Nigeria	45
Iran	330
Abu Dhabi	200
Indonesia	15
	1683

Export "unfavourable" areas

Libya	27
Venezuela	43
Middle East (ex. Iran and Abu Dhabi)	143
	213

Other areas

Australia	38
Latin America (excl. Venezuela)	36
Pakistan	16
	90
	1986

Source "Petroleum International", Jan. 1974, p. 22-4 updated using "Oil and Gas Journal", 30.12.1974.

At 500 cf of natural gas equivalent to 1 bbl of crude oil, the potential gas export reserves amount to some 400×10^9 bbl of crude equivalent, which is to be compared with the proven Middle East reserves of 403×10^9 bbls (BP Statistical review of the World Oil Industry, 1974, p. 4).

TABLE 2 - Let P_C be the price FOB in $/bbl of the marginal
crude

	$/bbl	$/ton	¢/therm
Crude (FOB Persian Gulf) per unit of furnace fuel oil	$1.2P_C$	$8.88P_C$	
Freight to US East Coast per unit of furnace fuel oil	1.62	12.00	
Built-up cost of cracking per unit of furnace fuel oil	2.03	15.00	
Value of furnace fuel oil	$1.2P_C+3.65$	$8.88P_C+27.00$	$2.07P_C+6.28$
Hence value of methyl fuel at US East Coast		$5.28P_C+16.1$	$2.07P_C+6.28$
(Less) Freight per unit of methyl fuel		(3.33)	
(Less) Manufacturing cost per unit of methyl fuel		(18.84)	
Netback per unit of methyl fuel		$5.28P_C-6.16$	$2.07P_C-2.42$
(Less) Fuel and loss			$0.89P_C-1.04$
Well-head value of gas			$1.18P_C-1.38$

leads to a manufacturing cost made up as follows:

	$/ton product
Capital charges (15% over life)	12.06
Maintenance	2.68
Operating costs	4.10
	18.84

6. 1 therm of methyl spirit requires 1.75 therms of natural gas.

The relationship:

Well-head value of gas = $1.18P_C - 1.38$

where P_C is the price of crude shows how the prices of gas and oil are necessarily connected. If Gulf crude is the marginal source, its price will determine that of gas. If some other process (e.g. LNG, Fischer Tropsch production of a naptha-like artificial gasoline) were cheaper than methyl spirit manufacture, gas would have a higher value. Alternatively, if gas were the marginal source, its price would determine that of crude oil – this price would be the upper price limit on crude. The World Energy Model can be used to discover under what conditions this might be expected to occur and which gas source could become the marginal supply.

4. THE CASE FOR BLENDING METHYL FUEL WITH MOTOR SPIRIT

There is now a considerable body of practical evidence to support the addition of up to 30% methyl fuel to motor spirit. Although pure methanol does not blend well with hydrocarbons, methyl fuel does. It has a high octane rating and is volatile. In traditional gasoline blends the straight-run front end components, which are good petrochemical feedstock have low octane ratings and require the addition of lead to raise the octane number to acceptable levels. A methyl fuel/hydrocarbon blend does not require lead, and this should reduce the necessity for constructing further octane improving and cracking plants.

Methyl fuel gives more miles/BTU than pure hydrocarbon mixtures. Extensive bench tests with a British Leyland Marina engine show that a 15% blend of methyl fuel, whilst improving power, gave almost the same mileage as with gasoline the need for adjustment of the engine. Even a 30% mixture, after optimum setting of carburettor and ignition tuning, reduces miles/gallon by only 8%. In the event that contraints are placed on the addition of methyl fuel to motor spirit, methyl fuel could be substituted for furnace fuel oil, again giving gas a higher value.

5. METHYL FUEL vs LNG

There are, in the literature, many calculations of the relative
merits of LNG and methyl fuel as means of transporting large
volumes of natural gas over large distances. But they suffer
from one or more of the following defects:

1. It is assumed that the price of gas is a necessary
    input to the calculation, and since this price can
    only be guessed at, the comparison is necessarily
    imprecise.

2. The possibility is ignored that the optimal supply
    routes for LNG and methyl fuel might be very different.

3. The comparisons are carried out on a calorific value
    basis.

The price of LNG or methyl fuel is fixed - at equilibrium - by
the price of the oil products for which they are substituting
(the energy system is unidimensional in price); the price of gas
is that price less the prices charged for conversion and transport.

With regard to supply patterns, either the LNG or the methyl fuel
route might be favoured in particular circumstances. For example,
methyl fuel manufactured in Algeria might be absorbed in Europe
and the oil products which it displaces exported to the USA, thus
saving on shipping costs per BTU. Such a route might not be available for LNG in view of the indigenous gas production in Europe
and the possibilities of importing pipeline gas from the USSR.
WEM's model, with its representations of alternative supply
possibilities and local demand patterns, permits us to analyse
these complex interactions.

6. THE WEM MODEL

The model is a complex representation of the international oil
and gas industry, based on linear programming to determine the
least cost strategy to meet a specified world demand, a technique
used widely in the petroleum industry (see "Understanding Energy"
for a fuller description).

Refining and marketing are represented as taking place in 22 locations throughout the world, crude oil production at 25 (more than
50 crudes are taken into account) and natural gas production at
19 locations. Oil and liquefied natural gas transport is included
using eight categories of ships and about 30 major pipelines are
considered.

Natural gas from the 19 locations is considered to be transported by one of the following routes:

- by pipeline to meet local gas demand and to exports
- by exporting as LNG, which is regasified to meet a distant gas demand
- by conversion to methyl fuel, which may either be used locally or exported for use as a component of motor spirit, or as a substitute for middle distillates and fuel oils.

The operating costs of these facilities and the investment costs pertaining to their expansion are supplied as input data to the model.

Using the reserves figures in Table 1, a reserves/production ratio of 25:1 was assumed. A price was chosen for light Arabian which was high enough to cause its production rate to fall to zero - an exaggerated and unrealistic case which could then be moved away from or towards a meaningful value.

With a stated future demand for petroleum, the model will determine the theoretical programme (in capacity and capital expenditure) for refinery construction worldwide, giving details of the type, size and location of plant, the optimal crude supply pattern and the necessary tanker construction by size category, to meet this demand.

With a specified price for light Arabian crude - the current marker crude - equilibrium prices for all other crudes and products and for natural gas at each location are determined. The model system also determines the cost of harbour and pipeline constraints and the equilibrium spot charter rate of vessels by class category. The results from running this model are given in Tables 3 to 6. This shows a time in the future when demand has increased and when oil refinery plant and tankers would need to be built if the methanol route were not used. With the high price chosen for light Arabian, its production falls to zero. Refinery capacity unused leaps to 35-40%: none is built. LNG is made in competition with methyl fuel up to the existing capacity of LNG tankers: no further tankers are constructed. Oil tanker construction is nonexistent in this case and some 19 m dwt of 25,000 ton vessels are left idle - stemming from the relative closeness of methyl fuel to the market. Some 640 mta*of methyl fuel is made in eight different locations, 470 mta being used as burning fuels. All motor spirit worldwide is at maximum methyl fuel content.

---

*
mta = 10^6 tons/annum

TABLE 3 - Methyl Spirit Production

	Quantity (10⁶ tonnes)	Disposal	Investment (10⁶ $)

	Quantity (10^6 tonnes)	Disposal	Investment (10^6 $)
SOLUTION A*			
Australia	35.2	18.8 local	2358
		16.4 Japan	
North Africa	115.1	25.6 local	7712
		89.5 France, UK, Spain, Italy, US	
Caribbean	50.9	8.3 local	3410
		42.5 US	
Persian Gulf	243.4	167.0 local + E. Africa	16308
		76.4 Japan, S. Africa	
E. Siberia	8.7	All to Japan	583
Alaska	35.0	22.4 to US	2345
		12.5 Canada	
Nigeria	42.6	16.6 local	2854
		25.9 US, Scand.	
Pakistan**	108.8	59.4 local	7290
		49.6 SE Asia	
W. Siberia	NIL	-	NIL
TOTAL	639.7		48369
SOLUTION B*			
Australia	3.1	All local	208
North Africa	64.4	2.5 local	4315
		61.9 UK, France, Spain, Italy, US, Benelux	
Caribbean	58.3	8.9 local	3906
		49.4 to US	
Persian Gulf	4.4	All local	295
E. Siberia	NIL	-	NIL
Alaska	22.1	14.6 to US	1481
		7.4 Canada	
Nigeria	1.9	0.5 local	127
		1.5 S. Africa	
Pakistan**	15.9	3.9 local	1065
		11.9 SE Asia	
W. Siberia	2.3	Scandinavia	154
TOTAL	172.4		11551

*In this and subsequent tables the solutions referred to are:
  A - Price of light Arabian set high enough to shut in produc-
      tion (Section 4).
  B - The minimax solution (Section 5)
**An error in the model gave a gas availability of 133 x 10⁹ m³.
The minimax solution B in fact used only 27 x 10⁹ m³.

TABLE 4 - Overall Disposal of Methyl Spirit

	SOLUTION A		SOLUTION B	
	$(10^6$ tonnes)	(%)	$(10^6$ tonnes)	(%)
To motor spirit	169.7	26.5	169.3	98.3
burning kerosene substitution	75.1	10.2	2.9	1.7
gas oil substitution	108.2	16.9	NIL	NIL
LSFO substitution	98.3	15.4	NIL	NIL
HSFO substitution	198.1	31.0	NIL	NIL
TOTAL	639.4	100.0	172.2	100.0

TABLE 5 - Refinery Plant Utilisation Existing and Planned
Capacity Left Idle

Process	SOLUTION A		SOLUTION B	
	$(10^6$ tonnes/year)	(%total)	$(10^6$ tonnes/year)	(%total)
Alkylation	22.8	81	28.9	100
Catalytic cracking	72.8	21	203.4	59
Crude distillation	896.3	36	432.2	17
Hydrofining	143.2	65	82.3	37
Catalytic reforming	96.2	33	37.8	13
Residue desulphurisation	20.0	100	3.1	15
Vacuum distillation	343.5	57	410.6	68
Hydrocracking	27.9	52	34.2	64

TABLE 6 - Total Capital Investment

	Solution A	Solution B
Pipelines	47.2	33.4
Shipping	NIL	NIL
LNG plant	2.5	0.4
Methyl spirit plant	42.9	11.6
Refining plant	0.0	0.2
TOTAL	92.6	45.6

Although this is not a realistic solution, since a cutback in OPEC production would never be allowed to fall entirely on one member, it gives us the starting point from which we can move towards the optimum price of light Arabian crude.

By reducing the price of this crude in stages, and running the model, natural gas liftings are found to decrease and crude liftings increase. The revenue to the crude producer, volume lifted multiplied by price, follows a saw-tooth curve passing through a global maximum (See Figs. 5 and 6). At this point, the producer's revenue is maximised and the consumer's total cost minimised. If the producer sets his price too low, the alternative sources will not be developed. The industry will build too much conventional refining and shipping facilities. Since alternatives take time to build and commission, there is a period in which the producer has no competitive alternative sources available in significant volumes, and therefore no upper limit to his price. As he capitalises on this situation, the consumer's bill increases and previously uneconomic alternatives are developed, and slack appears in crude refining and shipping capacities reflecting previous over-building.

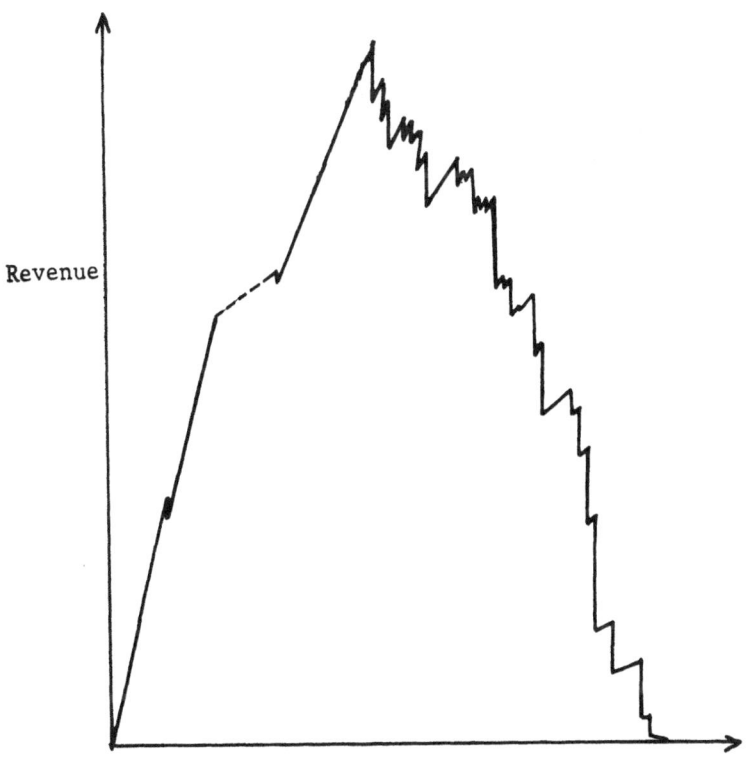

FIGURE 5 - Minimax solution - revenue vs price

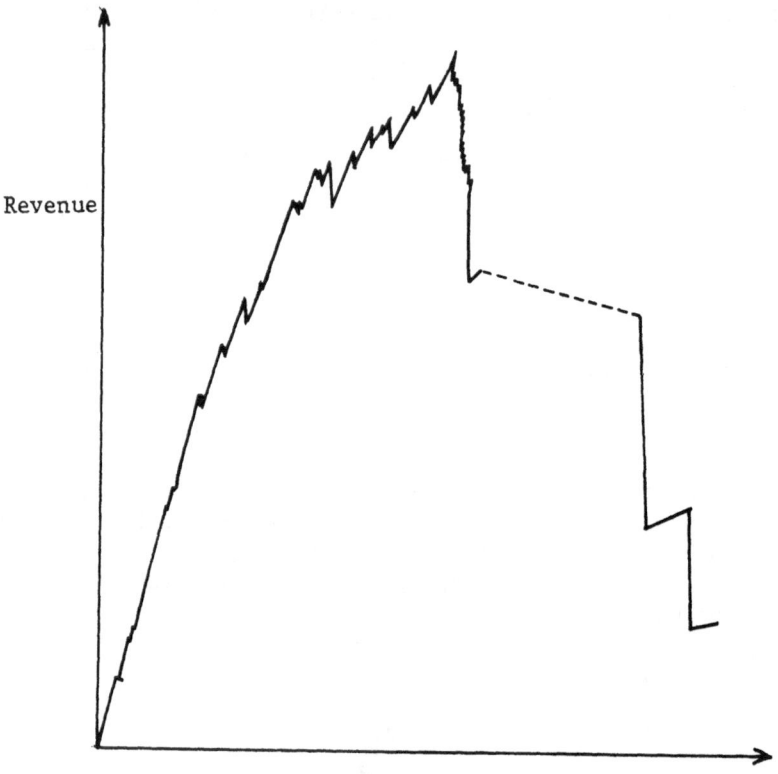

Revenue

Quantity of Marginal Crude

FIGURE 6 - Minimax solution - revenue vs quantity

This single demonstration gives the flavour of the technique and the logic behind it. To give a definitive price, further runs are required to cover a range of scenarios including variations in:

- possible OPEC prorationing schemes
- gas production rates and potential discoveries
- gas processing costs (LNG and methyl spirit plants, pipelines)
- demand assumptions
- for the longer term other forms of energy (coal, shale oil, nuclear, etc.) have to be included
- discounting over time (the current model is static).

7. CONCLUSIONS

It is in the best long term interest of both the producer and the consumer of oil that an economic price be established for crude oil. The producer should wish to avoid risking lower future revenues which would result from competition from substitution by other energy sources which would become relatively economic at higher prices. The consumer will wish to avoid being forced into uneconomic capital expenditure connected with these energy substitutions. WEM's model can help in the analysis of possible future consequences of today's political decisions.

It is possible that in the future a monopoly of gas producers (OGEC?) could replace the existing monopoly of crude oil producers. But this situation could only arise when energy demand has increased to the point where the crude producer's maximum revenue corresponds to the physical limits of production of the marginal source. The price of gas would then be subject to lower and upper limits - the cost of production and the cost of the next alternative energy source - in the same way as crude is now. The next alternative might then be coal, tar sands, shale oil, nuclear power, or some as yet undiscovered form of energy.

DISCUSSION

Deam's postulate that natural gas would be the future "marker crude" - that is to say the form of energy which clears the market and sets the price - caused many of those present to review their mental models of how the future price of energy would develop. Not least because Deam was implying a drastic reduction in energy price during the coming 30 years.

It was put to Deam that the inclusion of natural gas in the oil market would have the same effect as a new oil producer with a market share of 5%. OPEC would be able to squeeze this new producer out of the market. Deam replied that if the price leader, Saudi Arabia, were to freeze the newcomer out of the market, the Saudi's oil liftings would be reduced by 70% and this was very unlikely. There was very little opportunity of an OPEC + OGEC (Organisation of Gas Exporting Countries) cartel as oil and gas were competing fuels.

Deam asserted that for gas finds over 1000 km from the market, it would pay the gas producer to convert the gas to methanol at the well-head and pipe methanol to the market or transport it by sea. Methanol had a viscosity less than that of water, froze at low temperatures and was generally cheap to pump.

On the end use side, methanol could easily replace 2 fuel oil in electricity generation turbines. If waste heat from the turbines were used to decompose methanol to hydrogen and carbon monoxide, the thermal efficiency of the turbine could be pushed up to 80%. Hydrogen could have a premium use as a reducing agent in the metallurgical industries.

It was suggested that it would be better to consider the competition of methanol with gasoline than with # 2 fuel oil. Deam agreed; methanol increased the performance of the internal combustion engine. It had a very high octane value, was a pro-knock agent so lead could be dispensed with, and enabled the compression ratio to be increased to the region of 13:1 which implied more miles per gallon. The costs of converting cars to run on a mixture of 67% gasoline, 33% methanol were relatively small. In Brazil a 20% mixture was used without any change in the engine. Were methanol to compete on the gasoline market, it would reduce aggregate oil company profits and industrial opposition could be expected.

The question of cost escalation was raised. The example of oil extraction from tar sands had shown that estimates of $ 5/bbl in 1973 had grown to $ 20/bbl in 1979 and the technology was by no means perfected. Did we not risk the same sort of cost escalation in methanol production? Deam repeated that the technology was known and well-proven. The Russians had two large scale plants operating.

How safe was methanol technology? The potential risks from a fracture in an LNG ship were known to be enormous. Unlike LNG, however, methanol did not evaporate fast and was miscible in water. The major risk was thus in blinding fish.

Although there was general agreement on the time delays involved in the introduction of new technology into the market place, several speakers found it inexplicable that methanol was not already marketed. The oil industry was not monopolistic. Although the industry as a whole would lose money if the use of methanol became generally accepted, smaller oil companies - especially those with little cracking capacity - would have an interest in marketing a mixture of methanol + gasoline and increasing their market share. Deam maintained that inertia was the greatest reason for this delay, but that several companies (New Zealand was a prime mover) were becoming interested.

CHOICE OF MODELLING TECHNIQUE:
AN EXAMPLE OF A SIMULATION MODEL FOR WORLD OIL SUPPLY AND DEMAND

Leif K. Ervik

The Chr. Michelsen Institute
Department of Science and Technology
Bergen, Norway

ABSTRACT. This paper concentrates on general modelling issues. A world oil simulation model is used as an illustration, but many of the general observations are based on broad modelling experience and not just the oil model.

A limited attempt is made to explain the structure of the model and to relate it to the real world which it is designed to describe. The emphasis of the paper is centered on the development of the project itself.

By way of introduction, a little background from the real world is given.

1.   INTRODUCTION

Until several years ago, most oil companies believed that enough oil could be found to satisfy the world's demand for petroleum until the end of this century. Only a few geologists, such as King Hubbard of the US Geological Survey, voiced caution in interpreting the United States' energy reserves for the future, pointing out that discovery rates would decline as the major reserves had already been found in the US. It was commonly held that if the price of oil rose enough, companies would be able to make available close to unlimited reserves.

Lately, the situation has changed dramatically. After findings by some academic studies of the process underlying oil and gas discoveries and predictions of energy demand, such as the WAES study [1], oil companies began to reevaluate their data. Most of these reevaluations pointed to the possibility of a major oil supply gap before the end of the century. The likely form of the future oil production profile and timing of the "energy crisis" is still a matter of heated debate. The range spans from at one extreme predictions by Phillips Petroleum and the CIA, who foresee a serious imbalance between supply and demand already in the early 1980's to the more optimistic views, which place the levelling off in oil production somewhere at the end of the century.

Predictions of the future development are extremely sensitive to current events. The optimistic Mexico supply forecasts and the Iran supply cut backs have both influenced the common view of long-term price development.

Because of this extreme sensitivity, and for other reasons, most of these predictions are only partially relevant to a correct evaluation of future trends. They talk about a predicted gap between demand and supply, and translate this into a gap between productive capacity and consumption. Obviously, such a gap will never exist. The world cannot consume more petroleum in any given year that it produces (with the exception of tapping petroleum stockpiles which play a minor role in the overall picture). Both governmental action, and, in a free market situation, the price mechanism will tend to adjust demand to supply, thus reducing consumption until it equals available production. Most of the existing models are not able to formulate a simultaneous relationship between oil production, oil price, and oil demand, and have at least one of these variables as exogenous input.

Therefore, in 1977, The Chr. Michelsen Institute and SAGA Petroleum decided to develop a world oil model.

2.   THE PURPOSE OF THE MODEL

Generally, the possible objectives for a modelling effort may be summarized as:

    1. Assist clearly defined decision-makers
    2. Attain power for analysts
    3. Enlighten public discussion
    4. Pure (scientific) desire for understanding
    5. Communication/Education.

To tell the truth, we did not decide on the objective of the modelling effort in 1977. After the fact one can say that it started out as a tool for communication and will most likely end up as a tool for a set of clearly defined decision makers.

The purpose of the model is one of the key guidelines in model development. It is important in deciding how far to expand along a number of possible axes:

- ease of enrichment
- point-predictive capability
- ability to generate observed behaviour modes
- descriptive realism
- formal correspondance with data
- insight generating capacity
- transparency
- relevance to practical policy issues
- fertility

3. PROBLEM DEFINITION

The problem we are faced with is to structure and quantify the main determinants of the development of world oil price. We have also made an attempt at quantifying the uncertainty in these determining factors and the resulting uncertainty in world oil price.

Before we begin discussing the main determinants of oil price in the model and in the real world, I would like to state some of my biases:

- An inherent belief that the world is causal and that the important mechanisms can be perceived and represented. This is very different from those who believe behaviour results from stochastic processes, or from mechanisms which cannot be understood.

- A commitment to descriptive analysis rather than predictive analysis; to attempt to understand the behaviour of the system over time and the causes of that behaviour rather than a preoccupation with estimating future point values of a system element.

- A dedication to employ all the information that exists, being concerned about the quality of the data when interpreting the results, but not while choosing the parameters to be included in the model.

Given this preconceived notion of the world, simulation seems to be a natural choice of technique. In this particular study, we wanted to look at the long term behaviour of the oil market. Continuous simulation seems to be a good choice when dealing with aggregate descriptions of reality.

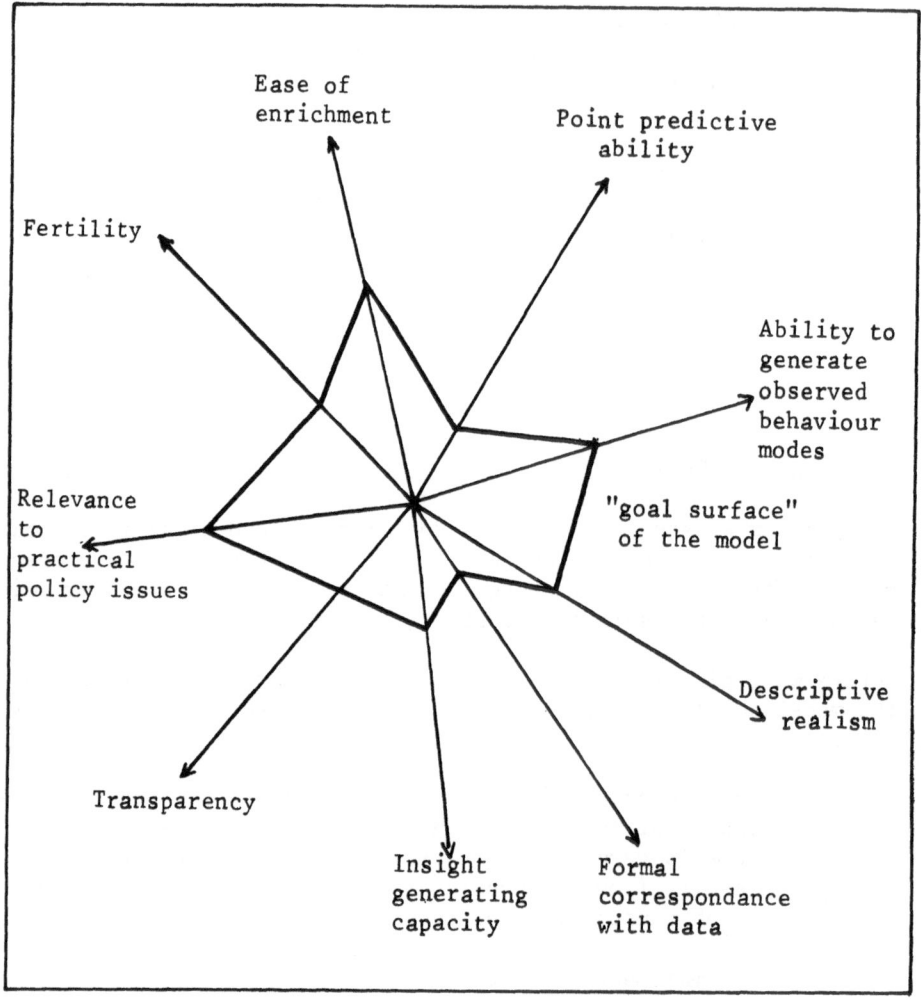

FIGURE 1 - The utility of a model is determined by how
well it satisfies the objectives selected as
important by the user. The cost increases
quickly when the model is pushed further along
any dimension.

For a given expenditure of modelling effort, there is a tradeoff between the degree of aggregation and the time horizon which
can satisfactorily be treated in a simulation model. If both a
more detailed model, and a model which simulates over a longer
time horizon is required, more effort must be expended (Fig. 2).

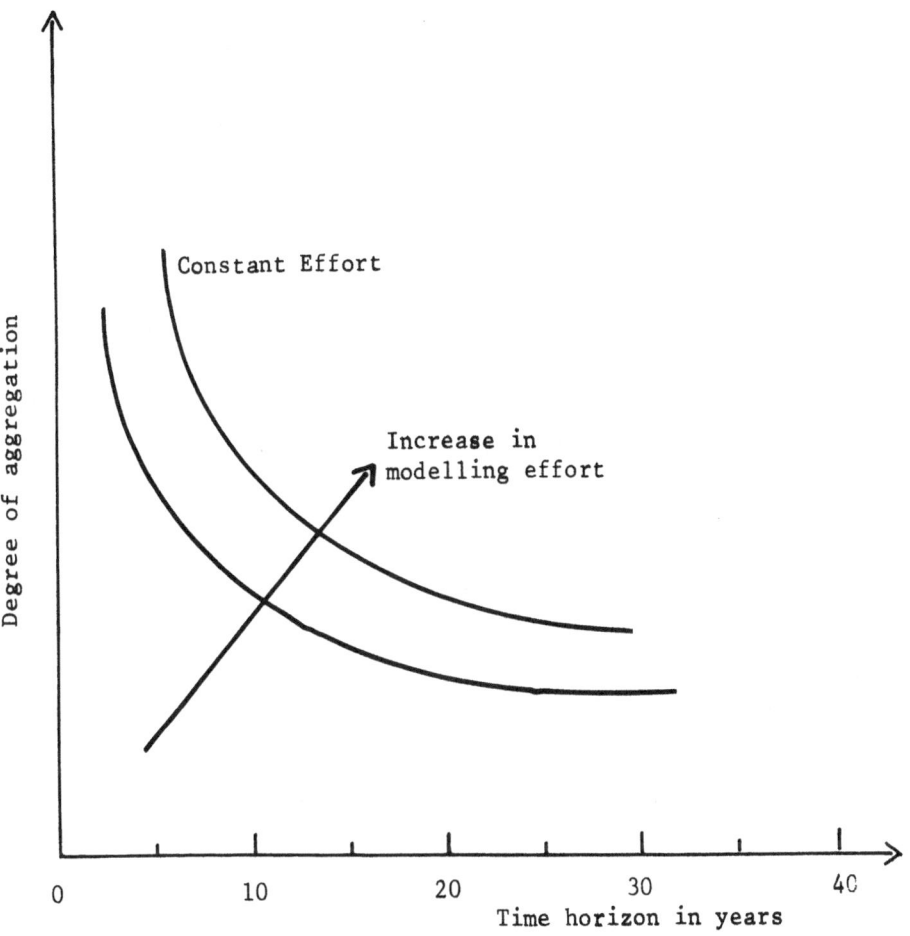

FIGURE 2 - Relationship between time horizon and degree
of aggregation

4.   MODEL DEVELOPMENT

The first oil model was a world oil production and demand model,
consisting of 9 regions grouped according to economic and political criteria (OECD, OPEC, developing countries, centrally planned
economies).

The choice of regions was made on the basis of importance (production vs demand) and on an expectation of future development.

Two regions that have reached the same stage in their development
and are expected to react similarly to forces might be grouped
together. Parallel with the first regional models, we developed

an aggregate world oil model which viewed the world as one sector. All the major behaviour modes in the development of oil price and production can be reproduced with this one-sector model.

A lesser detailed model has clear advantages when used for communication, education or public discussion. The aggregate model, however, is often mistrusted merely because it is too simple.

The aggregate model clearly benefits from the "audience" knowing that there exists a disaggregate model.

The initial models focused on development of oil price and production. Some of the results relating to oil production indicated a reduction in future demand for tanker tonnage. This result, coupled with the fact that we had a regionalized world oil model which simulated where oil was produced and consumed promoted the idea of studying oil transportation requirements.

As is well known, Norway has a very large tanker fleet. The future development of tanker freight rates, the so-called World Scale, is of major importance for shipowners and Norway's foreign exchange revenues.

To address this new set of questions the model was expanded. The new model has 16 fully formulated independent consumption and production regions. This was done mainly to separate the main oil consumption and production areas geographically, while still allowing the political or economical groupings of the previous model.

The oil export or import requirements of the 16 regions form the input for a linear programming model which distributes the amount of oil traded according to the formulated constraints.

OILTANK is probably the first example of a DYNAMO programme being combined with a linear programming model. The necessary software was developed at the Chr. Michelsen Institute. Linear programming is now a fully usable facility in the Institute's PRIME-computer version of DYNAMO. The main structure in OILTANK will be described in the following. Several LP models of the oil transportation system exist. They vary with regard to the number of regions and to the number of products.

We have not made a comparative study of these models, of which many are propriatory. We have been led to believe, however, that the long term accuracy of the description of the transportation pattern does not increase with more regions and more products. In any case, a 16 region one product model seems to be adequate for our purposes.

Typically, a LP model can be structurally complex, but dynamically simple. Since our LP is coupled to a dynamic model, dynamic properties are a little more interesting.

In order to generate the future freight rates, we coupled OILTANK with a model describing the supply of ships. The ship supply model had a ship building sector and all the various types of "capacity utilization", such as slow steaming and lay-ups. However, it soon became apparent that the feedback from freight rates and tanker availability to the consumption and production of oil was basically nonexistent. OILTANK is now simply used to prepare a consistent set of inputs to the tanker supply.

5.   OILTANK - AN OVERVIEW

OILTANK is a continuous simulation model integrated with a linear programming routine which calculates transportation requirements. It is also formulated as a Monte Carlo simulation model with uncertainty explicitly formulated in some 300 variables. The world is divided into 16 fully formulated consumption and production regions. The choice of regions is intended to separate the largest oil consumption and production areas geographically, while still allowing explicit representation of political or economic groupings of interest.

Fig. 3 shows the regions into which the model is divided. The world is split into four main categories:

OECD, OPEC, Developing Countries (except OPEC) and Communist Countries.

We will now take a closer look at the basic deterministic causal structure of the model.

The future development of oil price may be differentiated into three stages depending on which forces are dominant in determining oil price.

Fig. 4 describes the three stages as

-    OPEC stage
-    Scarcity stage
-    Alternative source stage.

It should be made clear that the same model describes all of the stages, but that different parts of the model structure are dominant in the different stages. Thus, it might very well be that the OPEC stage could last until the end of the simulation period, or that the scarcity stage becomes virtually nonexistent.

5.1 The "OPEC Stage"

This stage is characterized by excess supply capacity. The behaviour of the price curve will mainly depend on OPEC policy, i.e.
on the oil supply side. The demand parameters are less important.

It is assumed that the clearing price of oil is determined by
market forces. OPEC can influence price in the model only by
regulating production. We thus postulate that oil production
and oil price cannot be set independently of each other. This
is simply a consequence of observing that consumers of oil react
somewhat to price.

GROUP 10: OECD

    11   NORWAY
    12   WESTERN EUROPE (EX NORWAY)
    13   NORTH AMERICA
    14   JAPAN
    15   AUSTRALIA, NEW ZEALAND

GROUP 20: OPEC

    21   MIDDLE EAST
    22   NORTH AFRICA (LIBYA, ALGERIA)
    23   WEST AFRICA (NIGERIA, GABON)
    24   LATIN AMERICA (VENEZUELA, EQUADOR)
    25   SOUTH-EAST ASIA (INDONESIA)

GROUP 30: DEVELOPING COUNTRIES (EX OPEC)

    31   LATIN AMERICA (EX MEXICO)
    32   AFRICA
    33   ASIA, OCEANIA
    34   MEXICO

GROUP 40: CENTRALLY PLANNED ECONOMIES

    41   USSR, EASTERN EUROPE
    42   CHINA

FIGURE 3 - OILTANK model regions

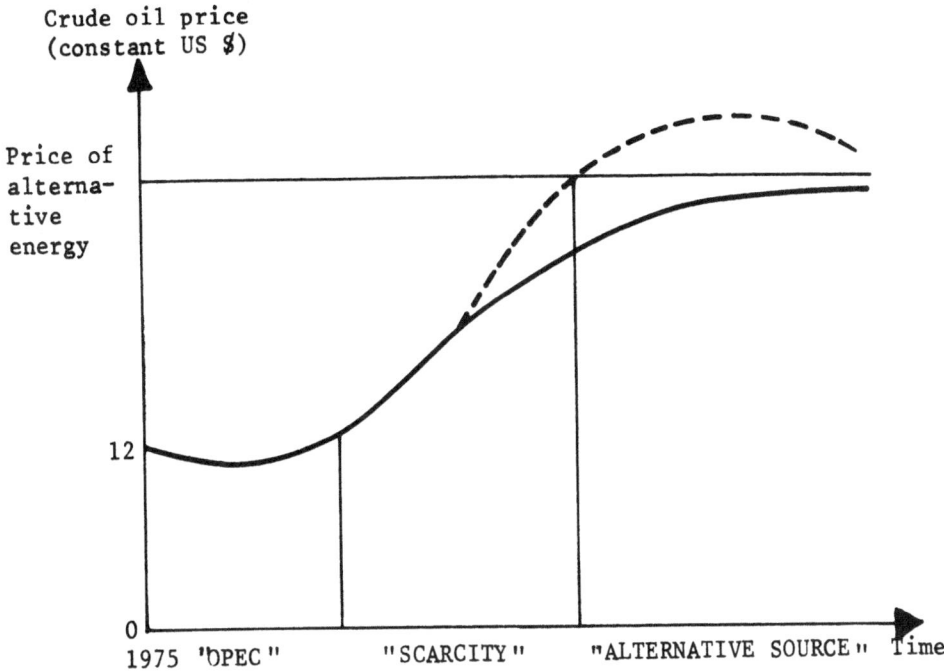

FIGURE 4 — The three stages of oil price development

Oil price determination

The rate of change of oil price is a function of the ratio between perceived need for oil and planned production. The perceived need for oil is, in turn, a function of expected oil consumption and inventory adjustment. Planned production is the production that would have prevailed if there were no cut-backs due to too large inventories.

The model is formulated such that a small excess demand leads to a large price increase while excess supply only marginally reduces price. In other words, the price is "sticky downwards".

OPEC production strategy

In our model, non-OPEC producers are just "price-takers" in their production policy.

The OPEC countries' production is modelled as a function of the following factors:

- Reserve production ratio (RPR)
- Actual oil price
- Desired oil price
- Exogenously formulated political factors

A high RPR indicates a situation with excess production capacity.
A high excess capacity makes it difficult to increase this excess
capacity any further.

When the desired oil price is higher than actual oil price, OPEC
tries to hold back production from the market. This is easy in
a situation with little excess capacity and difficult in a situation with large unutilized production potential. Production is
also cut-back when demand for OPEC production is low. The low
demand over a period will fill the inventories.

Fig. 5 shows the consequences for the development in price of an
OPEC-depletion policy. This policy involves two changes:

- Reducing the minimum Reserve Production Ratio (RPR)
  from 20 years to 15,

- making the RPR change-rate faster,

thus making the influence from OPEC-desired oil price less. The
result is a short term drop in oil price and a delay in the later
price increase of the order of 4-5 years compared with the standard run. This lower and delayed price is caused by increasing
depletion in OPEC, i.e. increasing production from OPEC members.

Depending mainly on the future demand for OPEC oil, the OPEC stage
might last longer than most experts have so far expected. High
prices will induce cut-backs and thus reduce demand earlier than
one previously could have expected. It is therefore quite conceivable that a surplus situation will still exist in 1985.

5.2   The "Scarcity Stage"

This stage is characterized by excess demand for oil products and
a stagnating or even declining crude production. Crude production
capacity will be fully utilized. Here oil price, together with
quantitative governmental restrictions, will be the main factors
regulating excess demand. The beginning of substitution of other
competitive energy sources may be observed.

The oil price in this stage will be determined by oil demand and
by medium term oil supply. Medium term oil consumption per region is determined by the variables:

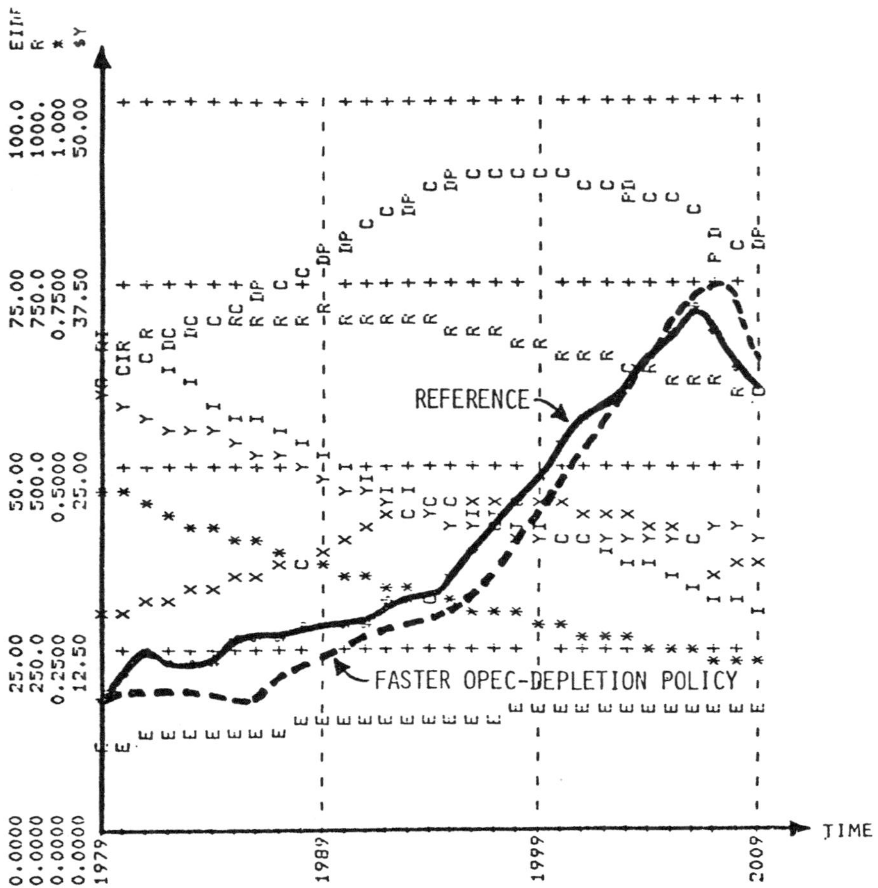

REGION:	World Total

ASSUMPTIONS:

Price tracks for standard (reference) assumptions and faster OPEC-depletion policy assumptions.

PRINTED VARIABLES:

P	PRODUCTION (P)	MB/D
D	CONSUMPTION (OC)	MB/D
X	EXPORT (MAX(0,NEX))	MB/D
M	IMPORT (MAX(0,-NEX))	MB/D
R	PROVEN RESERVES (PR)	BB
U	UNDISCOVERED RECOV. RES. (URR)	BB
I	DISCOVERIES (DIS)	MB/D
E	EOR-ADDITIONS TO PR (ADPR)	MB/D
Y	RESERVES/PROD.-RATIO (CRPR)	Y
*	UNDISC.REC.RES./EXPECTED ULTI-MATE RECOVERY-FRACTION (FR)	-
%	CUM.PROD./ULTIMATE RECOV. (CPUR)	%
$	CRUDE OIL PRICE (OP)	$/B

FIGURE 5 - Price tracks for standard (reference) assumptions and faster OPEC-depletion policy assumptions

258

- oil price
- availability of alternative energy
- economic growth
- demand elasticity
- income elasticity

The main causal loops are shown in Fig. 6.

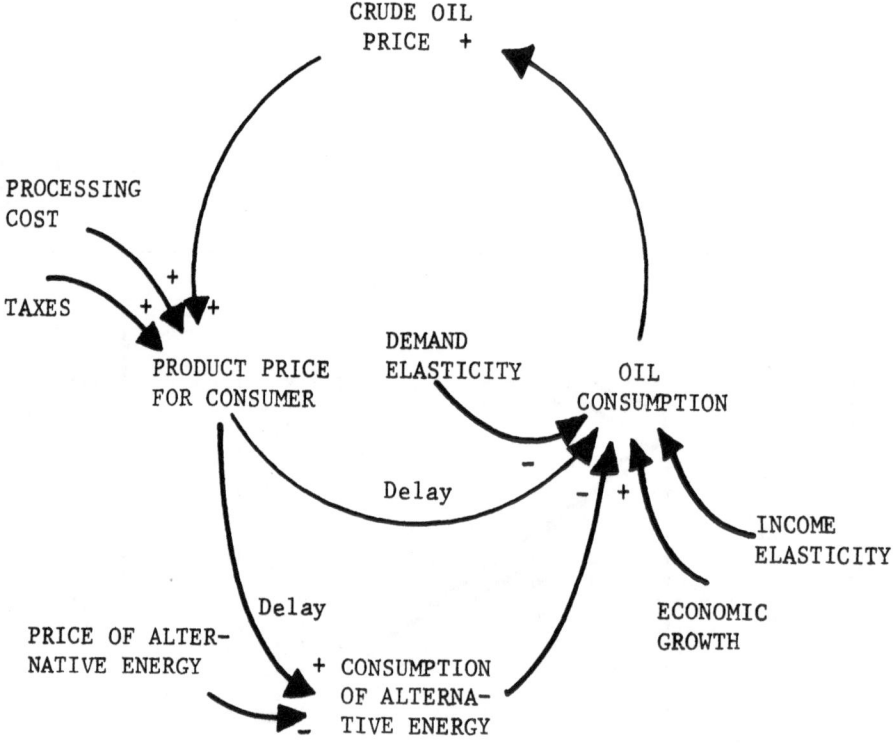

FIGURE 6 - The causal loops of oil consumption

Product prices

Although crude oil price is the basic model variable, we are forced
to use product prices on the demand side for formulating a sen-
sible price/demand relationship. The relationship between the
prices of crude and the prices of oil products which the consumers
are faced with is unclear. The relationship is disturbed by heavy
and frequently changing quantity and value-added taxes in the consumer countries. We should, in a model context, be aware of the
possibility of a correlation between changes in crude oil price

and tax rates. Quantity tax can be used to buffer the consumer from crude oil price changes.

A nonprogressive value added tax will increase the absolute change in consumer price, but keep the percentage change in consumer price equal to the change in the crude oil price. In 1970 the cost of crude oil composed 12% of the final consumer price. By 1975 this had increased to 34% as a world average. In our model, the tax-system in each region remains unchanged. The relationship between crude oil price and final consumer price is thus fixed. One should be aware that A. Al-Janabi, the chief of OPEC's economic department has expressed doubt that such a relationship exists at all [2].

To conclude: Consumer price is formulated by adding a fixed $/bbl cost for transport, refining, distribution and quantity taxes. This is then scaled up by a factor representing value-added taxes. These components are different for each region.

Consumption

The price/demand relationship is based on both an elasticity and a substitution concept. It also takes into account inherent time delays and the fact that natural oil will be increasingly substi-tuted when the price of alternative energy is approached.

Fig. 7 shows possible developments of crude oil price in this stage as a function of different assumed GNP growth rates in the key con-suming regions. The upper price curve shows the consequences of increasing the annual GNP growth rate in each region by one percent compared to the standard simulation.

The lower curve shows the consequences of reducing the growth rate by 1.5%. The middle price curve is the standard run price develop-ment.

The assumptions made on growth in the world economies is, as could be expected, of great importance to the medium to long term deve-lopment in crude oil price. Growth in GNP creates demand for oil. Oil demand forces the price upward until reduced consumption and increased production again clears the market.

Discoveries

In the model oil discoveries are a function of:

I. fraction of ultimately recoverable reserves remaining undiscovered. The assumed relationship simply states

260

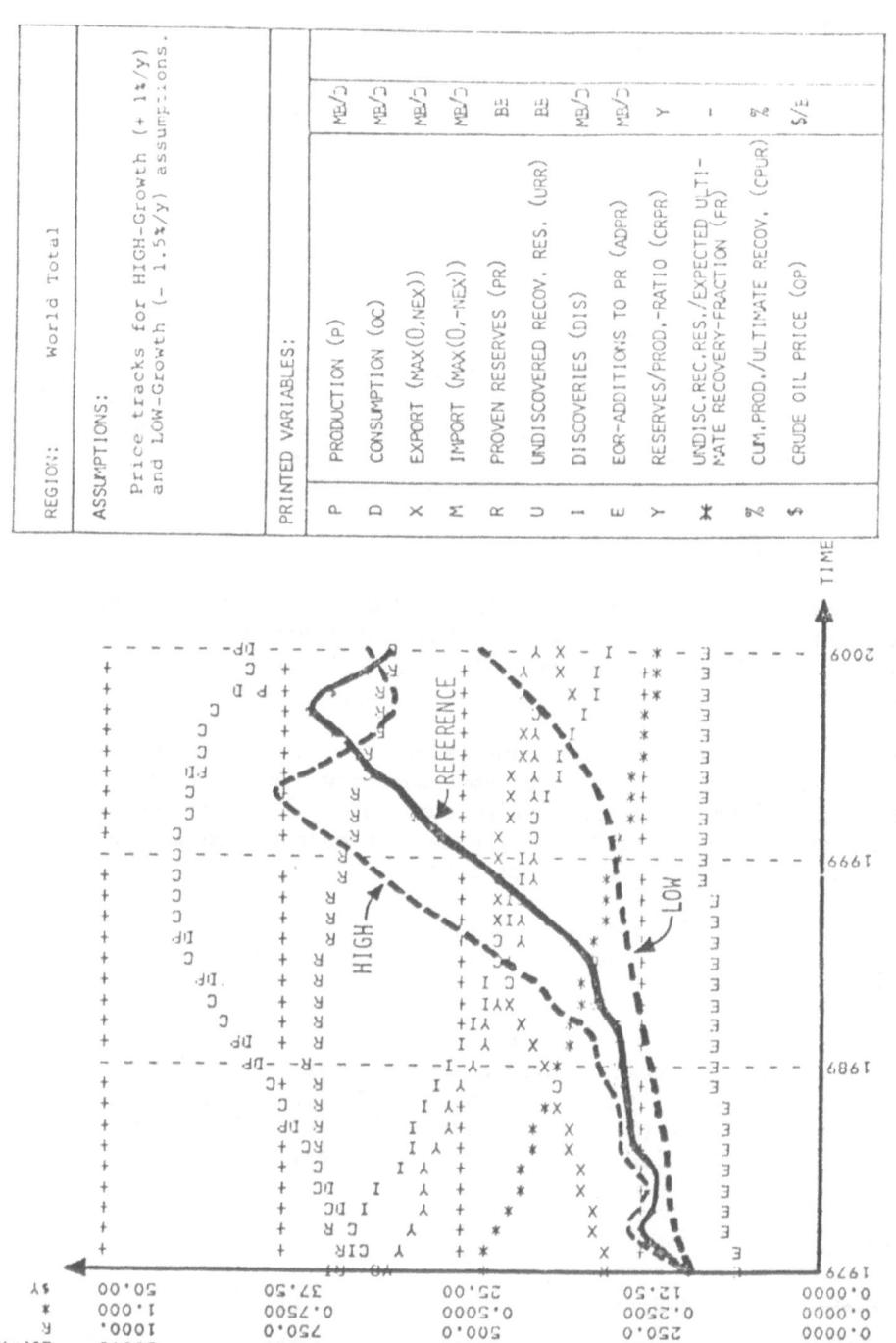

FIGURE 7 – Price tracks for different growth assumptions

that it becomes progressively more difficult to find
new reserves as the amount of oil remaining undiscovered
decreases. It also takes into account that it is difficult to find oil in completely unexplored areas.

II. net profit per barrel.

All other factors remaining unchanged, one tends to
drill more and thus find more, the lower the cost and
higher the price.

III. reserve incentive factor.

This variable reduces discoveries in mature areas with
excess capacity (OPEC).

IV. political incentive factor.

This is simply an exogenously formulated incentive on
discoveries.

There are new technologies to increase the fraction of oil recovered. These technologies are called enhanced oil recovery (EOR)
technologies. By increasing the fraction recovered from 0.3 to
0.6 we double the reserves in the world. In statistics, the contribution from EOR are often quoted together with discoveries. In
the model we have a separate formulation of enhanced oil recovery
in addition to discoveries.

EOR is formulated as additions to

- developed proven reserves
- undeveloped proven reserves
- undiscovered recoverable reserves.

For all three categories, enhanced recovery is assumed to be a
function of price and an exogenous growth in technology.

It is assumed that a fraction of the "old" fields can be subject
to enhanced oil recovery. "Old" fields are predominant in mature
production regions.

EOR thus plays a major role in determining supply in mature regions.
There is considerable disagreement amongst experts as to the future development of EOR.

*Enhanced oil recovery is the major source of uncertainty in
the development of future oil production.*

5.3   The "Alternative Source Stage"

Finally, oil price will approach the long term cost of alternative
energy. This stage will be characterized by a declining share of
natural oil in the total energy consumption picture. Alternative
energy sources are then the dominant factors in balancing oil demand with production. Compared to the "scarcity stage", price
will only change slightly. The price change will be larger if
an "overshoot" in the oil price in this stage is experienced.
This will basically depend on the length of the lead-time to
bring forth sufficient quantities of alternative energy.

Thus, the two main variables in this stage are:

- time to develop alternative energy sources
- cost of alternative energy.

The variable called "alternative energy development time" is designed to capture the time delay between an increase in crude oil
price and the actual supply of alternative energy. We hypothesize
that it will take a while before an oil price increase will be
used as the "expected price". The expected price is the price
that enters into profit calculations for alternative energy projects.

In addition to this delay come the actual planning, preparation,
and building delays. The sum of these delays amount to 15 years
in the standard run. Later, we examine the consequences of reducing this delay time to 8 years.

The cost of alternative energy is also a key factor in determining
oil price in this stage. In the standard run we have assumed this
to be equivalent to US $ 25/bbl (1975 $).

Fig. 8 shows the consequences of reducing this parameter to $20/bbl,
while keeping the delay time constant (labeled B). In Fig. 8 the
reference run is labeled A. If in addition to reducing the cost
of alternative energy to $ 20/bbl we reduce the delay time from
15 years to 8 years, we obtain the development labeled C in Fig.8.

We see that none of these changes have any effect prior to 1990.
The above two parameters are very important determinants of oil
price between 1995 and 2010. In this period, an overshoot in oil
price is clearly apparent. With 8 years delay time and $ 20/bbl
the overshoot is rather small, while 15 years and $ 25/bbl leads
to a significant overshoot in world oil price.

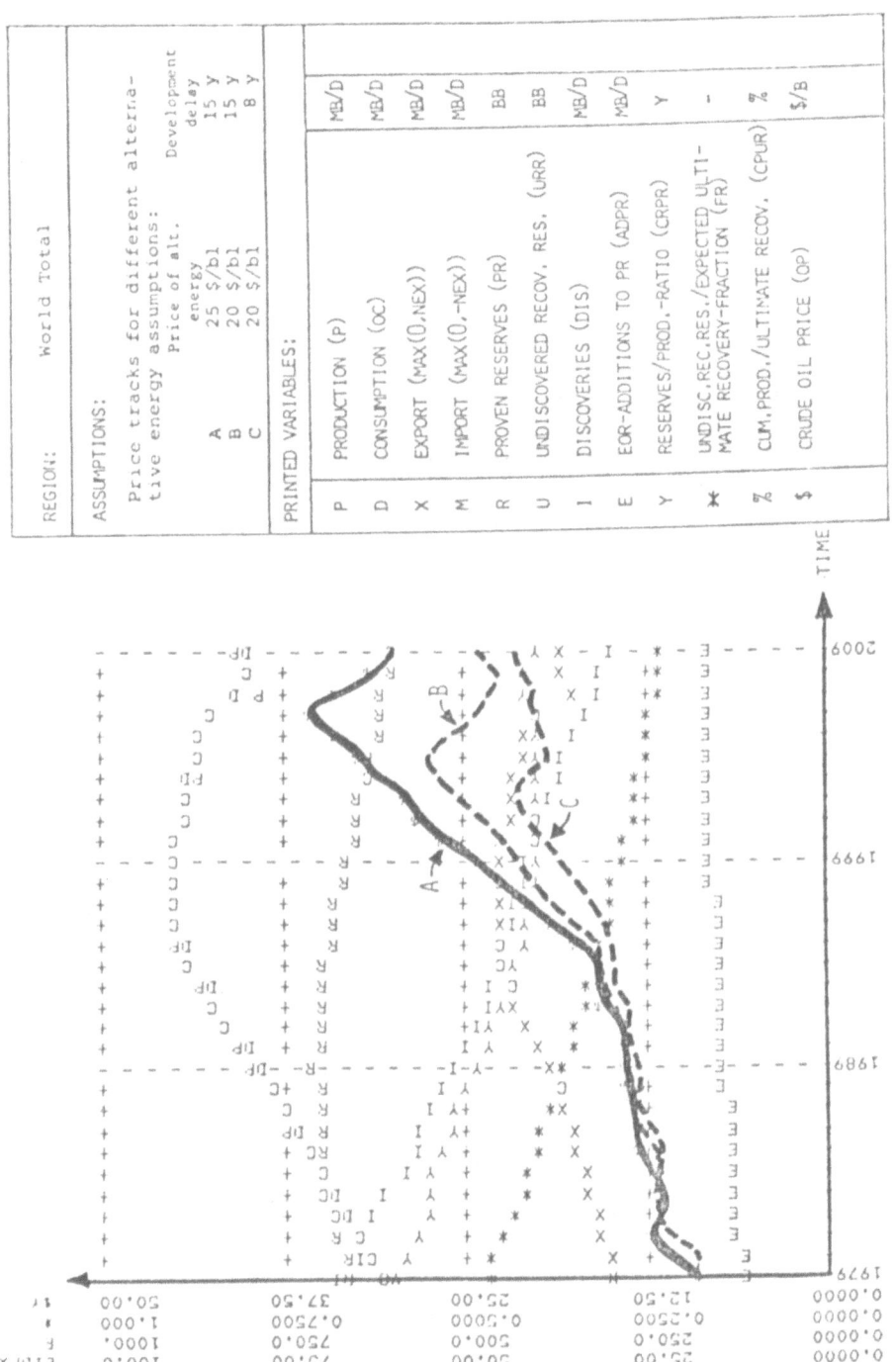

REGION: World Total

ASSUMPTIONS:

Price tracks for different alternative energy assumptions:

	Price of alt. energy	Development delay
A	25 $/bl	15 y
B	20 $/bl	15 y
C	20 $/bl	8 y

PRINTED VARIABLES:

P	PRODUCTION (P)	MB/D
D	CONSUMPTION (OC)	MB/D
X	EXPORT (MAX(0,NEX))	MB/D
M	IMPORT (MAX(0,-NEX))	MB/D
R	PROVEN RESERVES (PR)	BB
U	UNDISCOVERED RECOV. RES. (URR)	BB
I	DISCOVERIES (DIS)	MB/D
E	EOR-ADDITIONS TO PR (ADPR)	MB/D
Y	RESERVES/PROD.-RATIO (CRPR)	Y
*	UNDISC.REC.RES./EXPECTED ULTI-MATE RECOVERY-FRACTION (FR)	-
%	CUM.PROD./ULTIMATE RECOV. (CPUR)	%
$	CRUDE OIL PRICE (OP)	$/B

FIGURE 8 – Price tracks for different alternative energy assumptions

6.   QUANTIFYING THE UNCERTAINTY

The future price of oil is important for Norway's economy in several ways:

- extent of development of Norwegian hydroelectric
  potential

- development of North Sea oil fields

- planning of Norwegian export.

Everybody agrees that the future oil price is highly uncertain.
The decisions mentioned above must also take this uncertainty into
account. Nobody has so far managed to express the uncertainty
quantitatively. It is, however, possible to simulate the effects
of uncertainty by using the type of simulation model which we have
available. The model we have been talking about so far has been
deterministic. At the present time we are in the process of incorporating uncertainty into the model.

For overview, we can divide uncertainty into five groups:

   I.   uncertainty in the time development of exogenous
        variables

  II.   uncertainty in relationships in the model

 III.   uncertainty in parameter values

  IV.   random events during the simulation

   V.   experimental uncertainty (analytical solution impossible)

Exogenous variables

In OILTANK the growth rates for GNP in each region are exogenous.
There is considerable uncertainty in the numerical values for
these growth rates. We have formulated the uncertainty in the
growth rate for each region. However, there is probably enough
interdependence between the growth rates. This correlation has
to be formulated. In other words: Each "random growth rate" is
conditional on other growth rates.

This example illustrates the rule that formulating some of the
inputs as random variables forces one to consider relationships
that were left out of the deterministic model.

Uncertainty in the relationship in the model

The hypothesis is that there exists a relationship, but that its exact form is unknown. The relationship itself is then formulated as a bounded random relationship. Once a relationship is "drawn", it is maintained for the whole length of the simulation. An example here is the relationship between the Recovery Factor (RF) and oil price. Everybody agrees that RF increases as oil price increases (after a time delay). There is, however, substantial disagreement on the magnitude of this effect. This disagreement is formulated as an uncertainty.

Uncertainty in parameter values

This is the most common type of uncertainty. Such a parameter value might be the "initial recoverable reserves in Mexico". Today we do not know this figure with any degree of certainty. By plotting the different estimates that have been made, we have formulated a probability distribution for this variable. The information in that probability distribution is to a large extent subjective of course.

Random events during the simulation

This is the traditional type of uncertainty introduced in Monte Carlo (MC) models. In the future development of oil price production and transport, one must be prepared for several types of incidents. The development in Iran is an example. The Iran case illustrates another point: The effect of random disturbances depends on when it happens. The present Iran reduction in production is substantially compensated by increased production by other OPEC members. In 10 years time the possibility of other producers being able to compensate by increasing their production could be substantially more limited. The price effect of an Iran incident would thus be larger, the later it occurred.

Experimental uncertainty

The model is so large (several thousand equations) that an analytical solution is impossible. The experimental uncertainty comes from the fact that we can only run the model, due to computer costs, a certain number of times. This uncertainty is, however, under our total control.

Total uncertainty

In order to incorporate the first three types of uncertainty we
have formulated the model as a Monte Carlo simulator. We use 7
different types of distributions and have uncertainty in some 300
variables in the model. Fig. 9 shows experimental results for
crude oil price from this approach using 300 simulations. (In
order to get a fair representation of total uncertainty we would
need approximately 1,000 simulations.)

Fig. 10 shows the corresponding results for world oil production.
The uncertainty in price is quite large in the immediate future
until ca. 1985. In the period 1985-1995 uncertainty is somewhat
reduced, while the picture from 1995-2009 is one of increasing
uncertainty.

By following the individual price trajectories, the three stages
can be identified. The figures give an intuitive feeling for
the uncertainties involved. At any given point in time, a cross
section taken of the price profiles gives the probability distribution of oil price in that year based on today's best guess.

Fig. 11 shows the probability distributions for crude oil price
in 1985, 1990 and 2005. The figure indicates that there is a
very low (10%) chance that the price in year 2005 will be identical to the price in year 1990. This type of figure, however, does
not give any information about the simultaneous probability distributions. It might be that all the events that lead to low
price in year 2009 are also the same events that lead to low
price in 1985. Theoretically, therefore, there might be no event
that simultaneously leads to high price in 1985 and low in year
2009.

The "most probable" time-path is shown in Fig. 12. The "most
probable" is here defined as the time-path that goes through the
average price in each year. Fig. 12 also shows the "most probable"
development in world oil production. The production has a tendency to peak before the turn of the century. When reading these
curves, one should bear in mind the uncertainty involved. While
the expected price in year 1990 is approx. $ 15 per barrel, the
standard deviation in the same year is approx. $ 2.3 per barrel
(in 1975 $).

An inspection of the results together with the random parameters
that generated them will reveal how sensitive the model is to
specific parameter combinations. When one finds these sensitive
points, there are several courses of action that might be followed:

- The findings may indicate a fallacy in the model if it
  is clear that the real world behaviour being studied
  has persisted under a wide variety of circumstances

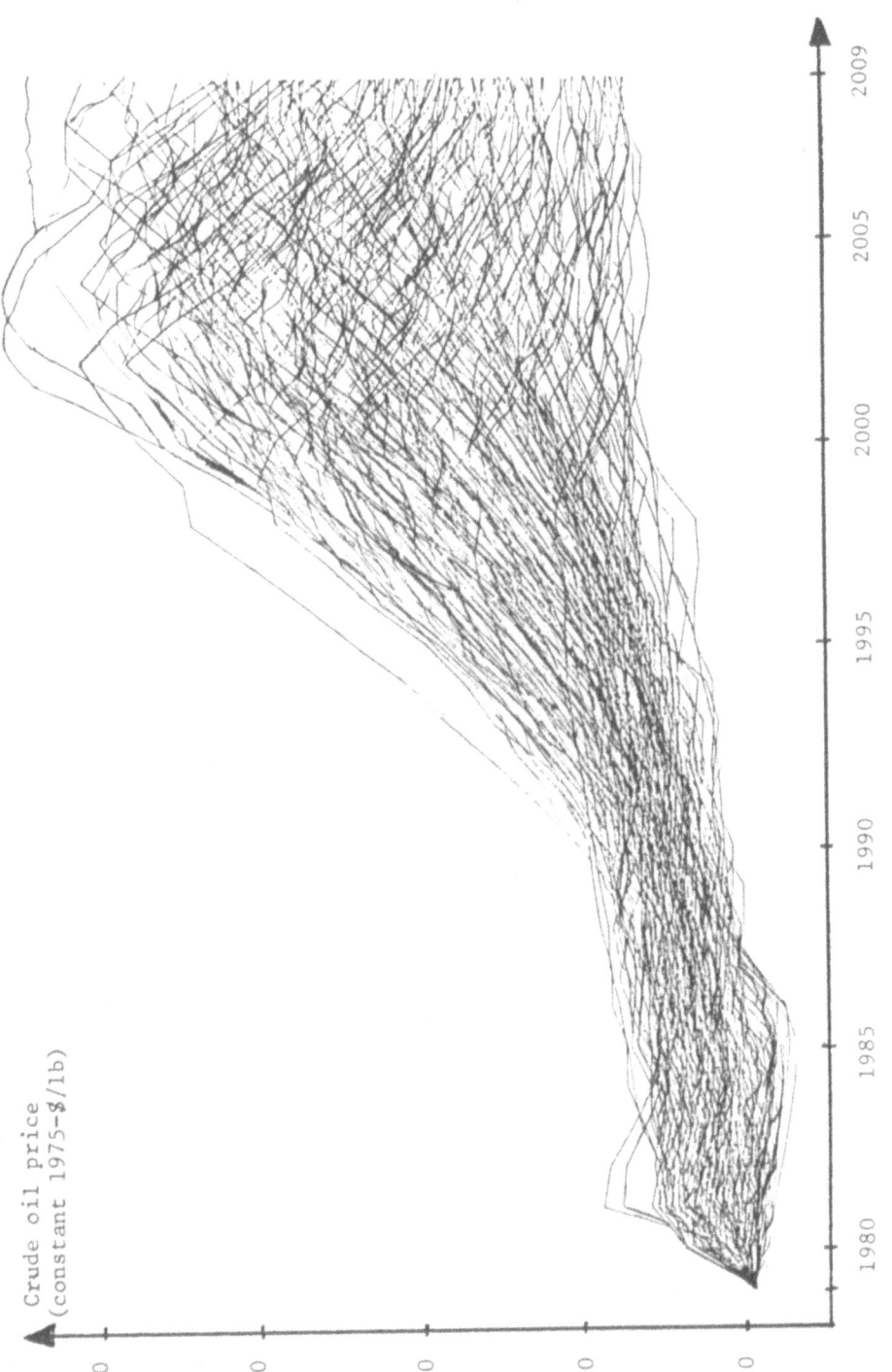

Crude oil price
(constant 1975-$/lb)

FIGURE 9 - Crude oil price tracks of Monte Carlo simulation runs

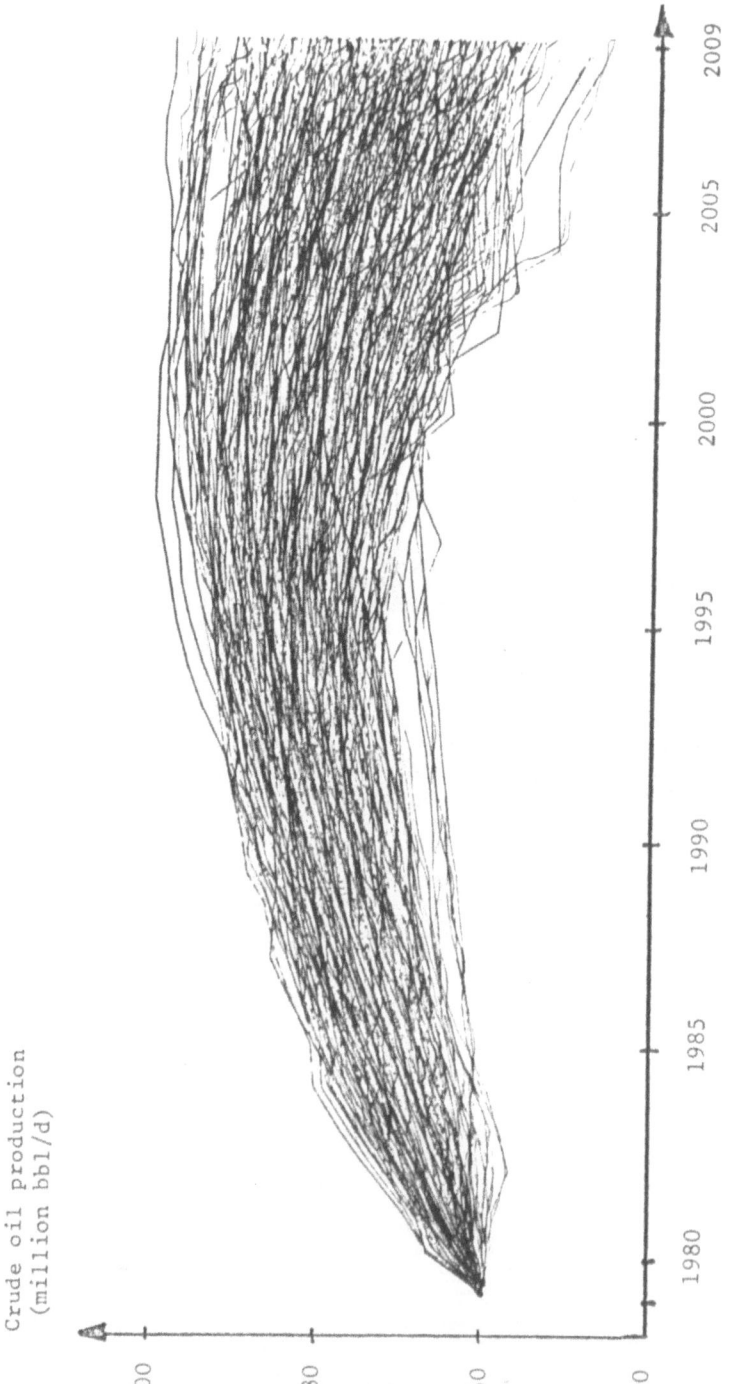

FIGURE 10 - Crude oil production tracks of Monte Carlo simulation runs

- The sensitive parameter set may become the focus of substantial empirical research so that its value can be estimated more precisely

- One may respond to the information on sensitive parameters by designing better policies in the real world so as to dampen their effects

- The client may also seize on the sensitive parameter and use it as a policy lever which can be manipulated in such a way as to produce the desired results.

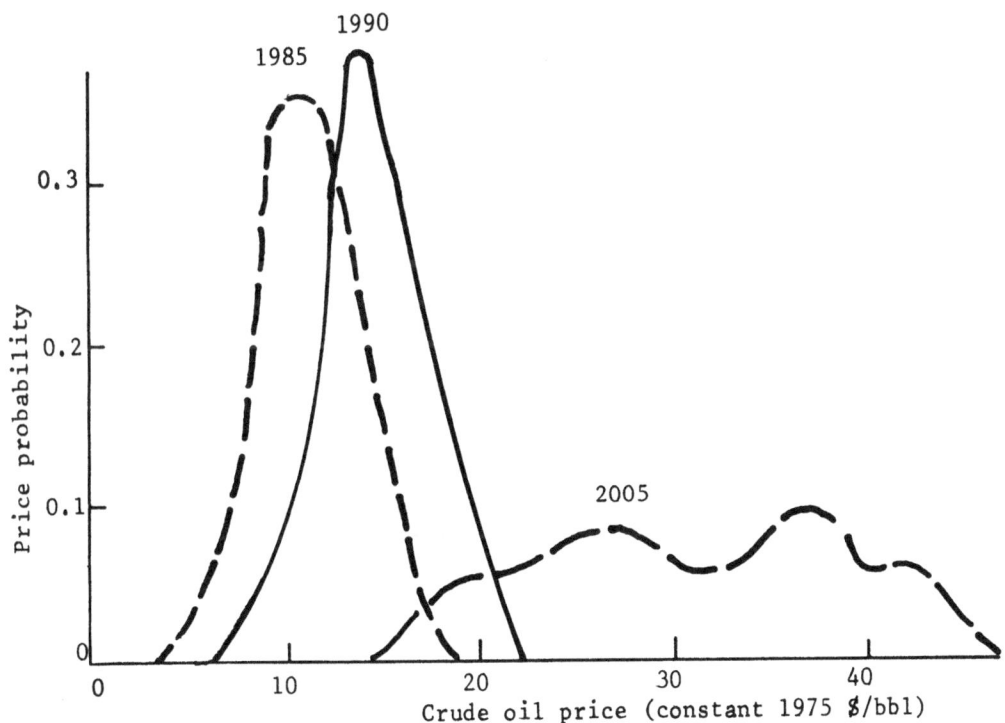

FIGURE 11 - Probability distributions for crude oil price in 1985, 1990 and 2005.

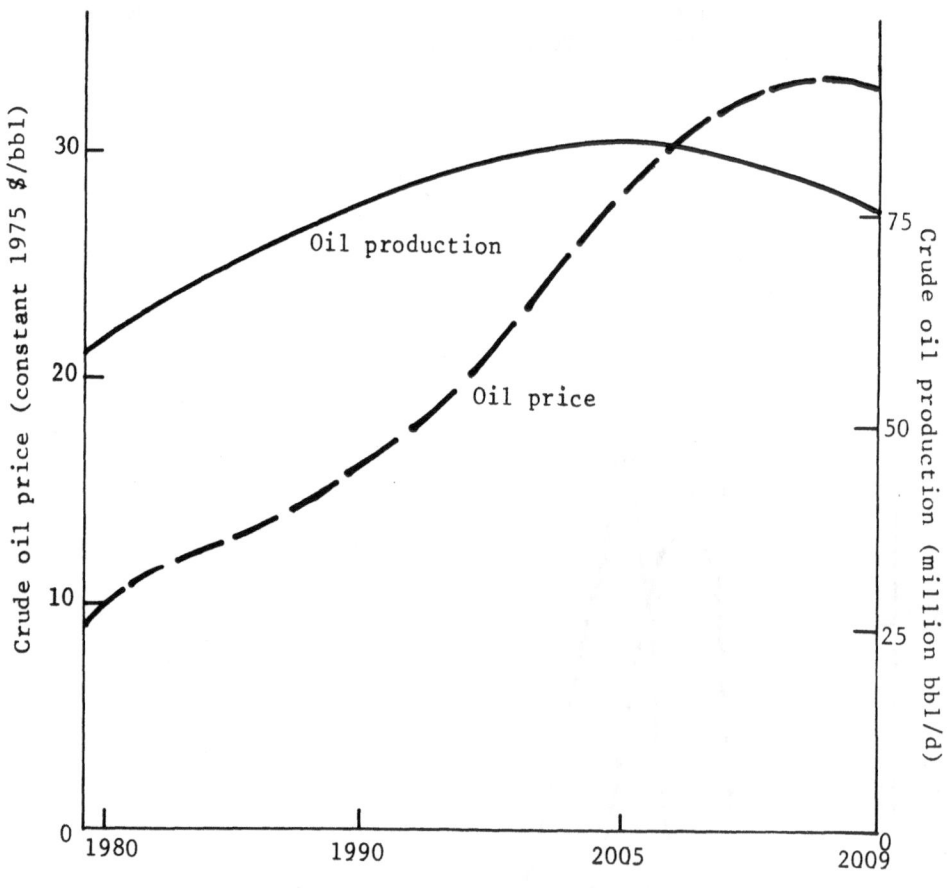

FIGURE 12 - "Most probable" trajectories for future
oil price and production

7. TESTING

No system analysis tool can provide new empirical data on individual components of the system. System analysis tools first become
useful in integrating this diverse data to obtain a comprehensive
picture of the system and to understand its probable consequences
for alternative policies. Information about determinants of future demand and supply of energy varies greatly in its comprehen-
sion and accuracy. The spectrum of information available is shown
diagramatically in Fig. 13.

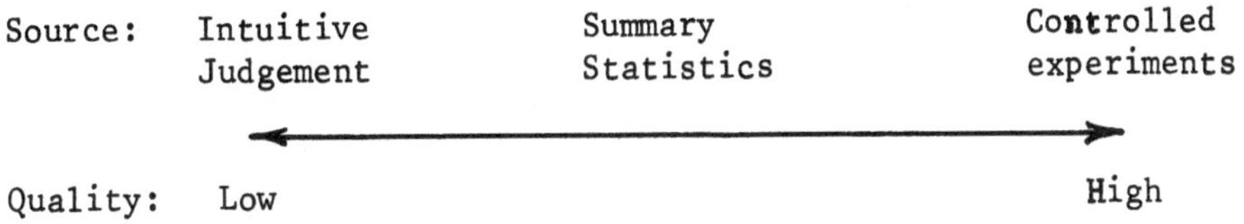

Source: Intuitive   Summary   Controlled
          Judgement   Statistics   experiments

Quality: Low   High

FIGURE 13 - The quality spectrum of information

In our work, we stress that each variable in the model should preferably have a real world counterpart. This "paradigm" helps in
using the whole spectrum of data in testing a model. The spectrum of information is clearly related to the question of uncertainty we just discussed. The "intuitive judgement" data typically are associated with a higher standard deviation than are
the data from controlled experiments.

Testing a model can be seen from at least three viewpoints. Each
viewpoint has different tests.

1. Types of Testing Appropriate to Modeller

   - statistical analysis
   - discovery (surprise behaviour)
   - comparison with real behaviour and explaining the
     difference.

2. Types of Testing Appropriate to Technical Evaluator

   - is boundary consistent with purpose?
   - does dimensional analysis check?
   - is behaviour independent of exogenous inputs?
   - is behaviour for different inputs reasonable?
   - has the model been run over a long enough time?
   - can the model evaluate multiple modes?
   - is behaviour plausible, given shifts in initial time?
   - if statistics are used, are techniques accurate?

3. Nontechnical Evaluator

   - is documentation complete?
   - are there verbal statements of the included theories?
   - does the model include/exclude theories you believe
     in?

We will now take a closer look at the three types of testing appropriate to the modeller:

Statistical analysis

OILTANK has not, as of yet, undergone a formal testing programme.
The demand side is based on results from other energy studies.
We are in the process of reformulating the whole demand side of
the model. There are very few data or analyses for the long run
development that can gainfully be used for our model. Even different studies of the short and medium term price response of oil
vary substantially.

On the supply side, however, we have cooperated with an international oil consulting firm. Their data and insight have been of
great value in testing the model.

Surprise behaviour

One possible behavioural test is provided when the model generates
a new or unexpected behaviour mode which is subsequently found to
exist in the real world.

In OILTANK, oil price development exhibits an "overshoot" followed
by damped oscillations before finally stabilizing at a level given
by the cost of alternative energy sources. The overshoot is caused
by delays in developing alternatives. Because of this delay, the
price can go far above the "cost of alternative energy". Unfortunately, we have to wait some 15-20 years before we can make a
comparison with the real world.

Real behaviour

The model does not endogenously explain the price jump in 1974 or
the Iranian price increase. There are models that do "explain"
the 1974 price rise. The essence in the explanation is "capacity
utilization" in OPEC as the key determinant. The fertility of
tuning a long term model to reproduce a single discrete event in
the historical development of the system is somewhat doubtful.
Our model does give a fair representation of the consequences of
the 1974 and 1978 production cut-backs.

From the original work with the first model came the surprizing
results (to us) that the single most important factor in deciding
the long term supply of oil was "enhanced oil recovery" (EOR).
EOR is the amount of oil brought forth due to increase in the
Recovery Factor we spoke of earlier. The work on EOR around the
world now seems to indicate similar conclusions.

Fortunately, improved long term policies can often be designed,
even in the absence of precise data, as long as the underlying

structure of the system - the complicated network of causal relationships - is generally understood.

8. CONCLUSION

The choice of modelling technique often follows from the problem
definition and the view of the world which the research worker
has. In this study we have applied

- continuous simulation
- linear programming
- Monte Carlo simulation.

When the modelling effort started, the changes in model focus
which are described in this paper were not planned.

The problems which the modelling effort has so far addressed have
not necessitated differentiating the various qualities of crude
oil. We have also managed to keep the number of regions down to
16. (Mexico was added in January 1979.) Clearly, aggregate,
transparent models have their advantages.

REFERENCES

1. C. Wilson, Energy: Global Prospects 1985-2000 , Report of
   the WAES: Workshop on Alternative Energy Strategies, McGraw-
   Hill, New York, 1977.

2. A. Al-Junabi, "The Determinants of Long Term Demand for
   OPEC Oil", Tripoli 1978.

DISCUSSION

Ervik's presentation stood in stark contrast to that of Deam.
The desirability of quantifying the uncertainty surrounding the
future development of oil price was accepted. Several speakers
did, however, question why none of Ervik's possible price paths
showed a decreasing price as Deam had described. Ervik admitted
that before Deam's lecture neither he, nor apparently any of the
other experts with whom he had consulted before assigning the mean
and variance values to the cost of alternative energy sources were
aware of Deam's methanol path. He would consider this on his
return and might well include it in his parameter estimates.

It was suggested that one reason why Ervik's future oil price did
not decrease was his assumption that the price was sticky downwards.

Ervik replied that this assumption did not affect the price level
necessary to clear the market, only the time taken to reach that
price level.

Ervik had asserted in his presentation that the development of
technology for enhanced oil recovery would significantly affect
the amount of oil available and the future production profile for
oil. He mentioned that in the model, the proven global reserves
were taken as only 1/10 of its quoted value, and the actual recoverable amount is taken to be a function of oil price. Several
speakers were sceptical of the possibility of significant improvements in the recovery factor. Debanne referred to a technology
assessment project in which he had taken part in which all the
nuclear engineers interviewed had placed great hope in enhanced
oil recovery, whilst all the petroleum engineers interviewed had
expressed similar hopes in nuclear technology.

The independent probability distributions of some 300 variables
was questioned: very few cases existed where there was no correlation between the random variables. For example, an increase
in oil recovery factor in one country would affect the technology
in other countries as well.

In response to a question, Ervik explained that each of the 300
runs was deterministic in itself. He also said that while equilibrium prices were endogenous, time delays were specified exogenously. It was pointed out that these could have crucial effects
on the model results.

PART IV

ELECTRICITY SYSTEMS MODELLING

A Comprehensive Approach to Forecasting Long
Term Regional Electricity Demand

Large Scale Integration of Wind Power in Power
Systems

Power Systems Planning Under Uncertainty in
Load Growth and Resource Availability

A COMPREHENSIVE APPROACH TO FORECASTING LONG TERM REGIONAL
ELECTRICITY DEMAND

S. Caldwell, W. Greene, T. Mount, S. Saltzman, R. Broyd

(Respectively) Departments of Sociology, Economics, Agricultural Economics, City and Regional Planning (2),
Cornell University, Ithaca, New York, 14853, U.S.A.

ABSTRACT. This paper describes a comprehensive system of models
which has been designed to forecast the long term demand for
electricity and other major sources of energy used in a region.
The important explanatory variables represent the socio-economic
characteristics of the spatial area being modelled as well as the
prices of various types of energy and the prices of certain other
goods and services. Aggregate forecasts of energy demand for the
residential, commercial, and industrial sectors of the region are
provided by a system of macro-econometric models which also includes
a broad range of aggregate economic and demographic variables. A
model for evaluating time-of-day rate structures and their impact
on peak load demand is driven by the aggregate electricity demand
forecasts of this economic/demographic/energy demand model. Other
outputs from this macro model are used to drive a micro simulation
model of households in the region in order to study the residential
demand for electricity on a disaggregate basis. A broader range
of issues and policy questions can be addressed using such linked
macro and micro models than would have been possible using a less
comprehensive approach in this application.

1. INTRODUCTION

The system of models presented in this paper was developed in response to a need to forecast the long term, regional (state) demand
for electricity. In the region under study, such forecasts are
used as annual inputs to a regulatory process from which must come
a determination as to whether or not there is likely to be sufficient
demand in the future to warrant the construction of new power plants
proposed for certification that year. It is clear, then, that such

long-term forecasts can impact on a wide variety of related economic, environmental, and social issues, especially for those living in the region.

A common approach used in the U.S. to forecasting the long term regional demand for electricity is based on single equation econometric models [3]. In these models, explanatory variables such as population, income, economic activity, etc., typically are assumed to be exogenous and, except in a general sense, independent of each other. Engineering or "end use" models represent another common approach used to forecast the long-term demand for electricity by a region's residential sector. These models also require assumptions about the region's future population and income levels, for example, which also are usually considered to be exogenous.

A more comprehensive approach to regional forecasting of electricity demand was recently developed* in an effort to avoid some of the problems inherent in the types of models described above. It consists of a system of integrated and linked macro and micro models in which many of the regional explanatory variables are treated as endogenous rather than exogenous variables.

The macro-econometric model treats the region on an aggregate basis in that virtually all of the variables represent the region as a single spatial unit. Each of the three major components or modules in the macro model (i.e., economic, demographic, and energy demand modules) consist of a system of multiple regression equations which are estimated from data from their own respective sample periods. The economic module treats the economy of the region within a consistent framework of regional gross product and income accounts. Variables are included for the components of final demand, value added, employment, and wage rates by industrial sectors, plus income and labor force. The demographic module disaggregates the region's population by race, sex, and nineteen age cohorts from birth to age 85 and over. Births, deaths, migration and household formation play important roles in the dynamics of this module. The energy demand module provides estimates of the demand for various

---

* The models presented here were developed in response to a need for fifteen-year forecasts of the demand for electrical energy in New York State. The support of the following organizations is gratefully acknowledged: The National Consumer Law Center, Boston, Massachusetts; the New York State Consumer Protection Board, Albany, New York; the New York State Public Service Commission, Albany, New York; and the Schenectady Community Action Program, Schenectady, New York. This paper draws on materials presented in the project's unpublished final report [2] and in a preliminary report [1].

types of energy such as electricity, oil, gas, etc. These demand forecasts are disaggregated by residential, commercial, and industrial users.

In order to forecast distributional patterns of household electricity consumption, the macro-econometric model has been connected to a micro-simulation model of the household sector. The combined macro-micro framework provides forecasts of aggregate electricity demand and also of the distribution of electricity use and expenditures by households with different characteristics. Since households are identified by income, size, appliance ownership, and other socioeconomic characteristics, a variety of distributional impacts can be examined. Moreover, the micro model treats households as actors whose behavior "adds up" to significant aggregate consequences for the region.* Thus, the micro model is used (a) to profile the distribution of electricity use and expenditures across households with different incomes and other different characteristics; and (b) to generate an aggregate forecast of household demand derived from "adding up" across the sample of households.

In theory, a household demand forecast derived by "adding up" demand across households could outperform a forecast derived from an aggregate model since the possibilities of efficient, unbiased parameter estimation are greater with the typically larger number of observations and variables available at the micro level [5]. In practice, however, problems of defining and measuring key variables, in obtaining price measurements, and in capturing changes over time can eliminate any advantage. By generating aggregate household sector forecasts from both macro and micro models we attempted to combine the distinct and complementary strengths of each approach.

Because there is a strong trend developing in the U.S. toward the use of time-differentiated-rates, the predictions generated by the use of time-differentiated-rates, the predictions generated by a load forecasting model should include possible changes in the design of electric utility rates. The forecasts from the electricity demand module are generated assuming that rates are designed in the forecast period as they have been in the past. These forecasts are modified in a simulation model in which alternative assumptions about electricity rates and elasticities can be evaluated.

---

*As contrasted with macro models in which the behaving unit is an entire economy, the behaving units in micro models are individuals, households, firms, and other micro actors. A sample of households is used to represent the regional population of households. Aggregates are obtained by "adding up" across micro-units. Since households of varying characteristics are represented, distributional questions can be addressed, in contrast to macro models which generate region-wide averages or aggregates [6].

2. THE MACRO MODEL: ECONOMIC AND DEMOGRAPHIC MODULES

Long-term demand forecasts for electricity are often based on the use of single equation regression models in which explanatory variables such as personal income, industrial output, employment, population, number of households, etc., are all considered to be exogenous. The hypotheses underlying the structure of the macro model described here is that these variables are actually endogenous. Therefore, an annual model in which these economic and demographic variables are determined simultaneously would be a more accurate representation of reality than a series of single equation models in which these variables are considered to be exogenous. The argument, in effect, is that the economic and demographic variables (i.e., the variables that determine the demand for energy) in a regional economy are interrelated to a high degree. Changes in the value of one variable which does not directly affect the demand for energy could have a significant impact on changes in other variables which do directly affect the demand for energy.

This section provides an overview of the economic and demographic equations in the macro model. Both sets of equations are fully integrated in the macro model in the sense that their endogenous variables are jointly determined. However, for convenience, each module is discussed in separate sub-sections. The endogenous variables (e.g., population, income, etc.) which provide the ties between the two modules are noted in the appropriate sub-section.

2.1 The Economic Module

The framework of regional product and income accounts provides an overview of the structure of the economic module of the macro model.* One of the standard methods for developing gross product estimates is based on value added by industrial origin. These usually include the nine major sectors of an economy: agriculture; mining; contract construction; manufacturing; wholesale and retail trade; transportation, communication, and public utilities; finance, insurance, and real estate; service and miscellaneous; and government.

Gross product can also be estimated by final demand components. These normally include the following economic activities: consumption; investments in business construction, producer's durable equipment, and housing; net inventory changes; net exports of goods

---

*The version described here is based on an earlier model of New York State [7]. The most significant differences between these two models are that the current version is annual rather than monthly, all national economic variables (e.g., GNP) are considered to be exogenous rather than endogenous, and the version is not multi-regional.

and services.

In addition to these variables plus personal income, the model also includes a number of other important variables disaggregated by major industrial sectors. These include employment, changes in wage rates, and wages and salaries. Each of these endogenous variables is determined by a behavioral equation, either linear or non-linear, which is developed from economic theory or the demand conditions in the relevant market. The current version of this module contains approximately 65 equations.

National variables in this module are all considered to be exogenous. These include, among others, gross national product, various price indices, interest rates, government purchases, etc. Demographic variables which are also considered to be endogenous, are described in more detail in section 2.2.

The following considerations entered into the development of the demand side of the regional product and income accounts. A simple aggregate consumption function based on classical consumer demand theory was formulated with income and a lagged consumption term as explanatory variables.* Typical explanatory factors such as interest rates, other costs, income, and a one period lagged term of the dependent variable, are incorporated in the appropriate investment functions. There is a relatively simple hypothesis for explaining regional and foreign trade in that the level of trade is assumed to be a function of the level of economic activity in the appropriate areas (i.e., outside the region for exports and within the region for imports). Government purchases of goods and services are assumed to be exogenous for the time being and the inventory adjustments are temporarily included in a statistical difference term which measures the variation between the two alternative estimates of gross state product. Because the micro model requires estimates of the annual number of dwelling units built in the region, equations are included for both the number of single family units and the number of multiple family units constructed.

Gross product originating in a sector was assumed to be determined by the demand for goods produced in that sector and the nature of the markets served rather than the state of technology implied by its production function. This approach was chosen because full employment was not generally observed throughout the sample period covered by the model and because of data constraints.

An economic base approach was used for the analysis of the

---

* Disaggregated demand functions for various sources of energy (e.g., electricity, gas, etc.) are included in the energy demand component of the macro model in Section 3.

demand and market characteristics of the major sectors of the region's economy. The markets for goods produced by importing and self-sufficient sectors are normally assumed to be local markets. Therefore, in addition to the lagged dependent variables in the value added functions for these sectors, the explanatory variables include some relevant local variables serving as measures of final and intermediate demand in the region. The following sectors were classified as being primarily importing or self-sufficient sectors: agriculture; mining; contract construction; retail trade; transportation, communications, and public utilities; service and miscellaneous; and government.

On the other hand, the demand for goods produced in exporting sectors is very likely to be explained by both local and national markets. Therefore, in addition to the lagged dependent variable, the other explanatory variables in such sectoral equations should include some applicable national and local variables. Manufacturing, wholesale trade, as well as finance, insurance and real estate are assumed to be exporting sectors serving both national and local markets with appropriate variables specified in these sectoral equations.

The sectoral employment functions are linearized forms of the constant elasticity of substitution (CES) production function, transformed so that employment is the dependent variable. The labor force participation rate is assumed to be a function of regional income and unemployment as well as a time trend variable which serves as a proxy for the increasing tendency for married women to work. Other demographic variables of interest in such a regional model such as migration, etc., are discussed in Section 2.2.

Regional price variables and relevant deflators are assumed to follow the paths of corresponding national variables which are exogenous to this model. Changes in sectoral wage rates are generally assumed to be influenced by the region's unemployment rate and the change in the consumer price level, both lagged one period, as well as the previous period's actual wage adjustment. These explanatory variables are lagged one period since current wage rates are often based on previous decisions which were influenced by values of these relevant economic variables. Equations for personal income as a function of gross state product as well as federal and state personal taxes as functions of income are straightforward.

2.2   The Demographic Module

Because of the long lead time necessary to put large capital facilities in place, energy-related investments are especially sensitive to population and household changes. Over the past several decades the growth in regional populations and the even faster growth of households have been major determinants of the growth of energy demand. Within the macro model, five components of population and household changes are explicitly represented: mortality, fertility, net migration, aging, and household status.

Because it is highly selective, mortality affects not only the size but also the composition of the population and household sectors. In this century steeply declining mortality rates have sharply altered age structures and household patterns.

Historical central death rates for the region can be derived for census years by combining vital statistics data on the number of deaths in the region with census counts of the region's population. The pattern of historical decline for the eight census years, 1900-1970, can be plausibly fitted to an exponential function of time with three parameters. Time in effect acts as a proxy for the underlying influences on mortality decline. The third parameter represents an asymptotic lower limit to mortality decline (6).

$$CDR_{it} = a_{i0} + (a_{i1} \, e^{a_{i2} * t} \, u_{it}) \qquad (2.2.1)$$

where CDR_{it} is the derived central death rate for the region

for age/race/sex

subpopulation i (i = 1,2...,76) in census year t (t = 1900, 1910,1920,...,1970)

a_{i0} = asymptotic lower limit parameter for subpopulation i

a_{i1} = intercept parameter for subpopulation i

a_{i2} = exponential rate of decline parameter for subpopulation i

u_{it} = error term for subpopulation i in year t

Assuming that the log of the error term is well behaved and in particular that adjacent census year errors are not correlated (an assumption supported by the Durban-Watson statistics), the parameters can be estimated with the historical central death rate data using an iterative, non-linear least square algorithm by transforming the equation into the following form:

$$\ln (CDR_{it} - a_{i0}) = \ln a_{i1} + a_{i2} \, t + \ln u_{it}. \qquad (2.2.2)$$

However, since for each of the 76 subpopulations only eight decennial observations were available with which to estimate three parameters, the asymptotic limit parameters (a_{i0}) were obtained from a different source. Hypothesizing that in the long run, the region's mortality rates will converge toward national rates, we drew on Social Security Administrations actuarial judgements concerning projected cause-specific declines in mortality by the year 2050. These projections for the year 2050 were used as the asymptotic limit parameter. The eight census year observations then were used to estimate the two remaining parameters.

$$\ln (\text{Diff}_{it}) = b_{i2} + a_{i2} + v_{it} \qquad (2.2.3)$$

where: Diff_{it} = the difference between the census year (t) central death rate and the actuarial lower bound estimate for subpopulation i

$$b_{i1} = \ln a_{i1}$$

$$v_{it} = \ln u_{it}$$

Of the 152 estimated parameters only seven failed to reach at least 5 percent significance levels. Since all 76 slope estimates were negative the model projects further declines in mortality rates over the next fifteen years. The actual rate of decline is different for different age/race/sex subpopulations.

Fifteen-year forecasts of energy demand are relatively insensitive to fertility projection since the effect of extra children in an existing household is small relative to the effect of extra households. For this reason, and also because fertility is a difficult, complicated process to represent endogenously, the module simply assumes plausible central birth rates for female subpopulations arrayed by age and race. Since national total fertility rates recently have been considerably below replacement levels of 2.1 children per female, and since the region's fertility rates have tended to be lower than national levels, we have assumed an eventual total fertility rate of 1.9, which is half-way between the low (1.7) and middle (2.1) Census Bureau forecasts. This of course, assumes that the region's fertility rates increase in the future as compared to present levels.

Births change only the youngest age groups and deaths change primarily the oldest age groups, but net migration to or from a region may substantially affect all age groups. Hence, accurate long term energy demand models require accurate submodels of net migration.

Time series measures of net migration rates for the region by race and age were derived indirectly from official population estimates by taking account of births, deaths, and aging in the region's official state population estimates and using the "residuals" as estimates of net migration. These "residual" estimates of net migration were checked against direct estimates of five-year migration patterns from the 1960 and 1970 censuses and also against recent estimates of 1970-75 state migration patterns.

These data were used to estimate the parameters of equations generating net migration rates for eight subpopulations. The independent variables aim to capture economic and social influences on net migration rates. For persons of working age, we hypothesized that net migration to or from the region is positively related to the region's income and employment levels relative to the income and employment levels of the rest of the nation. For younger persons we hypothesized that net migration rates are primarily dependent on the net migration rates of the parental generation and secondarily on employment and income levels. The latter variables serve as proxies for differential migration rates of persons with children vs. those without children. For the retired population, migration rates were hypothesized to be related to absolute per capita state income level (as a proxy for the economic ability to move) to the region's relative income position (as a proxy for cost of living) and to migration rates to younger groups (as a proxy for social ties).

These hypotheses guided the specification of equations generating net migration rates for eight separate race/age subpopulations. Estimated on time series net migration data for the region, these equations generate net migration rates which change endogenously in response to changes in state and national economic conditions.

To project annual changes in the size of, e.g., the 45-49-year-old white male subpopulation, we need estimates of both the number of 49-year-olds who will "age out of" the subpopulation and also the number of 44-year-olds who will "age into" the subpopulation. The simplest approach to this problem is to estimate the number of leavers as one-fifth of the 45-49 sub-population and the number of joiners as one-fifth of the 40-44-year-old subpopulation. We used a more realistic approach which draws on the relative sizes of neighboring subpopulations to make linear estimates of age structure. For example, the number of 49-year-olds can be estimated by assuming that (a) the number of 47-year-olds is one-fifth of all 45-49-year-olds; (b) the number of 52-year-olds is one-fifth of all 49-54-year-olds; and (c) the number of 49-year-olds falls on a straight line between these two numbers.

Once all 76 age/race/sex subpopulations are updated, household status (or group quarter/institutional status) is imputed to all persons over 14-years-old. Household status imputation equations were estimated using data from the 1960 and 1970 censuses for 12 northeastern states on the proportions in each age/race/sex subpopulation as a function of sex, age and the state's average income for persons over 14 years of age. Multinomial logit equations were estimated which serve to allocate all persons over 14 living in households exclusively and exhaustively among four statuses: (a) husband or wife in a husband–wife household; (b) "other" household head; (c) primary individual; (d) lives in a household, but is not a household head or wife of a head.

Before the logit equations allocate adults among the four house-hold statuses, a certain fraction of the population is first allo-cated to group quarter/institutions. The proportions of each age/sex subpopulation in group quarters or institutions in the 1970 census for the region are assumed to apply in future years. Once the group quarter/institutional population is determined, the remainder of the population is assumed to live in households and a particular household status assigned. Fifty-four equations (three household statuses for nine age groups for males and fe-males) were estimated with the regional data. For example, the three equations for 25–29-year-old males were as follows:

$$\ln\left(\frac{P1}{P4}\right) = f_1(\text{age, sex, mean personal income per adult})$$

$$\ln\left(\frac{P2}{P4}\right) = f_2(\text{age, sex, mean personal income per adult}) \quad (2.2.4)$$

$$\ln\left(\frac{P3}{P4}\right) = f_3(\text{age, sex, mean personal income per adult})$$

where: f_1, f_2, f_3 represent functions and

P_1 = proportion household heads, spouse present

P_2 = proportion household heads, relatives present but no spouse present

P_3 = proportion household heads, no relatives present

P_4 = proportion not household heads or wives of heads

The hypothesis underlying these equations is that the general trend toward smaller households is at least partly an "income effect". People seem to use increased prosperity to buy more separate living arrangements. Hence, future household growth is endogenous to the macro model. In prosperous times the number of households will grow more rapidly than in less prosperous times. In general, as real income increases, the number of males and fe-males living as household heads without spouses and as primary in-dividuals increases, and the number living as non-heads declines.

In summary, the macro model divides the population into 76 age/race/sex groups. The model starts from an initial distribution of the region's population among the 76 groups. Each group is first decremented by mortality using a central death rate generated by

an equation specifically estimated for that group. Births are
added to the population using central birth rates specific to
sixteen race/age groups for women. (Births are assigned a sex
stochastically, with slightly over half males.) Net migration
equations are used to project net migration rates. With the mortality, fertility and net migration rates and also the aging
algorithm, the 76 subpopulations are updated annually. Once the
updating has occurred, the population is allocated to group quar-
ters/institutions and among four household statuses.

3. THE MACRO MODEL: ENERGY DEMAND MODULE

The role of the demand equations for energy is to use predictions
of income, population, and economic activity to determine the
quantities of major fuels used. Three sectors of the economy are
identified, and separate sets of demand equations are specified
for each sector. The prices of fuels and some other related
measures, such as restrictions on the supplies of natural gas, are
used as additional exogenous variables.

3.1 Residential Sector

The residential sector deals with the direct purchase of fuels
used for space and water heating in homes, for cooking, and for
running other domestic appliances as well as the gasoline used
for private transportation. Since personal disposable income is
predicted by the macro model, the purpose of the demand model is
to allocate this income to different fuels and to other purchases.
The functional form chosen for the demand equations is a variation
of the multinomial logit model. The advantage of using this form
is that predicted expenditures are always positive and always
add to total expenditures. If demand equations are estimated individually for each fuel, particularly if a constant elasticity
form is used, there is no guarantee that predicted expenditures
are compatible with the predictions derived from the macro model,
and this can be a serious disadvantage if the model is to be used
for forecasting purposes.

An additional complication with modelling the residential sector
is that low expenditures on natural gas, for example, may reflect the fact that few customers can get access to gas pipelines
or that restrictions on new hookups are imposed by regulatory
agencies. For this reason, an additional model is specified
to allocate the population into four mutually exclusive categories:
(R1) use electricity for space heating; (R2) use natural gas for
space heating; (R3) have gas hookup, but not for space heating·
and (R4) do not have a gas hookup or use electricity for space
heating. Once again, this allocation can be treated using a

multinomial logit model, with the prices of fuels and income as regressors.

A dynamic version of the model is specified to account for the gradual adjustment of the allocation to changes in the explanatory variables. It is reasonable to expect that the rate of adjustment is quite slow because switching from one fuel to another is generally impractical unless the existing furnace needs to be replaced. This gradual adjustment mechanism is incorporated into the model by using a lagged dependent variable as a regressor. The coefficient of the lagged dependent variable can be interpreted as the proportion of total adjustment that takes place after the first time period, and therefore, must take on a value between 0 and 1. Given the dynamic specification of this model, it is relatively easy to deal with restrictions on new gas hookups by making the rate of adjustment in the model depend on the extent to which restrictions are imposed.

The allocation of consumer expenditures to different fuels is now made conditionally on the way that the population is divided into the four R categories. This is done by specifying the equation for each fuel as the product of the effect of income, prices, etc., and the corresponding proportion of the population purchasing the fuel. The behavioral part of the equation is dynamic to represent factors such as a gradual improvement in the overall efficiency of the stock of appliances, as older inefficient units are replaced by newer ones, in response to higher fuel prices.

The exact form of the estimated equations for the residential sector can be developed from the following representation of the multinomial logit model.

$$\frac{P_i Q_i}{M} = \frac{g_i}{\Sigma_j g_j} \qquad i = 1, 2, \ldots, 6 \qquad (3.1.1)$$

where $g_i = g_i (X_1, \ldots, X_N, \varepsilon_i)$ is a multiplicative function

of N regressors, such as prices and income, and a stochastic residual

P_i is the price of fuel i in real dollars for $i = 1,2,\ldots,5$

Q_i is the quantity of fuel i purchased per capita for $i = 1, 2, \ldots, 5$

M is real disposable income per capita

$P_6 Q_6 = M - \Sigma_{j=1}^{5} P_j Q_j$ is the real per capita expenditure on commodities other than fuels.

Expression (3.1.1) can be rewritten in a linear regression frame-

work by taking the logarithm of expenditure ratios to give the
following 5 equations

$$\ln \left[\frac{P_i Q_i}{P_6 Q_6}\right] = \ln g_i - \ln g_6 \qquad i = 1,2,\ldots,5. \qquad (3.1.2)$$

The five fuels identified in the model are: (1) electricity for
customer's heating, (2) electricity for nonheating customers,
(3) natural gas, (4) distillate oil, and (5) gasoline.

The basic form of g_i in (3.1.1) can be illustrated for natural
gas:

$$g_3 = \left[\frac{R2+R3}{RT}\right] \left[e^{\alpha_3} P_3^{\beta_3} M^{\gamma_3}\right] \left[\frac{P_3 Q_3 \times RT}{R2+R3}\right]_{-1}^{\lambda} \qquad (3.1.3)$$

where (R2 + R3) represents the number of customers with gas hookups,

RT represents the total number of customer of all types,

P_3 is the real price of natural gas,

M is real disposable income per capita,

Q_3 is the average quantity of gas used per capita,

α_3, β_3, γ_3 and λ are unknown parameters.

Expression (3.1.3) illustrates the economic rationale of the model.
The first term in brackets on the RHS represents the proportion
of the population that use gas. The second and third terms represent the use per capita by gas customers. The whole expression
implies that changes in the proportion of customers with gas hookups have an immediate effect on demand. Given this proportion,
however, demand responds gradually to changes of price and income,
reflecting the inevitable delays in the rate of which the existing
stock of gas appliances is replaced.

3.2 Industrial and Commercial Sectors

These two models are more conventional than the model for the
Residential Sector. The demand equations are defined on the
basis that there is an underlying production relationship linking
the output from each sector to the levels of inputs used. This
function is assumed to be linear homogeneous so that the cost minimizing combination of inputs is unaffected by the level of output.
The form of the cost function linking total expenditures on inputs to the prices of inputs is assumed to be a transcendental
logarithmic function. The demand equations that are derived from

this form of cost function can be written

$$\frac{P_i Q_i}{TX} = \alpha_i + \beta_{i1} \ln(P_1/P_N) + \beta_{i2} \ln(P_2/P_N) + \ldots + \beta_{iN-1} \ln$$

$$(P_{N-1}/P_N) + \varepsilon_i \qquad i = 1, 2, \ldots, N \qquad (3.2.1)$$

where P_i is the price of the ith input,

Q_i is the quantity of the ith input,

$TX = \sum\limits_{i=1}^{N} P_i Q_i$ is total expenditures on inputs,

α_i, β_{i1}, β_{i2}, $\ldots$, β_{iN-1} are unknown parameters,

ε_i is a stochastic residual.

The implication of (3.2.1) is that the estimated model is essentially a linear probability model in which the regressors are identical in every equation. However, the symmetry of the substitution elasticities in a production model of this type implies that constraints on the coefficients should be applied across different equations, implying $\beta_{ij} = \beta_{ji}$ for all i and j.

The allocation model can be made dynamic by introducing a lagged dependent variable in the same way that it is incorporated into the residential model. The implication is that expenditure patterns do not change immediately in response to price changes. Time is required to modify the existing stock of machinery and equipment.

In the commercial sector the following inputs are considered: (1) electricity, (2) natural gas, (3) heating oil (distillate and residual), and (4) labor. In the industrial sector, the inputs are: (1) electricity, (2) natural gas, (3) residual oil, (4) coal, (5) labor, and (6) capital.

There are additional problems with the natural gas equation because supplies of gas were curtailed to commercial and industrial customers in some years. An additional variable is incorporated into every equation to account for this effect, and it is defined as $(1-D) \left[\dfrac{P_2 Q_2}{D} \right]_{-1}$, where D is the proportion of gas requirements that are actually delivered. This variable can be interpreted as the anticipated amount of gas that will be curtailed in the current year based on the amount that would have been used in the previous year with no curtailments imposed. Hence, it is assumed that the value of D is announced, and that producers attempt to compensate for the anticipated curtailments by changing the mix of inputs. For example, it might be relatively easy to switch from natural gas to oil in some cases, but in other situations, jobs may be lost. This would imply a positive effect in the oil equation and a negative effect on employment.

4. TIME-OF-DAY PRICING MODEL

This section describes a simulation model for estimating the effects of time-of-day (TOD), or peak load pricing, on electricity consumption and peak load. The model consists of three basic equations: a demand equation which models the total energy requirements of the region, a substitution equation which models the allocation of consumption between the peak period and the off peak period, and a load factor equation which models the relationship between total energy requirements and summer coincident peak load. The demand and substitution equations are formulated as simply as possible in terms of relative prices. The demand equation predicts what consumer response will be to the observed change in the overall real price of electricity which attends the transition to time-of-day rates. The substitution equation assumes that consumers will allocate consumption between peak and off peak periods in accordance with the relative prices of consumption in the two periods. Unfortunately, economic theory provides no guidance in formulating the load factor equation. Load factor is determined more by <u>when</u> electricity is consumed than by <u>how much</u> is consumed. This equation is based essentially on observed phenomena. In particular, load factors have been observed to decline when real prices increase. However, load management and load shifting in response to peak load pricing should improve load factors. All elasticity effects in this TOD model are formulated to increase gradually over time. While individual customers may exhibit their full response in the course of a single year, the effect at the regional level will be dampened by the fact that implementation of peak load pricing is most likely to be gradual. Implementation of TOD pricing is assumed to be complete by the fifteenth year of the simulation.

4.1 Mathematical Formulation

The model will determine directly the following quantities:

Q_t = total consumption in GWH, in year t.

$W_t = Q_t^D / Q_t^N$ = the ratio of peak period consumption to off peak period consumption, in year t, and

α_t = ratio of summer coincident peak load in megawatts to annual energy consumption in GWH per year t.

The quantities Q_t^o, W_t^o, and α_t^o correspond to the above variables in the absence of TOD pricing. The various prices which appear in the model are as follows:

$\bar{P}^O_t$ = real level of non-TOD prices of electricity, averaged over all consumer classes and normalized to the value 1 in year t = 0.

P^D_t = real level of the TOD price of electricity in the day, or peak period in year t, relative to $\bar{P}^O_t$.

P^N_t = real level of the TOD price of electricity in the night, or off peak period in year t, relative to $\bar{P}^O_t$.

$\bar{P}_t$ = average of the day and night TOD prices, weighted by the equilibrium sales shares in year t.

Other variables, already determined are

S^D_t = sales share in the peak period in year t

$= W_t/(1 + W_t)$,

S^N_t = sales share in the off peak period in year t

$= 1/(1 + W_t)$, and

$\bar{P}_t = S^D_t P^D_t + S^N_t P^N_t$

K_t = summer coincident peak demand in year t

$= \alpha_t Q_t$.

Note that the ratio

$P_t = \bar{P}_t/\bar{P}^O_t$

gives the real level of the price of electricity under TOD pricing, relative to the case in which TOD pricing is absent. Like $\bar{P}^O_t$, $\bar{P}_t$ gives the real price of electricity relative to other goods and services in the economy; but $\bar{P}_t$ is not normalized to 1 in year 0.

The basic equations of the model are as follows:

Demand equation:

$$Q_t = Q^O_t \cdot [\bar{P}_t/\bar{P}^O_t]^{\epsilon_t} \qquad (4.1.1)$$

Substitution equation

$$W_t = W^O_t \cdot [P^D_t/P^N_t]^{\eta_t} \qquad (4.1.2)$$

Load factor equation

$$\alpha_t = \alpha_t^o \cdot [\bar{P}_t/\bar{P}_t^o]^{\sigma_t} \cdot [P_t^D/\bar{P}_t]^{\theta_t} \qquad (4.1.3)$$

The parameters ε_t, η_t, σ_t, and θ_t are elasticity parameters. The time dependence of each of these reflects the fact that the implementation of TOD pricing will take place over a number of years. Thus the full elasticity effect due to individual customers will be dampened at the regional level because (it is assumed) not all customers will face TOD prices until the final year of the simulation. Hence

Demand elasticity,

$$\varepsilon_t = \varepsilon_o + \varepsilon_1 t \qquad (4.1.4)$$

Elasticity of substitution

$$\eta_t = \eta_o + \eta_1 t \qquad (4.1.5)$$

Price elasticity of system load factor

$$\sigma_t = \sigma_o + \sigma_1 t \qquad (4.1.6)$$

Peak load effect of load shifting

$$\theta_t = \theta_o + \theta_1 t \qquad (4.1.7)$$

where t indexes time = 0,1,2...,15.

4.2 Input Data

Direct inputs to the model are Q_t^o and α_t^o. These will be, respectively, forecast energy requirements and the observed peak load to energy requirements ratio without TOD rates.

The base case, non-TOD, average price, $\bar{P}_t^o$, is determined using a forecast average rate of growth of the real price of electricity.

The levels of the time of day marginal prices are determined in a sub-model as follows: Ultimately two decisions will be made in the implementation of time-of-day rates: (1) the ratio of the peak period marginal price to the off peak period marginal price and (2) the real level of the marginal price of electricity in a regime of time-differentiated-rates. The first decision will be based upon the marginal cost of the utilities submitting the rates. However, the strict adherence to marginal cost pricing may be hindered by the revenue requirements aspect of rate making.

Under strict marginal cost pricing, the sales weighted average of the peak and off peak marginal charges may be considered higher than it is under the present, average cost pricing regime. It has been argued that the economically appropriate adjustment for the revenue requirements problem is by any of a number of schemes, all of which leave the marginal charges intact at marginal cost. If this policy is followed, then in addition to the real growth in the marginal charge for electricity, there may be a real growth component due to the transition to marginal cost pricing. If, however, marginal cost based revenues are reconciled to average cost based revenues by modifying marginal charges, this second component will be zero.

Given the assumed pricing policy scenario, the level of the TOD marginal prices are found as follows: the model is virtually insensitive to the value used for the proportion of total energy consumed in the peak period. One quarter, or .25, is probably fairly accurate if the peak is defined as most of the daylight hours of say April through October; we have used .25 here. Let the price ratio chosen by the Commission be r, and the additional price growth be i. Then the base price level for the TOD regime is

$$\bar{P}_t^* = (1 + .017 + i)^t.$$

(Note $\bar{P}_t^o = (1. + .017)^t$).

The marginal charges are determined from

$$.25 \ P_t^D + .75 \ P_t^N = \bar{P}_t^*$$

$$P_t^D / P_t^N = r.$$

These two equations yield values for P_t^D and P_t^N.

The qualitative effects observed through the simulation can be summarized in the following way. The growth in peak load is uniformly reduced by the introduction of TOD pricing. The larger the price differential, the lower the rate of growth. Under most of the scenarios, the peak load grows more slowly than the annual energy requirements, implying an improvement in the system load factor. A large rate of price growth coupled with a small peak/ off peak price ratio reverses this effect.

5. THE MICROSIMULATION MODEL

In order to forecast distributional patterns of household electricity consumption and as another approach to forecasting total household electricity demand, the macroeconomic model has been linked to a micro model of the household sector. This macro-micro framework provides forecasts of total electricity demand and also of the distribution of electricity demand and expenditures by households. Since households are identified by income, size, appliance ownership, and other socioeconomic characteristics, a variety of distributional impacts can be examined.

The microsimulation model begins from a sample of households representing all the region's households in an "initial" year. Each household record contains economic, demographic, housing, and other energy-related characteristics. The sample is "aged" from this initial year to the terminal year using assumptions about population, household and income growth; appliance distribution and saturation; unemployment; electricity consumption for households with varying characteristics; functional end use; and efficiency improvements by functional end use. Once the sample of households is aged to the terminal year, electricity use and expenditures are imputed to each household on the basis of the characteristics of the household and exogenously supplied prices. At this stage, functional end users are assessed and corresponding efficiency adjustments imputed. The final product is a detailed picture of electricity consumption and expenditures across households, as well as aggregate values for household sector consumption and expenditures.

5.1 Aging the Sample

Once the sample of households for the region was created for the initial year, the next step was to develop a set of control totals for the terminal year which would serve as inputs to the aging process. To a large extent we drew these control totals from the outputs of the macroeconomic model. Thus, economic, demographic, and energy demand modules provide most of the inputs needed by the microsimulation aging process. We will specify below which inputs were drawn from the macroeconomic model and describe generally how these inputs were used to age the sample.

In the microsimulation model, demographic aging is accomplished by reweighting the households and persons in the initial sample to conform as closely as possible to control totals for the terminal year supplied by the demographic module. One set of control totals gives the projected terminal year population of the region by age, race and sex including persons in group quarters and institutions. Another set of controls gives the proportion of the population by age, race and sex in group quarters or institutions. A third set of controls gives the projections of households by type, and sex and age of head.

The economic aging of the sample from the initial to the terminal year involves a two-step procedure, first supplying an unemployment rate for the terminal year which is used to adjust up

or down the labor force participators and income of selected
persons, and second adjusting the income amounts received by each
person by type of income. We derived multipliers for real income
growth by source by calculating the real rate of growth of employment income per 'employed person during the simulation period.
Rather than supply different multipliers for other sources of income (e.g., rent, aid to families with dependent children, pensions)
we judged it more prudent to forecast that all income would increase at the same rate over the period, rather than some kinds of
income increasing more rapidly than others. The state unemployment
rate for the terminal year was, of course, also taken from the
macroeconomic model base forecast.

Each household in the 1974 sample of households was identified
with a detailed set of housing characteristics. In the process of
adjusting to population and household weights during the demographic aging, the incidence of certain housing characteristics inevitably changes somewhat. For example, as the number of single-person
households increases, the number of households living in apartments
increases because single person households tend to live in apartments. For our purposes, the relevant fact is the critical relationships between housing characteristics and income and energy
demand are adequately perceived by this adjustment procedure.
Hence, we judged that no further housing computations were required. The assumption under which we operated in this case,
therefore, is that the relationship of housing characteristics to
other household characteristics will in the terminal year be
largely similar to these relationships as they existed in the
initial year.

In order to forecast the distribution of appliances in the terminal year, we first developed an accurate estimate of the distribution of these appliances across the region's households.
Then, using detailed cross tabulations of appliance ownership by
income, type of structure, and other variables, we fit log-linear
models to the cell counts to select the key variables. Once the
key variables and interactions were selected, the models were fit
by including only those effects which were statistically significant. The major effort was devoted to determining income effects
on appliance ownership.

The parameters are designed to capture distributions, not total
saturations. To assign total saturations we made a series of
judgements as to likely saturation levels and scaled the functions up or down by adjusting the intercepts of the log-linear
equations. This method has the nice property that it holds constant ratios of the odds of ownership for different classes to
each other.

5.2   Electricity Demand and Expenditure Imputations

Once the sample of households was fully aged to the terminal year in terms of its population, households, unemployment, income growth, housing changes and appliance imputations, it was prepared for the electricity demand imputation equation [4]. Once the quantity of electricity is imputed, the expenditure is determined using both price and quantity.

Each conservation measure affects only the portion of total electricity usage for a specific purpose, rather than uniformly reducing total annual usage of electricity. Therefore, a necessary preliminary step before the modelling of electricity conservation measures is the allocation of electricity usage by functional end use.

The allocation of imputed total electricity use (demand) to specific end uses was made for each household on the basis of the particular functions for which it used electricity and the national average usage of electricity for each function. The national values are used to calculate proportions of a household's electricity going to a particular use, given the overall consumption total.

A set of efficiency standards for appliances produced after 1980 were introduced into the model. These standards express, for example, post-1980 electricity usage of each appliance as a proportion of 1970 usage. Thus, all homes 13 years old or younger in 1993 would be assumed to have a new, more efficient water heater, stove, refrigerator, clothes dryer and dishwasher, if that appliance was owned. For older homes only those appliances replaced after 1980 would be new and therefore more efficient. However, expected appliance lifetimes make it reasonable to expect that by 1993 initially all appliances would be satisfying these efficiency standards. The only exceptions are water heaters. It was assumed that 93 percent of households with an income over $9,000 would have replaced their water heaters with a more efficient one by 1993 as compared to 70 percent of households with an income under $9,000. For all other appliances the new efficiency levels are assumed in the base forecast. Of course, sensitivity tests with alternative assumptions would be a useful exercise.

6.   SUMMARY AND CONCLUDING REMARKS

An economic, demographic, and energy demand macro model has been linked to a micro model of households and a time-of-day pricing policy model to provide a system for generating long-term forecasts of regional energy demand which are sensitive to a host of socioeconomic and price variables.* These forecasts of electricity demand are disaggregated by commercial, industrial, and residential sectors. In addition, detailed distribution impacts of consumption and expenditures can be provided for the residential sector through the use of the micro simulation model. The impact of time-of-day pricing policies can also be estimated. Because the demand for electricity is more realistically modelled within a framework of demands for other energy sources, the macro model also provides long run regional projections for other types of energy as well.

There are important advantages and some disadvantages in using this relatively more comprehensive approach to generating long range forecasts of regional demand for electricity. First, this approach to modelling regional demand is "data hungry" and can be much more costly in dollars and time than other less comprehensive approaches. On the other hand, the macro model also generates, in addition to total electricity demand, forecasts of demand for distillate oil, natural gas, and gasoline as well as forecasts of a wide range of regional economic and demographic variables. Thus, the integrated and linked models can be used in a more general way to evaluate alternative policy scenarios based on projections of a host of important socioeconomic and energy variables. Finally, this approach is generalizable to other regions and has the advantage of combining forecasts of aggregate and also distributional outcomes.

REFERENCES

1. S. Caldwell, W. Greene, T. Mount, S. Saltzman, R. Broyd, Forecasting Regional Energy Demand with Linked Macro Micro Models , Papers of the Regional Science Association XXV Annual Meeting, Chicago, Illinois, November 1978 (forthcoming).

2. S. Caldwell, W. Greene, T. Mount, S. Saltzman,"Forecasting Regional Energy Demand: The Case of New York State,"Technical Report submitted to the New York State Public Service Commission, 149(b) Proceeding, Case No. 27319, Albany, New York, January 1979.

3. Charles River Associates, Procedures for Independent Analyses of Electric Energy and Peak Loads by State Agencies, Vol. 1 and 2, Boston, Mass., 1978.

---

*The macro model is implemented on the TROLL system at Cornell University and now has over 400 equations. Long term (15 years) projections of alternative policy scenarios for New York State have been simulated with the macro and micro models [2].

4. Jill A. King, "Residential Energy Consumption by Functional End Use in 1975," Draft Final Report to Department of Energy, Appendix C, Mathematica Policy Research, Washington, D.C., Revised Nov. 30, 1978.

5. Guy Orcutt, Harold Watts, and John B. Edwards, Data Aggregation and Information Loss , American Economic Review, Vol. 58: 773-787, September, 1968.

6. Guy Orcutt, Steven Caldwell, and Richard Wertheimer, Policy Exploration Through Microanalytic Simulation, The Urban Institute, Washington, D.C., 1976.

7. S. Saltzman, and H.S. Chi, An Exploratory Monthly Integrated Regional/National Econometric Model , Regional Science and Urban Economics, Vol. 7: pp. 49-81, 1977.

DISCUSSION

Models as rich in detail as the one presented by Saltzman are by their very nature not transparent. It is difficult to trace the low demand forecasts produced by the model to particular piece(s) of model structure. It is not then easy to discern whether the main cause is saturation phenomena in the residential sector, or the assumptions on economic growth and energy price implicit in the model. Presentation of the direct-price and cross-price elasticities of energy use would, however, facilitate comparison of this model with others.

The technique would appear to be applicable both to other regions and to whole countries, especially to countries with imports and exports as a significant proportion of their GNP. (The similarity being that regions have a lot of economic "leakage" to their neighbours.) The stumbling block for extension of the model to other regions/countries would inevitably be the lack of data required by the model.

When using output from a demand model of this, or any other sort for that matter, as input to an energy supply construction program, the effect of variations in the exogenous imputs (price of competing fuels, economic growth rates, exports, etc.) should be quantified as much as possible. Quantifying the uncertainties involved would allow these to be accounted for explicitly in the further stages of the planning process.

Some doubt was however expressed as to the value of a "perfect demand forecast" at all. The investment programs which utilities companies had could be corrected as the program progressed at little extra cost. This criticism is, however, most relevant in the case of construction of many small facilities.

LARGE SCALE INTEGRATION OF WIND POWER IN POWER SYSTEMS

A. Traça

Departamento de Engenharia Electrotecnica
F.C.T.U.C., Universidade de Coimbra
Portugal

ABSTRACT.The use of wind power to generate electricity can make a positive contribution to solve the energy crisis. Wind power and velocity models are presented. The wind velocity model is generated from previous data records and is used to produce random values. These values allow the simulation of wind power connected to power systems. The use of hydro storage can provide an economical means to smooth wind power variations. Solar electric generation, although very expensive at the moment, is a complementary source to the wind, helping to reduce storage requirements.

1. INTRODUCTION

The sharp increase in fossil fuel prices as well as the awareness of the fast growing importance to preserve the environment have lead to an increased interest in renewable energy sources. To generate electricity, wind power offers the cheapest alternative which can be competitive in the near future. Wind-driven generators were used for the first time in Denmark in 1890, and since then thousands of small installations have been used to supply loads in remote places.

Wind power is a renewable energy source and has a reduced impact on the environment. Although wind power is free, the energy conversion equipment requires high initial investments. This equipment converts the kinetic energy of the air. The air density is very small; it is necessary to use a large area of interception to produce modest amounts of power. Wind has an intermittent nature which is highly variable and unpredictable. For a stand-alone system, this implies the need of storage, and for a diversified system, the need of back-up generating capacity. Unfortunately, the overall efficiency of wind-driven generators is poor (25-30%) and the wind variability does not allow high plant factors (15-40%), depending on the wind site characteristics. These features make wind power a challenge to modern technology to improve its competitiveness. The reward is attractive, the World Meteorological Organization estimates the availability of 10 to 20 TW of recoverable wind power. The likely spiral price increase of fossil fuels and its scarcity will favour renewable sources. The use of more strict safety and pollution regulations is also making conventional electricity generation more expensive.

2. WIND POWER DISTRIBUTION

Wind power is an indirect form of solar energy. The difference in heating of masses of air around the globe creates pressure gradients, causing the motion of the air. Wind power generation uses a fraction of the kinetic energy of the moving air. The power density per unit of area normal to the wind direction is given by

$$P = \frac{1}{2} \rho V^3$$

where V is the wind speed and ρ is the density. The expression above is derived considering the amount of air crossing unit area per unit time. Power variation with the cube of the velocity makes the careful selection of wind sites very important. Density is a secondary factor, decreasing by about 10% for each 1000 m above sea level and decreasing by about 3% for a temperature increase of 10°C around normal temperatures. The wind records available are normally from weather stations which are located in sheltered places and wind measurements are taken at low height (10 m). Wind data taken at a certain level above the ground can be converted to another level by an empirical formula

$$V_h = V_o (h/h_o)^a$$

where V_h is the wind velocity at height h and V_o is the measured wind velocity at the standard height (10 m). The exponent a changes with ground roughness. The value of a is 0.15 for coastal zones and open plains, increases to 0.30 in interior zones with medium size houses and trees, reaching 0.40 in cities with tall buildings. Topographic factors such as hills and valleys can produce acceleration in the lower air layers leading to a velocity increase.

Fig. 1 presents a block diagram of a wind generator, where A is the area of interception, that is the area of the circle described by the rotating blades, and $C_p(V)$, η_M, η_E, η_S are the efficiencies of the aerodynamical, mechanical, electrical, and storage subsystems, respectively. Transformer and transmission losses are not shown. The electrical energy output in a period T is

$$W = \int_0^T (1/2)\rho A C_p(V)\eta_M\eta_E V^3 dt.$$

The power coefficient $C_p(V)$ depends on the type of blades and on the tip speed ratio. The machine is operated from a cut-in velocity, which is enough to overcome friction losses, the power increasing with the cube of the velocity up to the rated speed where the maximum electrical output is reached. Above the rated speed, the blade regulating mechanisms keep the output constant and equal to the rated power up to the cut-off speed, when the machine is shut down to prevent mechanical damage. The cut-in and rated speeds are chosen according to the wind site characteristics.

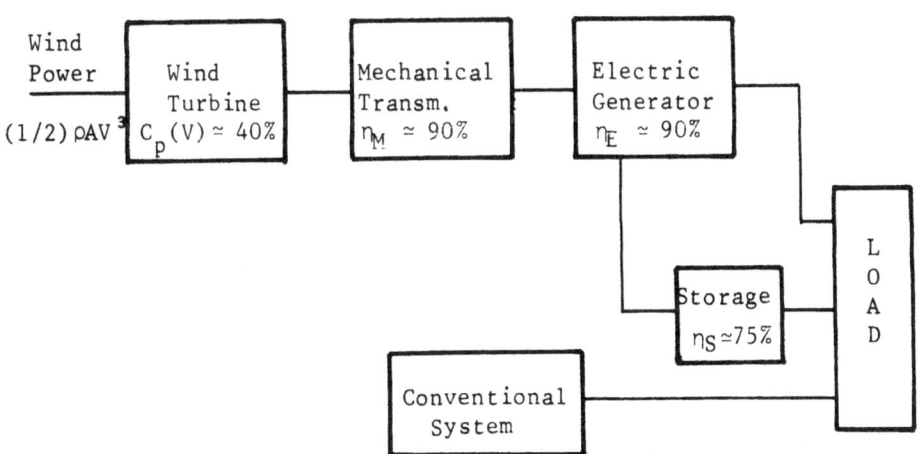

FIGURE 1 - Block diagram of wind energy conversion systems

To represent the wind power characteristics of a given site, several methods can be used. One method consists of ordering by decreasing magnitude the wind power to obtain the power duration curve (Fig. 2) in the same manner as the load duration curve is generated.

Another model uses the Weibull distribution (Fig.3) as the probability density function of the wind distribution, given by

$$f(V) = (K/c)(V/c)^{K-1} \exp(-(V/c)^K)$$

where c is the scale parameter and K is the shape factor. The average speed $\overline{V}$ is given by the first moment of the density function

$$\overline{V} = \int_{0}^{\infty} V\, f(V)\, dV.$$

The scale factor c is proportional to the average speed

$$1.11\overline{V} < c < 1.13\overline{V}.$$

and the shape parameter K for wind distribution is normally

$$1.3 < K < 3.$$

In order to determine the Weibull parameters the cumulative distribution function is used

$$(V) = 1 - \exp(-(V/c)^K)$$

rewritten as

$$1 - F(V) = \exp(-(V/c)^K)$$

and applying twice the logarithm on both sides, to find the expression

$$\ln(-\ln(1 - F(V))) = K \ln V - K \ln c$$

which can be seen as the expression

$$y = ax + b$$

where

$$y = \ln(-\ln(1 - F(V)))$$
$$a = K$$
$$x = \ln V$$
$$b = -K \ln c$$

Using linear regression analysis for n pairs of values (x,y), a and b are obtained giving c and K as

$$K = a$$
$$c = \exp(-b/K).$$

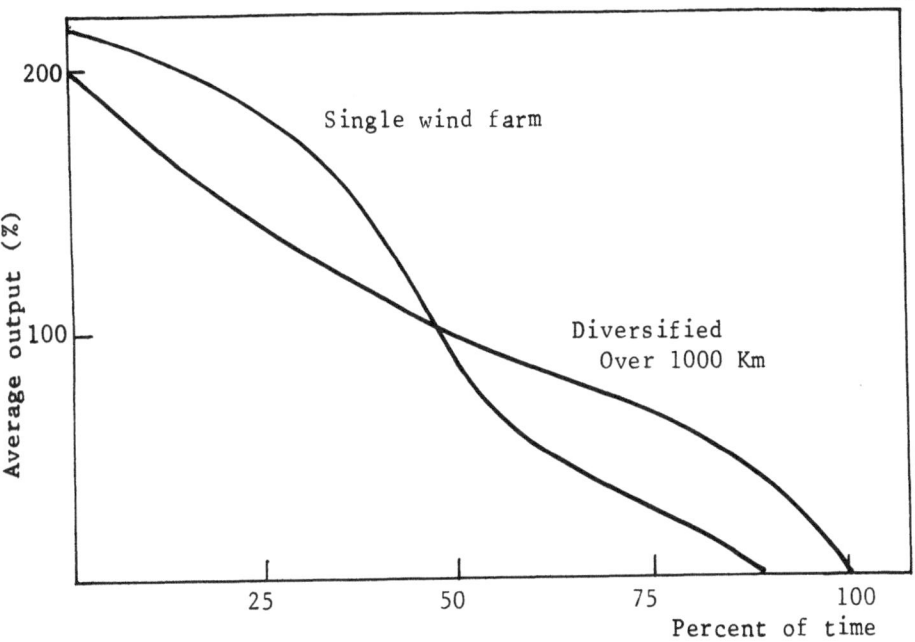

FIGURE 2 - Reduction in windpower variability using site
diversity

The knowledge of wind distribution can be used to select the wind
driven generator which is appropriate for the site [3]. Then,
the energy output, the average power and load factors can be forecast. The average electrical power will be given by

$$\bar{P}_e = \frac{1}{2} \rho A \int_{V_c}^{V_R} C_p \eta_M \eta_E V^3 f(V) dV + \frac{1}{2} \rho A C_p \eta_M \eta_E V_R^3 \int_{V_R}^{V_o} f(V) dV$$

where V_c, V_R and V_o are the cut-in, rated, and cut-off speeds.
The load factor is calculated by dividing the average power by
the rated power.

Although the wind distribution models previously described are
helpful in choosing the aerogenerator characteristics, they do not
provide enough information about the cyclic changes of wind speed,
which are mainly dependent on the time of the year, the hour of
the day and the existence of a front system. To evaluate the contribution of wind power systems, the daily and seasonal variation
must be known, as demand changes with time of the year, day of the
week and hour of the day.

Fig. 4 shows the hourly wind speed variations during four months
of different seasons at Porto (Portugal). It can be seen that a
pattern of variation where wind power levels are generally higher

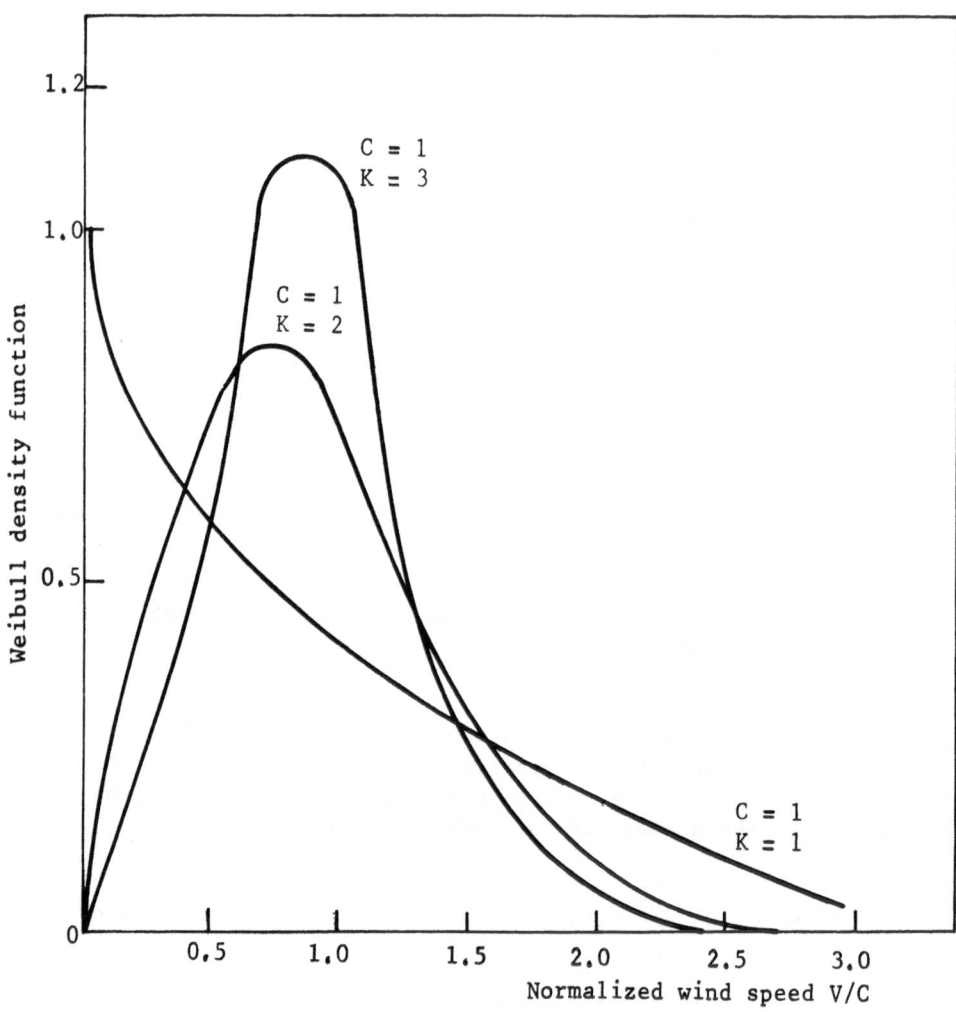

FIGURE 3 - Weibull distribution for c = 1 and various
values of k

during daylight, usually reaching peak value around 14-15 hours.
This effect is more pronounced in the summer and is due to heating
of the ground, causing turbulence which brings energy from faster
moving upper layers of the atmosphere. For each period of the
year, a month or two weeks according to the degree of wind speed
variation, an hourly model can be used. Due to the shape of the
curves of Fig. 4, it was found that a trigonometric sum could provide a suitable model. Having 2N + 1 arguments of a data function,
the trigonometric sums which fit those points is given by

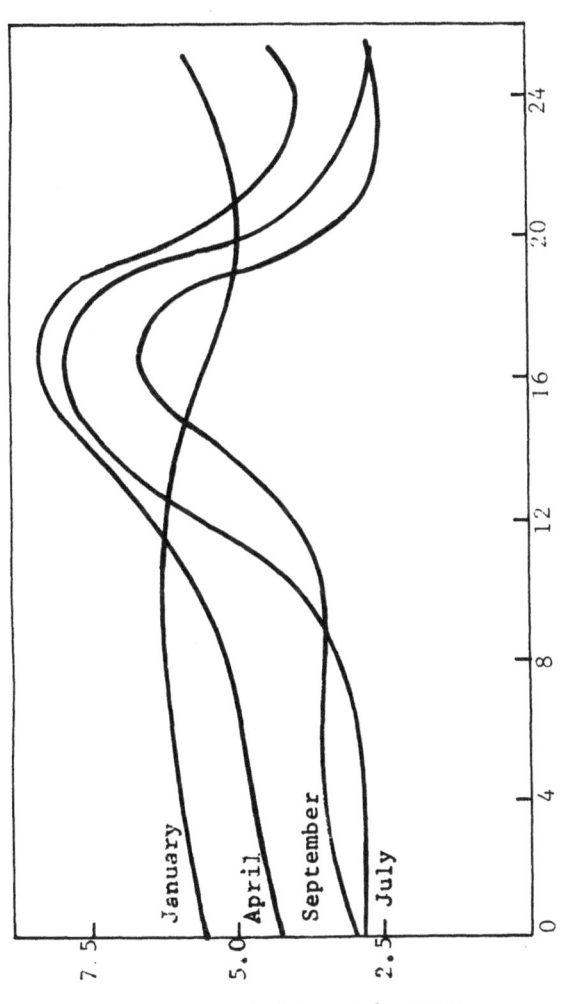

FIGURE 4 – Average daily wind speed variation, during the months of January, April, July and September, Porto (Portugal), 1957–1966

$$y(x) = \frac{1}{2} a_o + \sum_{K=1}^{N} (a_K \cos \frac{2\pi Kx}{2N+1} + b_K \sin \frac{2\pi Kx}{2N+1}).$$

The coefficients a_K and b_K can be calculated as

$$a_K = \frac{2}{2N+1} \sum_{x=0}^{2N} y(x) \cos \frac{2\pi Kx}{2N+1} \qquad K = 0,1,\ldots,N$$

$$b_K = \frac{2}{2N+1} \sum_{x=0}^{2N} y(x) \sin \frac{2\pi Kx}{2N+1} \qquad K = 0,1,\ldots,N.$$

An even function can be used as an extension of the month hourly distribution producing a more convenient expression, which is

$$y(x) = \frac{1}{2} a_o + \sum_{K=1}^{N} a_K \cos \frac{2\pi Kx}{2N+1}.$$

It was found that only the first six terms of the trigonometric sum contained relevant information. Each month hourly average wind speed distribution can thus be represented by six parameters. The curves of Fig. 4 represent average daily wind speed variation over the period 1957-1966. For each one hour period, the wind speed has an average speed taken from the model curve and a standard deviation calculated from several years data. If the hourly standard deviation is approximately constant for several hours of the day, one value is enough, otherwise a model similar to the one previously described can be used to represent its variations. With the average speed and standard deviation, the wind speed random values can be generated, allowing the evaluation of wind generator performance.

3.   ECONOMIC EVALUATION

Wind power can be used to supply loads in remote places, where the cost of conventional systems is too high. Wind power can also be used connected to a utility system, competing with other sources of energy, having to satisfy economic criteria and not compromising the reliability of the system. Other aspects must be considered such as the low impact on the environment, the absence of health and safety hazards, energy supply security, and the effects on the balance of payments.

3.1 Use of Wind Power as Fuel Saver

Wind is not a source of firm power and has a highly unpredictable behaviour. The system installed capacity must be large to cover the low wind periods. However, if the wind generators are spread over a large area, the different weather conditions give rise to a smoothing effect on the wind duration curve [5]. Fig. 2 shows the effect of having a dispersed system. Even in these conditions, although less likely, the wind can have a reduced output for short periods of time.

The wind generators used as fuel savers inject power into the system, allowing the power output of conventional units to be reduced, saving fuel. The most expensive fossil units are the first to be replaced. Due to its low running costs, it is not economical to unload base load generation, such as nuclear units. The start-stop cycle of fossil units consumes power, needs staff, and

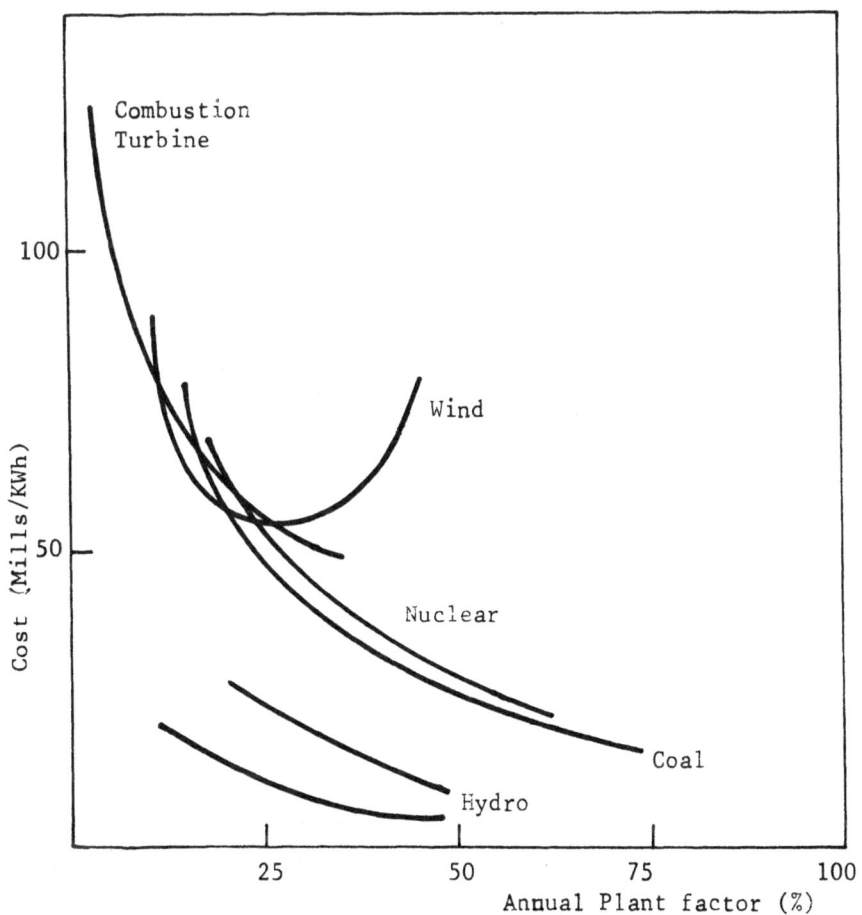

FIGURE 5 - Cost of generated energy as function of plant
factor

increases maintenance requirements and thus should not be used
frequently.

Fig. 5 shows typical capacity costs of several types of units.
The value for the windmill conversion system does not include
transmission and storage. Fig. 6 represents the generating cost
assuming that the wind system has hydro storage [4]. To know the
breakeven price of wind generators, the value of fuel saved must
be evaluated.

$$\text{Fuel Saved (mills/KWh)} = \text{Fuel Cost (mills/}10^6\text{BTU)}$$

$$\times \text{ Incremental Heat Rate (}10^6\text{BTU/KWh)}$$

The incremental heat rate is higher on peaking units, and must be
changed accordingly with fossil units which have its output reduced.
Each generator with a capacity factor of K produces a value per KW
of installed power given by

$$\text{Value Produced (\$/KW)} = 8760 \times K \times \text{Fuel Saved}$$

The cost of operating wind plants is given by

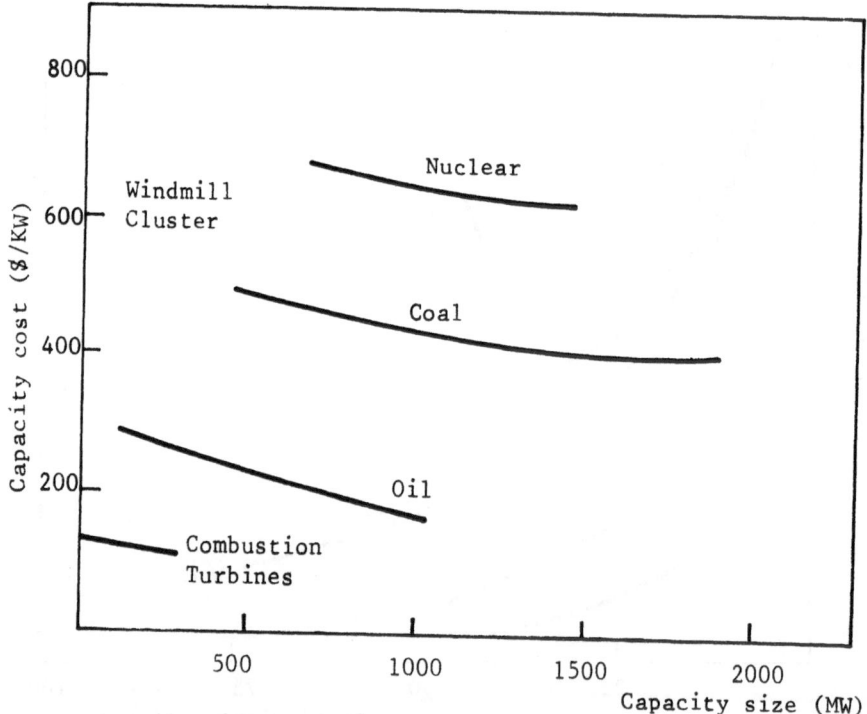

FIGURE 6 - Typical capacity costs

Plant Cost ($/KW) = Fixed Charge Rate x Unit Capacity Cost

($/KW) + O&M ($/KW)

where O&M are the operation and maintenance costs. Recent experience [6] indicates that O&M is a small fraction of total cost. Making equal the expressions relating the Plant Cost and Value Produced, both per unit of capacity installed, the breakeven price for Wind Unit Capacity Cost is given by

Unit Capacity Cost ($/KW) =

$$= \frac{8760 \times K \times \text{Fuel Costs} \times \text{Incremental Heat Rate}}{\text{Fixed Charge Rate}} - \text{O\&M}$$

As O&M is small compared with other costs the competitiveness of wind generators is proportional to the capacity factor, fuel costs and incremental heat rate, and decreases with the fixed charge rate.

The capacity factor depends on the wind site and machine characteristics. Sites with high average and steady speeds are more attractive. Fossil fuel costs are increasing at an alarming rate. At the writing of this article (May 1979), oil prices are touching $20/Barrel. This will favour the renewable sources, and in particular, wind power. The incremental heat rate depends on the type of fossil unit and the load operating point. Figs. 7 and 8 show that for most of the time wind power is in phase with consumer demand, both on an annual and a daily basis. For this reason, peaking plants having higher incremental costs will be replaced. To assess these potential savings, a simulation study is being made using average daily wind speed model previously described, together with the time load variations and the utility generating characteristics (Fig. 9). Load variations can be simulated using average values from previous years corrected for load growth, by extrapolation or correlation.

The fixed charge rate includes depreciation, rate of return, insurance, and taxes. In a capital intensive investment, the cost of money plays an important role. Insurance will cover the risk of failure due to storms. Whatever speed is adopted for design purposes, there is always a probability, however small, that it may be exceeded in a storm of exceptional violence. As wind generators will be located in exposed sites, the design must include an adequate safety factor. The unit capacity cost includes the aerogenerator, land, erection, and connection to the grid. To be competitive, that cost would have to be in the order of 600-700 $/KW. Only mass production of 1-2 MW machines, together with government incentives, can bring the costs to that level.

312

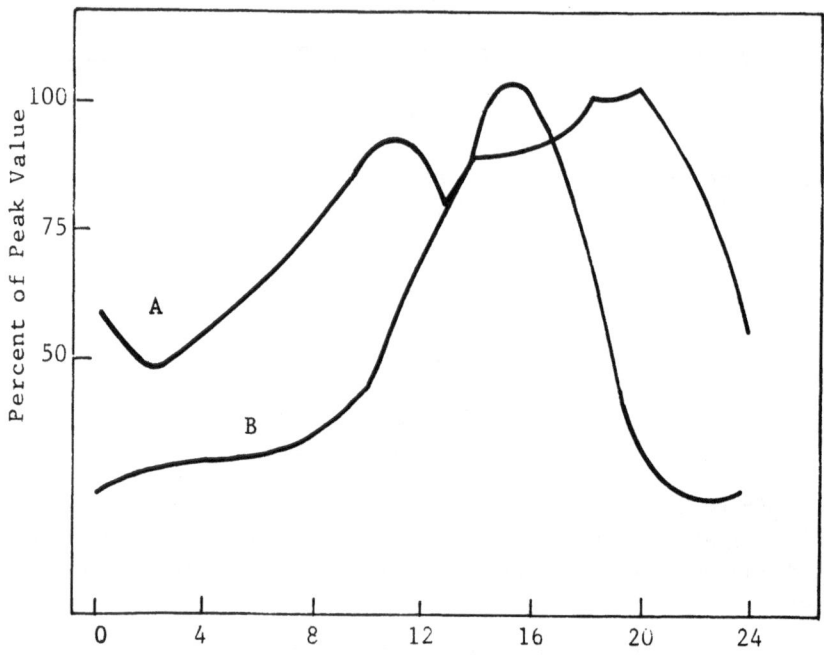

FIGURE 7 - A) Typical daily load curve, Portugal
          B) Average daily wind power curve, Porto

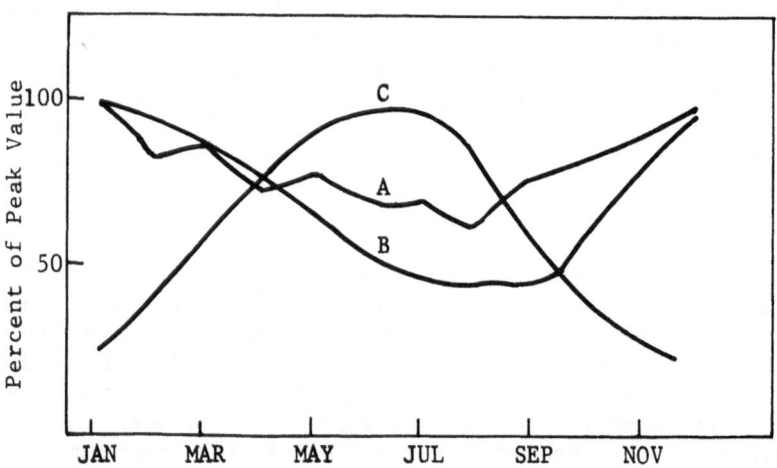

FIGURE 8 - A) Annual load variation, Portugal
          B) Average annual wind power curve, Porto
          C) Average annual solar power curve, Porto

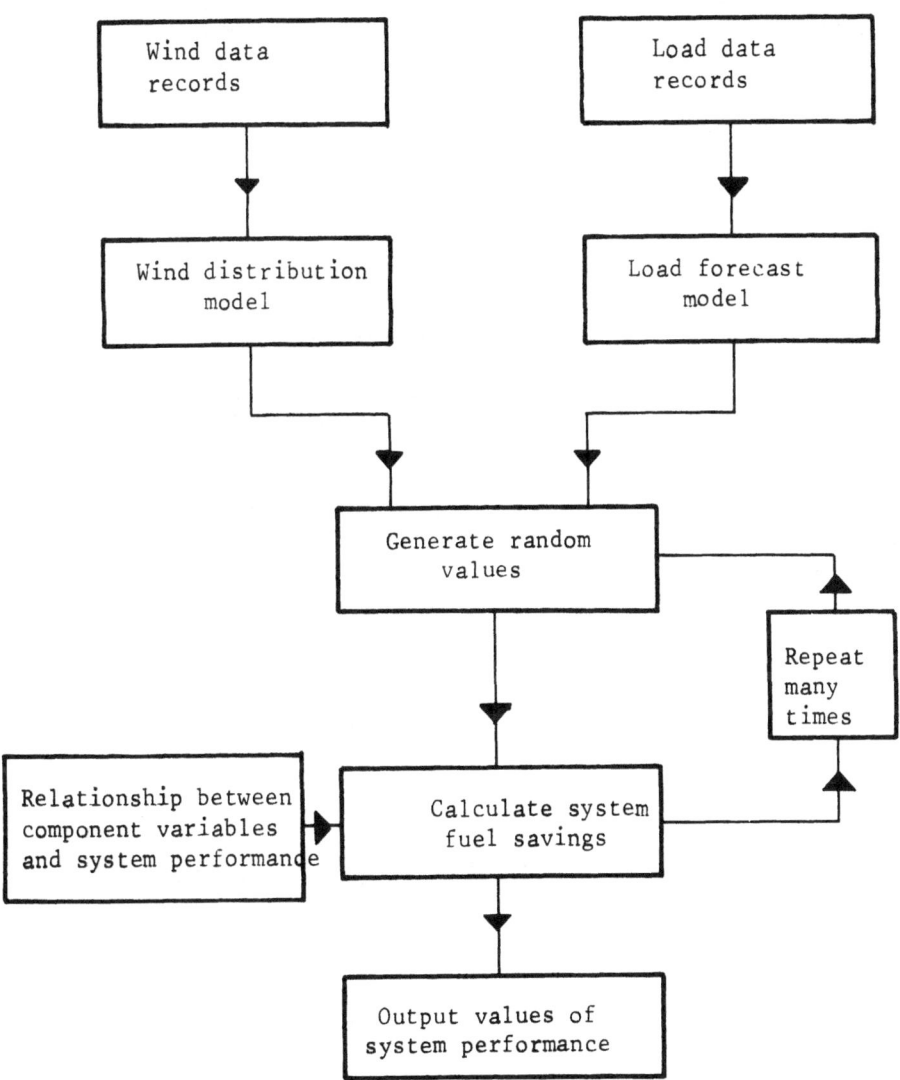

FIGURE 9 - Flow chart of Monte Carlo simulation model
applied to obtain the approximate distribution
of fuel savings

The connection of significant wind power to the utility can give rise to some technical problems if it is above 10-15% of total capacity [5,13]. As wind power is not controllable, the overall system stability and frequency control can be reduced, spinning reserve must be kept at fairly high levels, and during high wind speeds, transmission can be overloaded. The experience with the Clayton (USA), 200 KW generator, representing 15% of off-peak power demand showed that the system is able to absorb wind turbine power fluctuations caused by gusting.

3.2 Wind Power Combined With Hydro Storage

In order to smooth wind power fluctuations, either pure pumped storage or river flow hydro storage can be used. Utilities having already a high amount of hydro storage are in a better condition to take advantage of wind power. If wind is blowing water is pumped to the upper reservoir or accumulated in the river flow reservoir. If wind and rainfall are not in phase, the utility is in even better conditions, as surplus generation appears at different times of the year. Pumped storage is nowadays the most economical means of storing electrical energy. Typical costs are at 250-350 $/KW per unit of installed capacity having an overall efficiency of about 75%. These plants also present a non-polluting way of operation and a quick time of response.

Using storage for peaking duty, the cost of peaking energy will be given by

Peaking Energy (mills/KWh) =

$$= \frac{\text{Capital cost of storage (\$/KW) x Fixed charge rate}}{K \times \eta_S \times 8760}$$

$$+ \text{O\&M (mills/KWh)}$$

where K and η_S are the load factor, and overall efficiency, respectively. If at any time the reservoir is full, the wind power is dumped into the network giving rise to fuel savings. With an additional investment and with a power loss, hydro storage improves the quality of the energy supplied by wind power. Using the average daily wind speed distribution, simulation studies can be carried out for several wind and storage capacities to check the frequency of outage, that is, the periods when the wind is not blowing and the reservoir is empty.

3.3 Wind Power Combined With Solar Power

In some regions of the world, solar power is a complementary
source in relation to wind power. Fig. 8 shows the annual solar,
wind and load variations at Porto, Portugal. Combining the two
sources, a supply with much less fluctuations and which can follow
the load, may be obtained. The short term variations can be
evened with a reduced storage capacity, as wind blows stronger
when solar power is weaker, and vice-versa. Solar electric generation is still very expensive, solar thermal generation being
about two to three times dearer than wind power, but the potential
is huge.

4. CONCLUSIONS

As energy plays a fundamental role in our daily life, in a period
of short supply it is important to analyse the different alternatives. Wind power, although unpredictable, can be described by
statistical models which help to evaluate the effects of its widespread use.

REFERENCES

1. E.W. Golding, The Generation of Electricity by Wind Power,
    London Halsted Press, 1976.

2. P.C. Putnam, Power from the Wind, Van Nostrand, New York,
    1948.

3. G.L. Johnson, Economic Design of Wind Electric Systems ,
    IIEE Transactions on Power Apparatus and Systems, Vol. PAS-97,
    No. 2, March/April 1978.

4. G.E. Jorgensen, M. Lotker and R.C. Meir, Design, Economic
    and System Considerations of Large Wind-Driven Generators,
    IEEE Transactions on Power Apparatus and Systems, Vol. PAS-95,
    No. 3, May/June 1976.

5. C.J. Todd, R.L. Eddy, R.C. James and W.E. Howell, Cost-
    Effective Electric Power Generation from the Wind , Wind
    Engineering, Vol. 2, No. 1, 1978.

6. T.W. Reddoch and J.W. Klein, "No Ill Winds for New Mexico
    Utility", IEEE Spectrum, March 1979.

7. R.L. Sullivan, Power Systems Planning, McGraw-Hill, 1977.

316

8. R.I. Harris, Energy Supplies for Remote Communities, Wind and Wave Conference, Dublin, April 26, 1978.

9. A.E. Allen, Potential for Conventional and Underground Pumped-Storage , IEEE Transactions on Power Apparatus and Systems, Vol. PAS-96, No. 3, May/June 1977.

10. B. Sorensen,"Direct and Indirect Economics of Wind Energy Systems Relative to Fuel Based Systems", International Symposium on Wind Energy Systems, Cambridge, September 7-9, 1976.

11. J.P. Molly,"Balancing Power Supply From Wind Energy Converting Systems", ibidem.

12. E.E. Johanson, and M.K. Goldenblatt,"An Economic Model to Establish the Value of WECS to a Utility System", Second International Symposium on Wind Energy Systems, Amsterdam, October 3-6, 1978.

13. G.H. Bontius, A.H.E. Manders and T. Stoop,"Implications of Large Scale Introduction of Power from Large Wind Energy Conversion Systems Into the Existing Electric Power Supply System in the Netherlands", ibidem.

14. S.T.Y. Lee and C. Dechamps, Mathematical Model for Economic Evaluation of Tidal Power in the Bay of Fundy , IEEE Transactions on Power Apparatus and Systems, Vol. PAS-97, No. 5, Sept/Oct., 1978.

15. J.W. Andrews, Energy Storage Requirements Reduced in Coupled Wind-Solar Systems , Solar Energy, Vol. 18, No. 1, 1976.

16. W.G. Pollard, A General Method for the Evaluation of Possible Systems for Electric Generation with Solar Energy , IEEE Transactions on Power Apparatus and Systems, Vol.PAS-97, No. 5, Sept/Oct., 1978.

DISCUSSION

In the discussion that followed, the point was made that the stability of the integrated power system would be low and there would be voltage drops according to the wind speed variations. Traça replied that AC generators do not take the full benefit of wind power; instead, variable speed generators could be used. Also, to increase the stability of the integrated system, hydraulic pumped storage systems can be used. With 5% additional cost, reversible turbines could do this job. This type of storage would

increase the amount of firm power obtained from the integrated system.

In response to another question, Traça attested that wind turbines do have economies of scale but above 1 MW, stress increases beyond acceptable range. Also, noise is another problem.

It was recommended that instead of basing costs on power, it would be more rational to make comparisons on the basis of energy costs in order to take into account the availability of wind power. It was reported that average generating cost/KWh from wind (storage included) would be about 4-5 ¢/KWh.

It was also pointed out that in order to integrate wind power, hydro and solar power, more comprehensive studies were needed. Simulation models, for example, could take into account the load curves of the demand, and variations in wind speed and solar intensity.

POWER SYSTEMS PLANNING UNDER UNCERTAINTY IN LOAD GROWTH
AND RESOURCE AVAILABILITY

Arun P. Sanghvi*

ICF Incorporated, 1850 K. Street
Northwest, Suite 950
Washington, D.C., 20006
U.S.A.

ABSTRACT. This paper describes a decision analysis framework and
methodology for analyzing the economic impacts of alternative
electrical capacity expansion strategies to cope with some key
uncertainties that permeate the system's planning environment.
In particular, uncertainties vis-a-vis load growth forecasts and
hydro energy availability are directly addressed in this paper.
An application of our model to the Pacific Northwest region of
the United States is presented.

1.   INTRODUCTION

This paper discusses some problems that arise from the increasing
degree of uncertainty that permeates the generation capacity planning environment. It goes on to describe a decision analysis
framework to analyze the economic impacts of alternative expansion
strategies to cope with some of these uncertainties. In particular, the methodology that is developed allows for the explicit
consideration of the reality that even under the most carefully
designed expansion plan, projections of electricity demand and
generating capacity will exceed or fall short of actual demand
and capacity. Therefore, one way or another, society will have
to bear the costs of such deviations. These costs can be very
high, since on the one hand, a potential excess capacity situation will impose unnecessarily high fixed costs of generation.**

* The author is grateful to Mike Cobereley and George Fegan for
  their insights and assistance in preparing this paper.

**The electric utility industry is one of the most capital inten-
  sive industries in the United States. Estimates of the investment required in the next decade for providing additional generating capacity range from $ 215 billion to $ 323 billion (1).

On the other hand, potential shortfalls can lead to adverse economic and social impacts. With so much at stake, it is important that every investment policy, i.e., a capacity expansion plan, be evaluated vis-a-vis the societal costs incurred in the event of a shortfall and in the event of an excess capacity outcome. The methodology developed in this paper does this by computing for each expansion plan, the expected delivered price of electricity in a selected future target year. These costs are determined by the interaction of the corresponding market demand and supply curves that represent the fixed and variable cost of the generating capacity, and the societal cost of power shortfalls, if any. The costs of any slowdown in the given expansion plan are also reflected in the installed cost of the generating plant, and therefore in the supply curves.

The methodology developed is valuable in helping choose the best capacity expansion rate given a probabilistic load growth rate forecast. It is also valuable to policy makers in the selection of a socially desirable capacity expansion rate when confronted by divergent load growth forecasts. We illustrate this procedure with an application to the capacity expansion decision for the Pacific Northwest (PNW) region of the United States. A key uncertainty in the PNW, the availability of hydroelectric energy, is also considered in the analysis.

Section 2 describes how the economic impacts of certain issues, that are key to the capacity expansion decision, are incorporated in the analysis. Section 3 presents an overview of the methodology. A detailed discussion of our results is contained in Section 4.

2. MAJOR ISSUES IN THE CAPACITY EXPANSION DECISION

The nature of the electric utility planning environment has changed considerably in the recent past. Capacity planning used to be a relatively mechanistic and simple process. Key variables that impact the expansion plans were known or could be predicted with considerable confidence. As an example, load growth was quite predictable. Fuel availability was not a major issue. The lead times in obtaining siting and construction permits and satisfying a host of other regulatory requirements were short and more or less certain. The possibility of procedural delays and extended litigation by state or federal regulatory agencies, legislative bodies, environmentalists, and consumers' groups did exist, but seldom materialized. In contrast, the capacity planning process is now characterized by the availability of a large number of generating technologies, rising fuel and construction costs, changing patterns in end-use consumption, more "difficult" financing and capital availability position, increased uncertainty in fuel availability, and in the siting and certification approval process, and the higher probability of regulatory and procedural delays and extended litigation. Other issues related to the capacity expansion decision are: different load growth forecasts, uncertainty in forecasts, availability of hydro energy, capacity and demand for exports, capacity for and availability of imports, price of exports, price of imports, costs of power shortages, plant construction time, ability to accelerate/delay plants, nuclear vs. coal vs. combustion turbine plants, environmental considerations, regulatory requirements, and political and social climate. The decision theoretic model that is developed in this paper incorporates, to varying degrees, a majority of the issues described above, and also the uncertainties associated with some of them. This section discusses some of these uncertainties, and how their impacts can be measured.

2.1   Uncertainty in Load Growth

The "need for power" in the United States has increased at an average rate of 7 percent per year in the past twenty years. For the most part, the actual growth rate was stable, seldom deviating significantly from the average. In contrast, the utility planning environment now is characterized by increased uncertainty about the future demand for electricity. Nationally, for example, load growth forecasts through 1995 range from 2.83 percent* to 6.38 percent.** Within this range there are over two dozen different forecasts. In the Pacific Northwest (PNW) the degree of uncertainty about the future is no less. The West Group Forecast [6] expects firm load to grow at 4.4 percent through 1990. In contrast, the Northwest Energy Policy Project (NEPP [5]) forecasts for low, medium and high growth scenarios are 1.4 percent, 2.98 percent, and 4.4 percent, respectively. Finally, the Pacific Northwest Utilities Conference Committee (PNUCC) has forecast load growth through 1989 to be in the range of 3.5 percent to 5.2 percent, with an expected growth of 4.4 percent.

Differences in load forecasts may be due to the forecasting model/ technique used and/or due to differences in the estimates of judgemental variables, such as the prices of competing fuels, degree of mandatory or voluntary conservation in the future, etc. Irrespective of the reasons, uncertainty in the forecast of future demand for electricity can significantly impact the ability of utilities to meet the electricity demand of consumers. The economic impact of over- or underestimating load growth can be illustrated by means of Figs. 2.1 and 2.2. Fig. 2.1 depicts the

---

* Edison Electric Institute, low growth scenario.

**Federal Energy Administration, electrification scenario.

interaction of the supply and demand curves for electricity in a
given year in the future. The demand curve represents the quantity of electrical energy demanded for a given price. The supply
curve represents the average cost of supplying a given quantity of
electrical energy. The U-shaped nature of the supply curve is a
consequence of the fact that at low levels of system utilization,
the fixed costs have to be allocated over fewer units of electricity. At very high levels of system utilization, on the other
hand, the variable costs rise since in a fixed capacity system
higher demand levels must be met by bringing higher cost plants
on stream. The intersection of the two curves gives the quantity,
Q, of electrical energy that will be supplied and consumed, at the
prevailing market price of P. The market equilibrium point defined
by P and Q is a key element in quantifying the impact of alternate
capacity expansion decisions as shown in Fig. 2.2. This exhibit
contrasts the impact of over-building and under-building on electricity rates in the illustrative target year of 1988. Supply
curves corresponding to three different capacity expansion rates
are shown. Point b represents the market equilibrium if the utility was fortuitous enough to predict the true load growth outcome
of 4 percent. The impact of building over and under is contained
in the prices implied by the points c and a, respectively. In this
illustration, over-building results in under-utilization of system
capacity as evidenced by the fact that point c is located on the
"left" side of the corresponding supply curve. Analogously, under-
building results in over-utilization of system capacity as evidenced by the point a lying on the "right" side of the corresponding supply curve. In this illustration, point a lies above point
c; i.e., over-building happens to be better than under-building.
The situation can as well be the other way around, as shown in
Fig. 2.3. The point is that the economic impact of over- and
under-building can be measured by studying the interactions of the
corresponding supply and demand curves. It is true that the electricity market is regulated and therefore not likely to behave as
a perfect market. However, we assume that a one year period is
sufficient for the demand supply interaction to achieve equilibrium,
even through the Public Utilities Commission's rate setting procedure.

2.2   Availability of Hydro Energy

The Pacific Northwest has a relative abundance of hydroelectric
power. Approximately 84 percent of the named plant generating capacity is hydro. Actual energy capability of the hydro system,
however, also depends upon the amount of run-off in spring and
early summer. This run-off can vary considerably over the years.
Consequently, the energy capability of the hydro system typically
ranges from around 12,000 MW in a critical water year (the hydro
system's firm load carrying capability - HFLCC) to about 18,000 MW

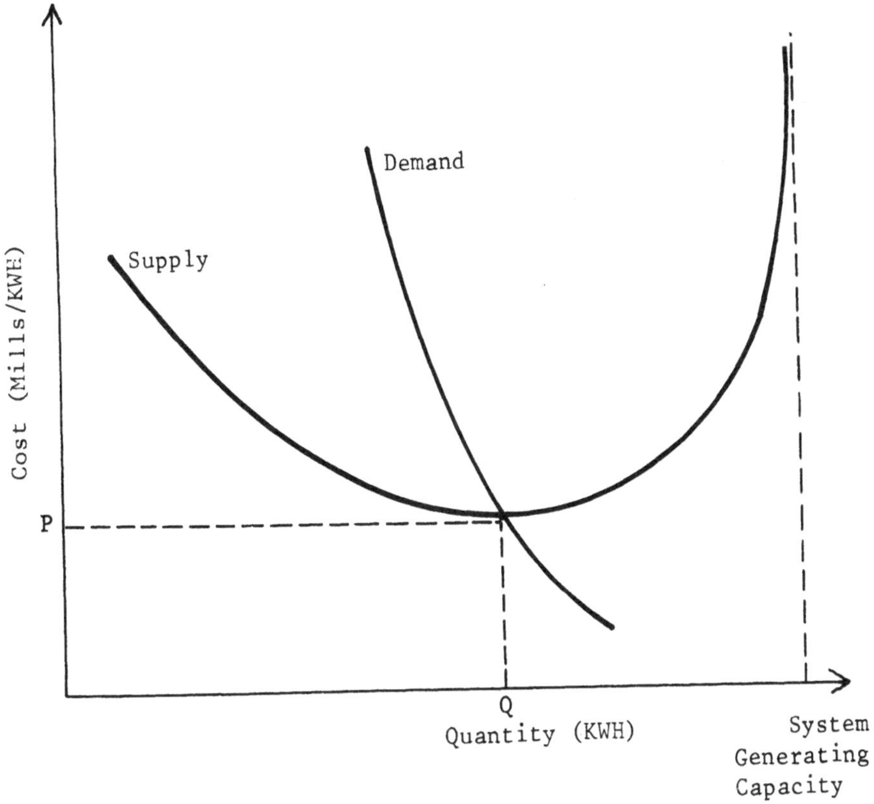

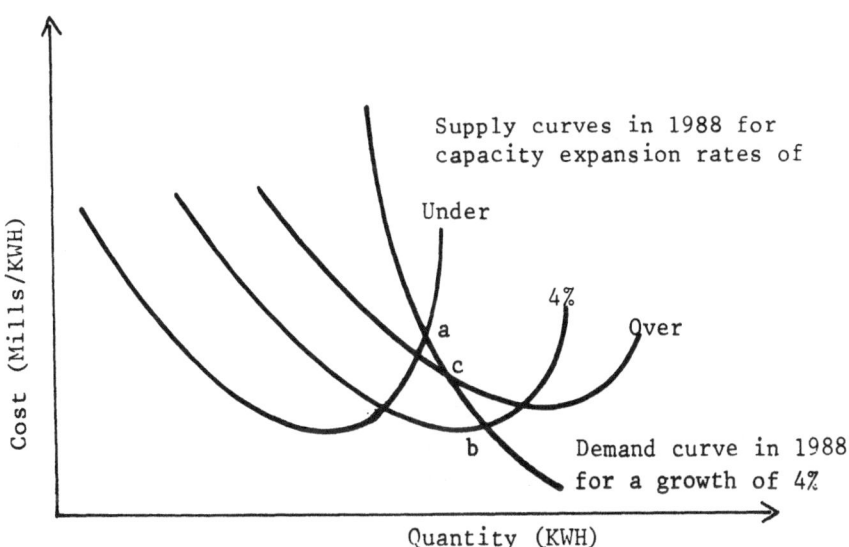

.FIGURE 2.1 - Relationship among supply-demand and
capacity

FIGURE 2.2 - Illustration of the impact of over-building
and under-building on the market price of
electrical energy

324

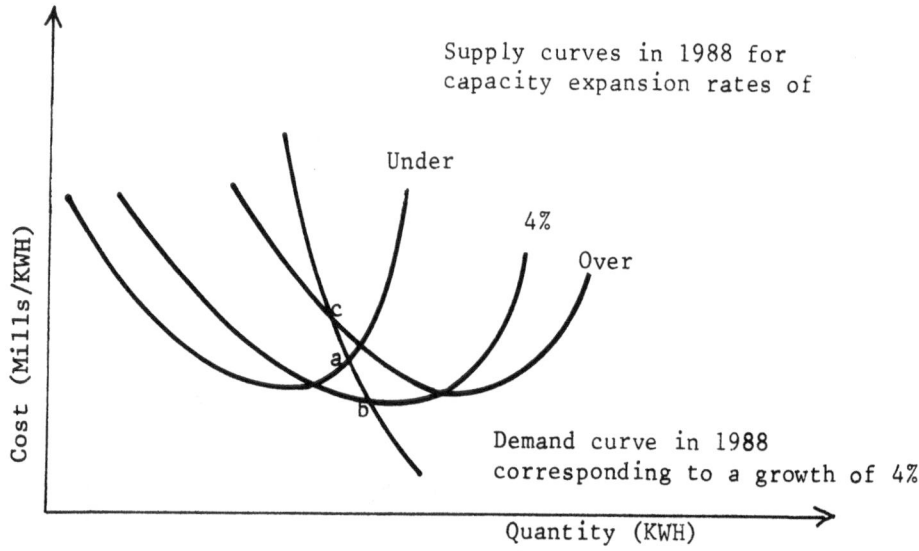

Supply curves in 1988 for
capacity expansion rates of

Under

4%

Over

c

a

b

Cost (Mills/KWH)

Demand curve in 1988
corresponding to a growth of 4%

Quantity (KWH)

FIGURE 2.3 - Illustration of a situation where under-
building is better than over-building

in the best water years. The abundance of hydro in a good water
year, naturally, reduces the costs of meeting a given load, in
contrast to adverse water conditions. This fluctuation results
in a different supply curve for each water year and is depicted
in Fig. 2.4 for three different water years. As the amount of
hydro energy available increases, the cost curve shifts downward
and to the right to reflect the increased generating capability
and the fact that the variable costs of hydro generation are the
lowest of all generating technologies. Fig. 2.4 also reveals
that the market equilibrium is a function of the water year outcome. If the three water years in question are equally likely,
then the expected market price can be determined by averaging the
prices corresponding to points a, b, and c.

2.3   Costs of Power Outages

Uncertainty in load as well as in generation availability make it
a virtual certainty that at times consumers will be denied power.
The social costs of such outages must be considered in deciding
the optimal generating capacity. This is easier said than done.

The cost of an electrical outage to a consumer is a function of
the time of day and duration of outage, the activities impacted,
the nature of and degree to which the activities impacted are

dependent upon electricity, the availability of a backup power source, the ability to resume the impacted activity normally after power is restored, the frequency of the outages, and a host of other determinants. Consequently, the cost of an outage is different for each consumer. To some it may be the mere inconvenience of being stranded in an elevator. To another it may involve loss due to fire, burglary or vandalism. To yet another, it may be the loss of production, spoilage of in-process inventory and of equipment. Information about the costs of outages is vital for several reasons. Determination of the optimal amount of generating capacity of a utility should be based upon a careful balancing of the associated costs and benefits of each investment alternative. As system capacity is decreased, the associated direct costs decrease, but at the expense of increasing costs of energy denied, due to the more frequent generation failures.

Assessing the cost of potential electricity shortages is also important from the standpoint of a utility's decision to over-build or under-build. Consider a scenario where a utility faces an under-build situation, say three years down the line; it has the option of adding combustion turbines and restoring the proper reserve margin, or relaxing the reliability requirement.* If it decides to relax the requirement, to what extent should it do so? These issues beg the question: What is the societal cost of an outage? Currently, such decisions are primarily based upon value judgements of certain individuals. In doing so, however, an implicit cost of outage figure has been assigned. Such value judgements invariably lead to inconsistent decisions over time. We believe that the cost of outages is a key exogenous variable that should be explicitly used in the capacity expansion planning process. Only then can a consistent evaluation of alternate expansion plans be achieved. Such an approach affords an analysis of the sensitivity of the optimal capacity expansion rate to the costs of outages. This flexibility is especially important, since current estimates of the costs of outages vary widely in the range of .60 $ to 16 $ per KWH denied [4].

---

*Most major utilities strive to achieve a loss of load probability (LOLP) of almost one in ten years. This typically translates to a reserve margin of 20 percent. The use of a LOLP of one in ten is, unfortunately, arbitrary, and most likely not optimal in a social cost-benefit sense. In fact, some recent studies (e.g., see [9]) argue convincingly that substantial economic gains, with virtually no perceptible social cost, by reducing current reserve margins by only a small amount.

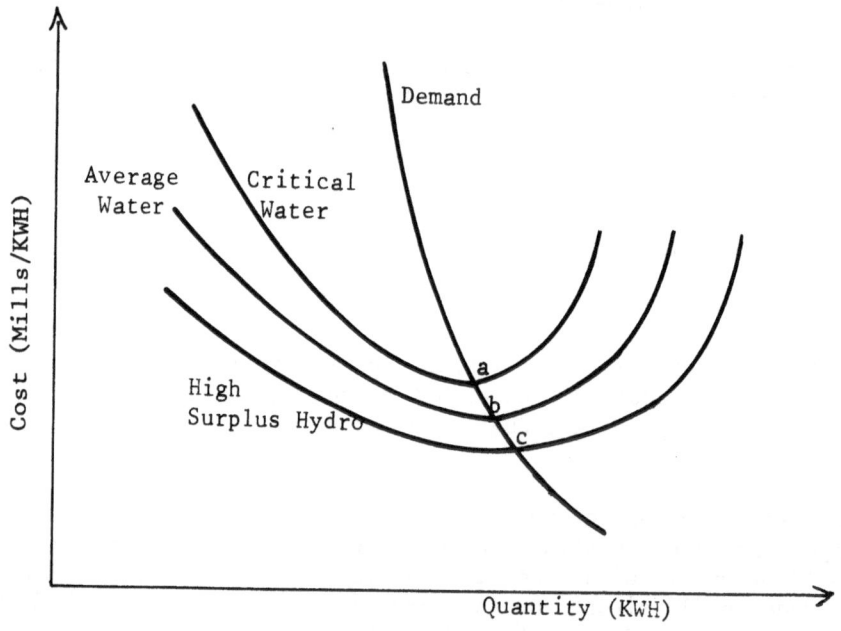

FIGURE 2.4 - Shifts in supply curve because of surplus
            hydro

3.   OUTLINE OF METHODOLOGY

This section provides a general description of the model that was
used to assess the economic impact of being over- or under-built
and of alternate over-build and catch-up strategies to meet load
in the selected target year of 1988. Fig. 3.1 is a schematic
view of the model and the directional linkages among the various
modules. Fundamentally, the model attempts to compute the market
price of electricity in the target year of 1988, given different
capacity expansion rates, load growth outcomes, and growth scenarios. The capacity expansion rate determines the plant expansion
program to be undertaken, starting in 1977, so that sufficient
generating capacity is available, under current planning procedures
in 1988.* The growth scenario specifies the year in which the

---

*Under the present planning process, the capacity required in any
year is determined by the projected load and a reserve requirement that is equal to half of the increase in "utility type" loads
from that year to the next. In this report, whenever we speak
of a capacity expansion rate of 3 percent, for example, we mean
that capacity is expanded at a rate sufficient to meet the load
and reserve requirements, as defined above, corresponding to a
load growth rate of 3 percent.

eventual load growth outcome is discovered. This determines the exact nature and extent of any capacity adjustments made between 1977 and 1988. If discovery of the actual load growth rate is made prior to 1988, then capacity adjustments, either additions or delays, are made to ensure that there is adequate generating capacity available in 1988. If the true load growth rate is only discovered in 1988, then the capacity expansion plan initiated in 1977, with no interim adjustments, will generally lead to an over- or under-built situation.

On the supply side, the model determines the average cost (to consumers) curve of electricity for a specified load growth rate and growth scenario. This is achieved by estimating the cost of meeting a number of different load levels, and then using a curve fitting routine to determine a continuous supply curve. Such cost curves are determined for each of eight representative water years. On the demand side, the load growth rate outcome and a specified demand elasticity are used to determine the demand curve for electricity in the PNW in 1988. Finally, an equilibrium routine determines the market equilibrium price and consumption quantity that will prevail under each of the eight water year conditions. The expected price and quantity under the specified load growth rate and growth scenario are then computed as simple averages of the corresponding eight values that are determined by the equilibrium routine. These and other results are then printed by the report generator.

In this study, we examine the costs of adjustments in energy capabilities rather than in the capacity of the generating system. A hydro dominated generating system, such as the Pacific Northwest, typically offers large capacity and therefore flexibility in meeting peak loads. This is not to say that a capacity shortfall or excess will never accompany an energy shortfall or excess. In such situations, an assumption implicit in our model is that any adjustments made to resolve an energy shortfall or excess situation will automatically take care of the capacity situation.

3.1   Capacity Adjustment Model

The ultimate load growth outcome is generally different from the load growth rate used to plan capacity expansion. This outcome may be known as early as 1982, at which time construction of additional generating capacity on-line by 1988; or in 1985, at which time additional combustion turbines can be added, if necessary; or as late as 1988, in which case, we have an under-built situation. On the other hand, knowledge of the true load growth outcome in 1982 or in 1985 may require a slowing down of existing construction to delay bringing one or more plants on-line by 1988. Such a discovery made in 1988 will result in an over-built situation. Additions and/or delays of the type just described, are

328

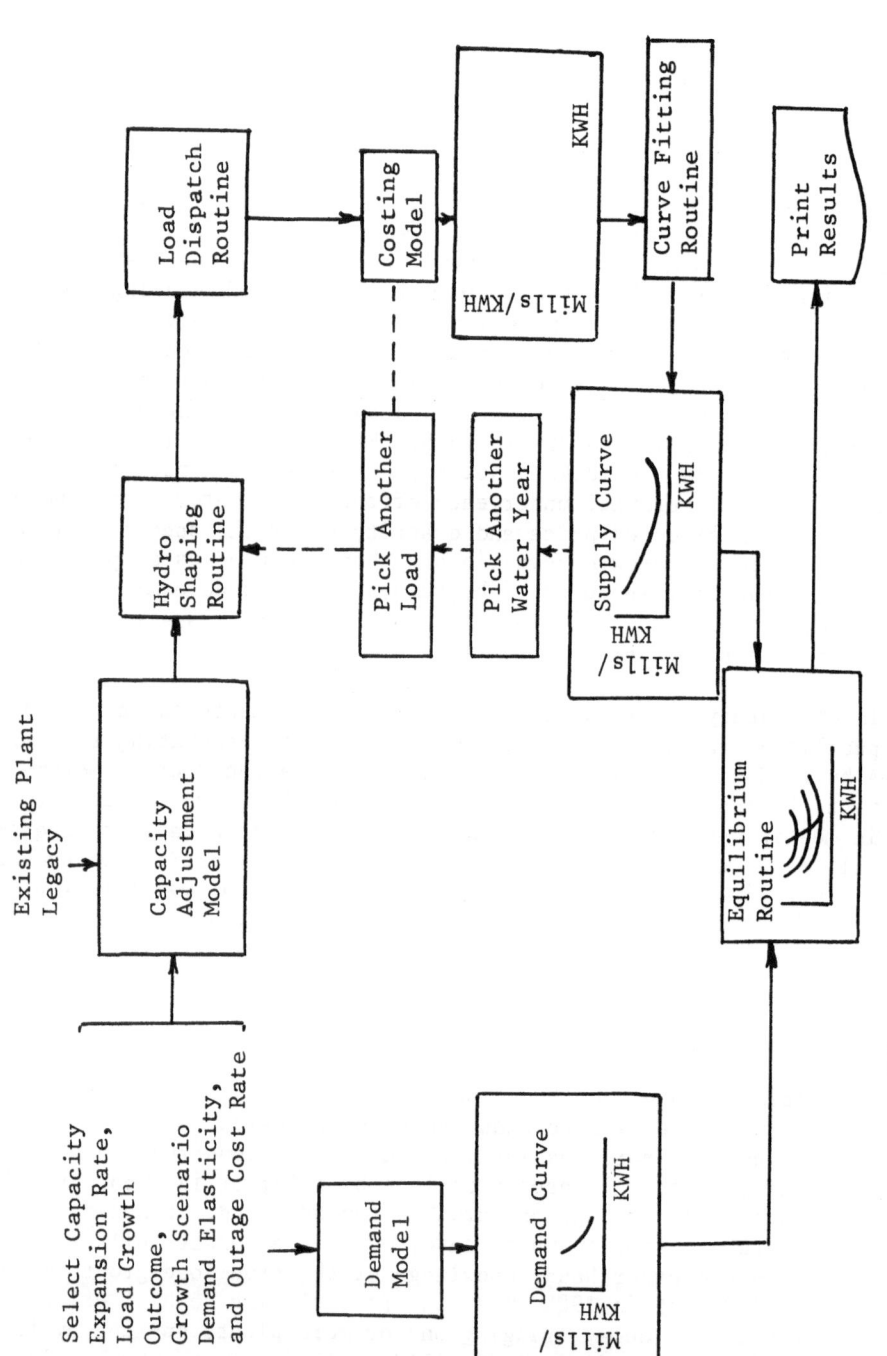

FIGURE 3.1 — Overview of the Model

performed by the capacity adjustment model.

3.2 Hydro Shaping Routine

Most of the hydro and thermal generating facilities in the PNW are located on or along the four major rivers that are interconnected; Columbia, Snake, Willamette and Pend Oreille. Generally, these generation sites cannot be operated in isolation. Downstream effects of upstream generation, hydro as well as thermal, must be considered. In other instances, a spilloff may be necessary just to provide sufficient cooling water for a downstream thermal plant. A grand plan is necessary to effectively operate the hydrothermal system, and to prevent the myopic optimization policies that are likely to be pursued by managers of individual generating facilities. Such a plan exists in the PNW and goes by the name of "Agreement for Coordination of Operations Among the Power Systems of the Pacific Northwest", or, in short, the coordination agreement. This agreement is a legal document that spells out in detail the operating procedures, obligations and entitlements that are binding on all the major utility transmission companies, the Corps of Engineers and the Bonneville Power Administration. The coordination agreement not only specifies rules for the determination of the Hydro system Firm Energy Load Carrying Capability (HFLCC) for maximum advantage, but permits the shaping of this firm capability from month to month within a water year and from one water year to the next. This shaping is governed by the specification of the Energy Content Curve that determines the maximum amount of stored energy that can be drafted at any point in time. An exception is made to the extent that some provisional energy can be borrowed from the future.

The Hydro Shaping routine is a model of the operating rules that govern the use of hydro energy in the PNW. For a given load level and thermal generating capacity, this model shapes, within the constraints of the Coordination Agreement, the HFLCC into three seasons in a manner that offers the best odds of meeting the load. These seasons are:

Early drawdown season, September-December; reservoirs are drawn down, and no forecasts of runoff are available.

Late drawdown season, January-April; reservoirs are still being drawn, but runoff forecasts might make more storage available for use.

Refill-hold season, May-August; the spring runoff allows filling, and most thermal scheduled maintenance is performed in this season.

Next, for the pre-selected water year being simulated, the program determines the natural streamflow capability from historical data of actual runoffs and flows in the PNW. The surplus hydro (i.e., energy in excess of HFLCC) is then computed and dispatched to meet load, by regulating the system within the specified policies. Details of this program are contained in [3]. In short, the output of the hydro shaping routine specifies for any given load, the availability of hydro energy, as a function of the water year being simulated, and by the season within the water year.*

3.3   Load Dispatch Routine

The variable costs of supplying a specified load is primarily a function of the generating mix dispatched to meet the load. The load dispatch routine attempts to do this dispatch in the most inexpensive manner. Specifically, for a given load, the following dispatch merit order is used: hydro, nuclear, coal, existing small thermal plants, and new combustion turbines. The only exception to this merit order ensures that under no circumstances is the utilization factor of nuclear plants lower than .70. The maximum utilization factor of nuclear plants is .75. This constraint of not under-utilizing nuclear capacity stems from the concern of utility planners that disturbing the nuclear fuel cycle now has ripple effects on future fuel cycles. The resultant cost increases in all future cycles, it is generally believed, will exceed any savings in the current cycle. The decision to under-utilize nuclear plants in a good water year should be a function of the amount of surplus hydro available, and the prospects of surplus hydro in the next year or two. If the current surplus is high and prospects for the future also look good, then benefits are likely to outweigh the cumulative cost increases of future fuel cycles. The development of such a model was beyond the scope of this study. Moreover, it is extremely unlikely that information about future hydro surplus will ever be available.

A typical load dispatch proceeds as follows. For the given load, output from the Hydro Shaping Routine specifies the amount of hydro energy available. First, 70 percent of the nuclear nameplate capacity is dispatched. Following this, as much of the hydro energy is dispatched as is necessary or available. Any unfulfilled load is then met by dispatching energy from minemouth coal plants, followed by energy from unit train coal plants and

*From historical records of stream flow data for 40 years, eight ersatz "water years" were constructed. Each water year is characterized by the amount of surplus hydro energy that is available in that year. Furthermore, each such water year was constructed so as to be equally likely.

so on, down the merit order. This process culminates with specification of the amount of the load met by various generating
sources. A feature not included in this study is the consideration of the load duration curve in dispatching the generating
plant. A load duration curve analysis is particularly critical
for predominantly thermal systems that are peak-constrained. In
contrast, the PNW, because of the abundance of hydro energy, is
energy-constrained for the most part. However, the incorporation
of the standard mathematical programming approach to load dispatching would be straightforward to incorporate, given the modular
component nature of our model.

3.4   Costing Model

The costing model calculates the unit cost to consumers (Mills/
KWH) of supplying a specified amount of electrical energy (KWH)
associated with a given growth scenario. In several growth scenarios analyzed, capacity adjustments involve a slowdown of construction in progress. Under current regulatory procedures, only
a small portion of construction work in process is included in
the rate base, with the entire investment in new generating plant
being included in the rate base starting with initial operation
of plant. Consequently, the fixed cost of a plant that is delayed so that it comes onstream after 1988 will no longer appear
in the 1988 rate base, but will slow up as a higher cost in a
later year. This would pose no problem in a multi-target year
study. In a single target year study, such as this, the use of
current rate making procedures would not adequately show the impact of delaying construction in progress. Therefore, our costing model creates an artificial rate base. This base in 1988 includes a portion of the fixed cost of generating plant, still under construction in 1988, in proportion to the fraction of the
plant completed as of that year, vis-a-vis the extended construction schedule.

3.5   Demand Model

A demand curve is a functional relationship between the amount of
electrical energy consumption as a function of the selling price
of the energy. The nature of the demand curve is equal   fundamentally to the capacity expansion decision as is the supply curve.
Together they determine the market equilibrium and provide a
measure of consumer welfare.

This study assumes the following relationship between the demand
and price of electricity in the target year of 1988:

$$Q = KP^e \qquad\qquad (1)$$

where:

Q = Quantity of electrical energy consumed in 1988 (10^9 KWH)

P = Price of electrical energy in 1988 (Mills/KWH)

e = The short term price elasticity of demand in 1988, i.e., the percentage change in consumption for a unit percentage change in price

K = A scaling constant.

4. RESULTS

The model described in Section 3 was used to determine the impact of being over- and under-built assuming different load growth rates and growth scenarios. We also studied other growth scenarios to assess the impact of deliberate "over-build" and "catch-up" strategies. Fig. 4.1 displays the growth scenarios analyzed in the base case.* Identical scenarios were run for the deliberate over- and under-build cases of 5.5 percent, and 2.5 percent, and 3.5 percent respectively. Each stop node in Fig. 4.1 is defined by the load growth outcome and a generating mix that evolves as a consequence of the plant expansion program undertaken in 1977 and adjusted appropriately. For each such stop node, the entire model, as depicted in Fig. 3.1, is run to determine the expected market price of electricity in 1988. Furthermore, each of the cases was run under different assumptions on the capability to import and export power into and from the PNW. However, in this paper, we only report the results for the case of full import and export capacibility.** Furthermore, we report the results for those growth scenarios where the true load growth outcome is discovered in the years 1985 and 1988 only. The presentation and analysis of results in all other scenarios and import export capability combinations is beyond the scope of this paper. Such a comprehensive analysis is contained in [7].

Our results for the case where the true load growth outcome is discovered in 1985, are presented in the Economic Impact (Pay Off)

---

* This case corresponds to the expansion plan currently being implemented in the PNW.

**After accounting for existing firm contracts, these capabilities are, respectively, 2,213 MW, and 4,813 MW.

Matrix displayed in Fig. 4.2. The comparable Pay Off Matrix
for the static case* is displayed in Fig. 4.3. The planner
has a choice of adopting a deliberate under-build strategy by
building for a load growth of 2.5 percent or 3.5 percent, a deliberate over-build strategy of planning for load growth of 5.5
percent, or planning to match the load growth of 4.4 percent. In
each case, one of four possible outcomes will result. The economic impact of the particular decision and outcome is measured
by the expected price of electricity in 1988, as measured in current Mills/KWH. As an example, if capacity expansion was planned
for a load growth of 2.5 percent and the eventual outcome is 4.4
percent, then the cost is 53.3 Mills if no capacity adjustments
were made in the interim (Fig. 4.3). If, on the other hand, a
capacity-adjustment could be made in 1985, then the cost is only
40.8 Mills (Fig. 4.2). The higher costs in the former case
stem from a combination of factors; the increased utilization of
higher variable cost generating equipment, and the economic cost
of power shortfalls, valued at a dollar per KWH denied.** The
results in Figs. 4.1 and 4.2 support the following conclusions.
The expected costs of ending up in an under-built situation are
significantly higher than being in an over-built situation. This
follows since the entries above the diagonal of either pay off
matrix are larger than the corresponding entries symmetrically
located below the diagonal. These differences are obviously less
when capacity adjustments can be made to match load growth. Another conclusion is that whereas adjustments in capacity can only
be done at some cost, the net impact can be beneficial. This
conclusion is supported in those instances where off-diagonal
entries in the matrix in Fig. 4.2 are smaller in magnitude than
the corresponding entries in Fig. 4.3. A third conclusion is
that if the load growth outcomes were known a priori, then the
optimal capacity expansion decision is to plan to match this growth.

Fig. 4.4 reproduces the matrix in Fig. 4.2, with the additional information about the probabilities of the various load
growth outcomes displayed at the bottom. The probabilities do
not represent official forecasts. However, this distribution has
an expected value of 4.4 percent per year, the current "high" forecast of load growth in the PNW and is typical of the type of distribution D^H that was discussed in the introduction. The expected
costs of each decision alternative are displayed in the column to
the right of the matrix. The decision that minimizes the expected

---

* I.e., the case where true load growth is only discovered in 1988.
This implies that no intermediate capacity adjustments are made.

**The general nature of conclusions in this paper do not change
when outage costs as low as .50 $/KWH denied were used. Results,
using outage costs other than 1 $/KWH denied are contained in [7].

334

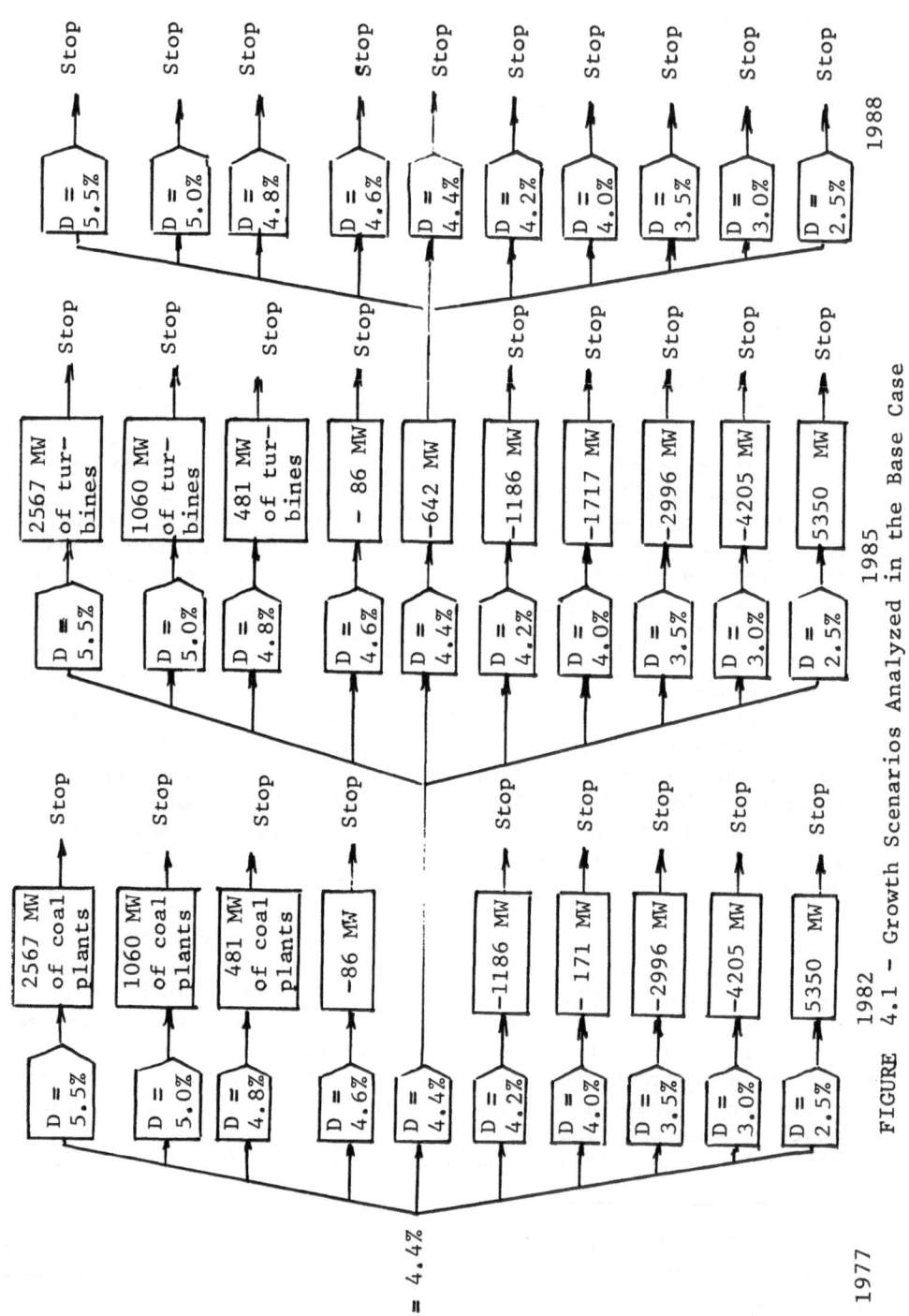

FIGURE 4.1 – Growth Scenarios Analyzed in the Base Case

cost is to expand at 4.4 percent. The information that is contained in Fig. 4.2 can also be used to shed light on the capacity expansion decision when one is faced with divergent probabilistic forecasts. Suppose that another group's forecast of the odds of the four load growth outcomes of 2.5 percent, 3.5 percent, 4.4 percent, and 5.5 percent is .6, .3, .08, and .02, respectively. Under this "low growth" forecast, the expected costs of the four decision alternatives, in the presence of capacity adjustments, are 30.6, 31.8, 35.3, and 36.7 Mills/KWH. In contrast to the "high growth" forecast, under the "low growth" forecast, the optimal expansion decision is to plan for a load growth of 2.5 percent. The question now arises as to which capacity expansion rate should be used. Fig. 4.5 displays the alternatives and expected costs under each outcome. The maximum strategy is to expand at 4.4 percent. The minimal regret strategy is to expand at 2.5 percent. Perhaps a more meaningful way to look at the decision problem is to compare the economic impacts of choosing the 2.5 percent expansion rate with the outcome being determined by distribution D^H and choosing the 4.4 percent rate with the outcome being determined by distribution D^L. The numbers in Fig. 4.5 show that these costs are, respectively, 40.4 Mills/KWH and 35.3 Mills/KWH. Consequently choosing the incorrect rate results in a cost of 5 Mills/KWH. This, conservatively, translates to an expected savings in PNW's consumers' electric bills of approximately 750 million dollars in the year 1988.* Under this criterion, the strategy is to expand at 4.4 percent.

A "second order" expected value analysis can be performed on the 2 x 2 payoff matrix in Fig. 4.5. However, this does not clarify the basic tradeoffs further. Members of the regulatory body who accept or "lean heavily" towards distribution D^L will in all likelihood favor an expansion rate of 2.5 percent. By the same token, those who accept or lean heavily towards distribution D^H will favor the 4.4 percent rate. The swing votes may lie with those members who truly cannot make up their minds about accepting D^L or D^H. Many of these members are likely to use the reasoning laid out in the previous paragraph and choose the 4.4 percent rate. Still other undecided members may assign** likelihoods to the occurrence of D^L and D^H. This immediately implies an unconditional distribution that lies "between" D^L and D^H. Such distributions will favor a compromise solution, i.e., an expansion rate between 2.5 percent and 4.4 percent. Often, however, such "compromise"

---

* This estimate assumes that the 1988 consumption level will be at least 150 billion KWH.

**If not specific values, in a fuzzy sense.

Load growth outcome

Decision alternative	2.5%	3.5%	4.4%	5.5%
2.5%	26.9	32.0	53.3	140.8
3.5%	29.5	31.0	35.6	70.9
4.4%	33.8	33.4	34.1	44.1
5.5%	37.3	35.8	35.5	36.0

FIGURE 4.2 - Economic impact matrix
Scenario: Load growth outcome discovered in
year 1985

Load growth outcome

Decision alternative	2.5%	3.5%	4.4%	5.5%
2.5%	26.9	34.2	40.8	47.1
3.5%	31.1	31.0	37.3	43.8
4.4%	35.6	34.9	34.1	39.3
5.5%	36.8	36.7	36.6	36.0

FIGURE 4.3 - Economic impact matrix
Scenario: Load growth outcome discovered in
year 1988

Decision Alternative	Load growth outcome 2.5%	3.5%	4.4%	5.5%	Expected Cost (Mills/KWH)
2.5%	26.9	34.2	40.8	47.1	40.4
3.5%	31.1	31.0	37.3	43.8	37.4
4.4%	35.6	34.9	34.1	39.3	35.6
5.5%	36.8	36.7	36.6	36.0	36.5
Probability of given load growth rate outcome	.05	.2	.5	.25	

FIGURE 4.4 - Economic impact matrix
Scenario: Load growth outcome discovered in year 1985

338

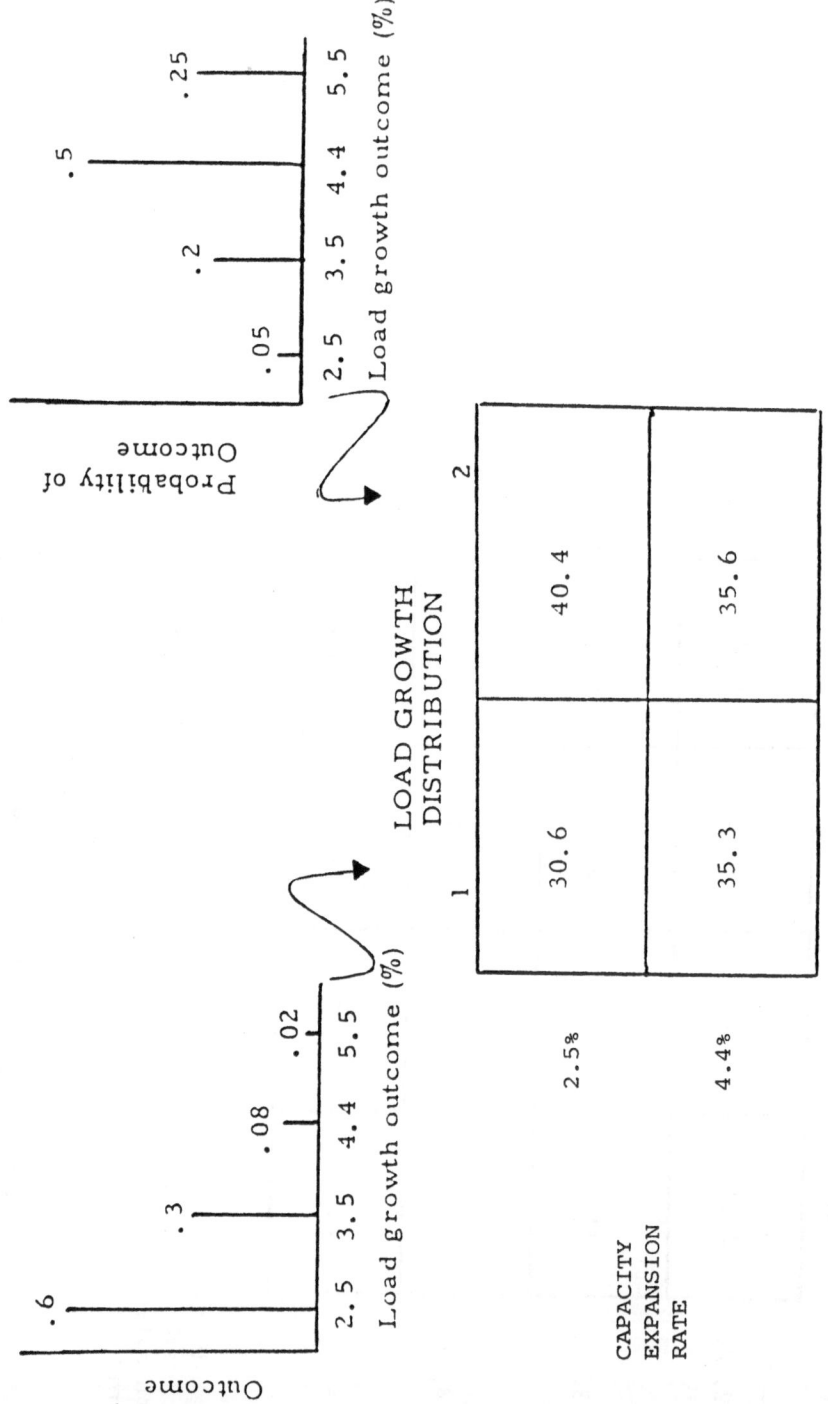

FIGURE 4.5 — Capacity expansion decision under different probabilistic load growth forecasts

solutions may not be feasible, since the basic decision is discrete in nature - grant or deny a permit for an additional plant.

In the final analysis, the decision to accept one expansion rate
over the other rests upon the agency with the mandate to do so.
Our decision analysis purposely falls short of recommending this
final step, since the capacity expansion rate decision must also
incorporate other socio-political considerations. However, this
analysis provides an objective assessment of the economic and environmental impacts of each decision alternative.* The decision
analysis approach developed here forces a logical basis and structure to an otherwise informal reasoning process. It therefore
provides an effective medium for communicating the reasoning that
underlies the final recommendation.

REFERENCES

1.   Federal Energy Administration, "National Energy Outlook",
      1976.

2.   Gordian Associates, "Optimal Capacity Planning Under Uncer-
      tainty in Demand", Report submitted to the Federal Energy
      Administration, November 1976.

3.   D. Lewis, R. Duncan, and M. Schultz, "Energy Reserve
      Planning Model", Progress Report, Northwest Power Pool,
      Portland, Oregon, 1975.

4.   Mitre Corporation, "Need for Power Study: An Assessment of
      the Adequacy of Future Electric Generating Capacity", Report
      MTR-7549, 1977.

5.   NEPP, "Northwest Energy Policy Project, Supply Module",
      1977.

6.   PNUCC, "West Group Forecast of Power Loads and Resources,
      July 1978-June 1989", March 1978.

7.   A.P. Sanghvi, and D.R. Limaye, "Planning for Generation Ca-
      pacity Expansion in the Pacific Northwest: A Decision Analy-
      sis of Over- and Under-Building", Mathematica Report sub-
      mitted to the Portland General Electric Company, Portland,
      Oregon, August 1978.

---

*The environmental impact of various strategies is reflected in the
analysis to the extent that the costs of scrubbers or other pollution abatement devices mandated by legislative action are
reflected in fixed and variable costs of each generating plant.

8. Stanford Research Institute, "Decision Analysis of California Electrical Capacity Expansion", Report submitted to CERDC, February 1977.

9. M.L. Telson, "The Economics of Alternative Levels of Reliability for Electric Power Generation Systems", Bell Journal of Econ. and Management Sci., No. 16, No.2, Autumn 1976.

PART V

METHODOLOGY IN MODELLING

Regional Network Based Modelling as a Tool in
Energy Planning: An Evolutionary Exposé

Applications of Multiple Objective Linear
Programming to Energy Modelling

An Open Dynamic Linear Activity Analysis Model
of Production

REGIONAL NETWORK BASED MODELLING AS A TOOL IN ENERGY PLANNING:
AN EVOLUTIONARY EXPOSE

Joseph G. Debanné

Department of Management Science
University of Ottawa
Ottawa, Ontario, Canada

ABSTRACT. The evolution of regional network based energy supply
and distribution models from a single commodity pure trans-shipment case to a multi-commodity generalized trans-shipment case
with investment options is described. The generalized version
featuring convex production functions and demand functions for the
various commodities, producing regions and demand sectors in the
respective energy consuming regions are also modelled, thereby
allowing the determination of equilibrium prices in the market
place, given a price leader's base price for oil. The superior
computational efficiency of certain network flow optimization algorithms renders economical the modelling of very large and detailed regional energy supply-distribution systems. The problems
and techniques of interfacing a network optimization model with
highly nonlinear peripheral models such as the petroleum exploration and production function are also discussed in the paper.

1. INTRODUCTION

1.1 Historical Background

The first regional energy supply distribution model featuring the
various oil producing regions and refining centers of North America was published in 1969 [2]. In this model, Alaska's North Slope
which was then newly discovered was assigned the oil price leader's
role in North America under the assumption that import quotas and
barriers would continue to insulate the North American oil producing industry for some time. This one-product model led to several valuable results and conclusions, not only about the possible
impact of Alaska's North Slope discovery, but also to the conclusion

that the value of this type of exercise could be considerably enhanced if other energy commodities, in particular natural gas, could be concurrently taken into account.

This study led to a second paper in 1971 which proposed Model II, an Oil, Gas, and "other" model [3], patterned along similar principles and methodology as the single commodity model. In this second generation model, "other" was a proxy for whatever other energy source may be available in a particular region (e.g., coal and/or nuclear power). Improvements and innovations introduced in the multi-commodity model were made partly in the light of the experience gained with the one-commodity model and partly because of some fundamental differences between a one-commodity economy and a multi-commodity economy, notably the substitution effects between commodities in the market place, price interactions between different commodities and the effect of prices and demand levels in the various producing regions on investments in production and transmission capacity.

An expanded version of this model, referred to as Model IIa, features coal as a specific commodity in addition to oil and gas [4]. Considering that coal can be transported by conventional means, i.e. railways, barges and ships as well as in coal slurry pipelines, the Model IIa version of the regional energy supply-distribution model can represent these means of transportation and has the capacity of designing and expanding coal slurry pipelines.

1.2   The Various Functions Within Model II and Model III

The following agents of "functions" interact within Model II and Model III:

Regions

The various producing (or potentially productive) regions in North America are represented in Model II. "Regions" here means the industry and the state or provincial government representing together a regional interest and entity within the model. Every region determines its investments in oil and gas exploration and development. Because of the abundance of coal in North America, exploration for coal was ignored. States or groups of states, e.g. Western Europe constitute the "regions" of Model III.

Price leader

Overseas oil is treated as the "price leader" producing region with an "infinite" capacity in Model II, while any commodity in any energy producing region can be treated as the price leader in Model III.

Other energy

Model II features "other" forms of energy, where "other" is a
proxy for any other source of energy made available at each demand
center at a higher price than oil, gas or coal, the three commodities explicitly represented in Model II. Accordingly, "other"
is consumed only if it is impossible to meet energy demand requirements with oil, gas and/or coal in Model II.

Other energy commodities could be featured within the Model II
network without major algorithmic changes so long as this does not
entail transformation processes and energy losses. For example,
the various refined petroleum products could not be featured in
this supply-distribution model together with oil. Fixed proportions of refined products dictated by crude oil composition and
refinery technology cannot be represented in a standard network
flow model. This limitation imposes the restriction that all petroleum liquid products including oil are indistinguishable in the
model. Likewise, conversion of petroleum, natural gas or coal
into electricity, or of coal into liquid fuels cannot be modelled
in Model II as such processes feature substantial energy losses
which require the use of a generalized network flow optimization
algorithm. Note that gas transmission losses are already modelled
with generalized networks in a regional supply-distribution system
of natural gas [1].

Model III has been developed around a generalized network optimization code, designed for this purpose [8], to represent among
other things commodity transformations and transmission losses.
Model III features for instance the uranium mining, enrichment
and consumption processes as well as technological options with
fixed set up charges.

The capability to correctly feature "flow split" processes such
as petroleum refining where the input stream is split into a number of output streams according to specified proportions would
require yet another special network algorithm, to be developed
and which will eventually be incorporated in a Model IV version.

Transportation

The transportation sector of Model II comprises oil, gas and coal
transmission by pipeline; oil and NGL by tankers and can also accommodate coal by railways (and water ways). Pipelines are designed,
costed and expanded within Model II with due regard to transmission
requirements, the type of terrain traversed by the pipeline, the
cost of steel and the mean winter temperatures prevailing throughout
the regions where the pipeline passes. Tariffs are set for all
pipelines according to simplified but reasonably representative

tariff making rules. Railways and sea routes are treated with less detail and merely feature an estimated port-to-port transportation cost per TBTU.[*] Model III is less elaborate in terms of detail regarding the transportation sector, being a more aggregate model.

Energy demand

Energy demand in Model II can be specified by regions, e.g. PAD districts in the US and provinces in Canada. Each demand region is characterized by coordinates (latitude-longitude) of the center of gravity of its population, and forecasts of minimum demands for oil, natural gas and coal, as well as of total energy demand. Period by period within the planning horizon, the model uses a modified Ford-Fulkerson primal-dual algorithm (9) to determine equilibrium prices and to optimize flows in the supply-distribution network, i.e., to allocate oil, gas, coal and "other" commodities throughout the North American market in order to satisfy all demands at minimum total cost. Since oil, gas and coal are cheaper than the "other" energy, this "other" source is called upon in Model II to satisfy total demand in any region only if oil, gas and coal cannot satisfy total demand in this region.

Specification of nonsubstitutable energy demands by commodity and by region may not be obtainable from most regional energy demand models. Conversely, price demand elasticities by product and sector are easily obtainable from most energy demand models. In order to facilitate the "plugging in" of existing regional energy demand models, Model III is designed to accept price-demand elasticity information and return the minimum cost or equilibrium allocation of supply to demand as well as commodity demands and prices at every iteration and time period.

This feature of Model III will be discussed further and in greater detail in the appropriate sections of this paper.

Government

The last and increasingly important agent in the energy sector is government. Energy policy is implemented in Models II and III by subsidies or surcharges per unit commodity produced, imported or consumed and by constraints to "flows" e.g. import or export quotas. Likewise, environmental protection constraints can be taken into account by featuring pollution control costs in the appropriate arcs of the network. Models II and III are designed to facilitate the specification of policy parameters by providing considerable flexibility in network modelling.

---

[*]Trillion Btu

Externalities

To the above six agents interacting within Model III it would be
appropriate to add an "externalities" function which links this
regional energy supply and distribution model to its economic and
physical environment. It may be desirable in some scenarios e.g.
of the world energy system to feature the interactions of the
energy sector with the world economy and ecosystem, here referred
to as the "peripheral" models.

1.3   Comparative Overview of Models II and III

Figures 1a and 1b respectively describe the computational schemes
and information flows in Model II and Model III. Essentially,
equilibrium between supply and demand as well as between price
demand elasticities and the product mix "consumed" by the model
is endogenously attained in Model III while demands that drive
Model II are in effect exogenously specified period by period.

More specifically, the following differences in treatment are
noted:

- Own price-demand elasticities by product and by sector
  are generated by the demand model of Model III and are
  mapped onto the supply-distribution network to determine
  the equilibrium production rates, prices, and demands.
  Total energy demands by market or demand center are exogenously specified, period by period in Model II, and
  effectively drive the network flow allocation model under the assumption that total energy demand is inelastic
  with respect to price. Minimum nonsubstitutable demands
  by product must also be specified in Model II, thereby
  allowing the allocation model to meet the specified total energy demands with some measure of flexibility as
  to the choice of product.

- Investments in production capacity are determined in the
  course of the network allocation of Model III as this
  allocation yields the equilibrium production rates,
  prices, and demands throughout the model for the period
  of interest. This is made possible by mapping the cost-
  supply production functions onto the network, at the
  beginning of every period as pictured in Figures 1b and
  2. Production capacities are also determined at the
  beginning of every period in Model II; however, production capacity is assumed to be inelastic with respect
  to price in this model.

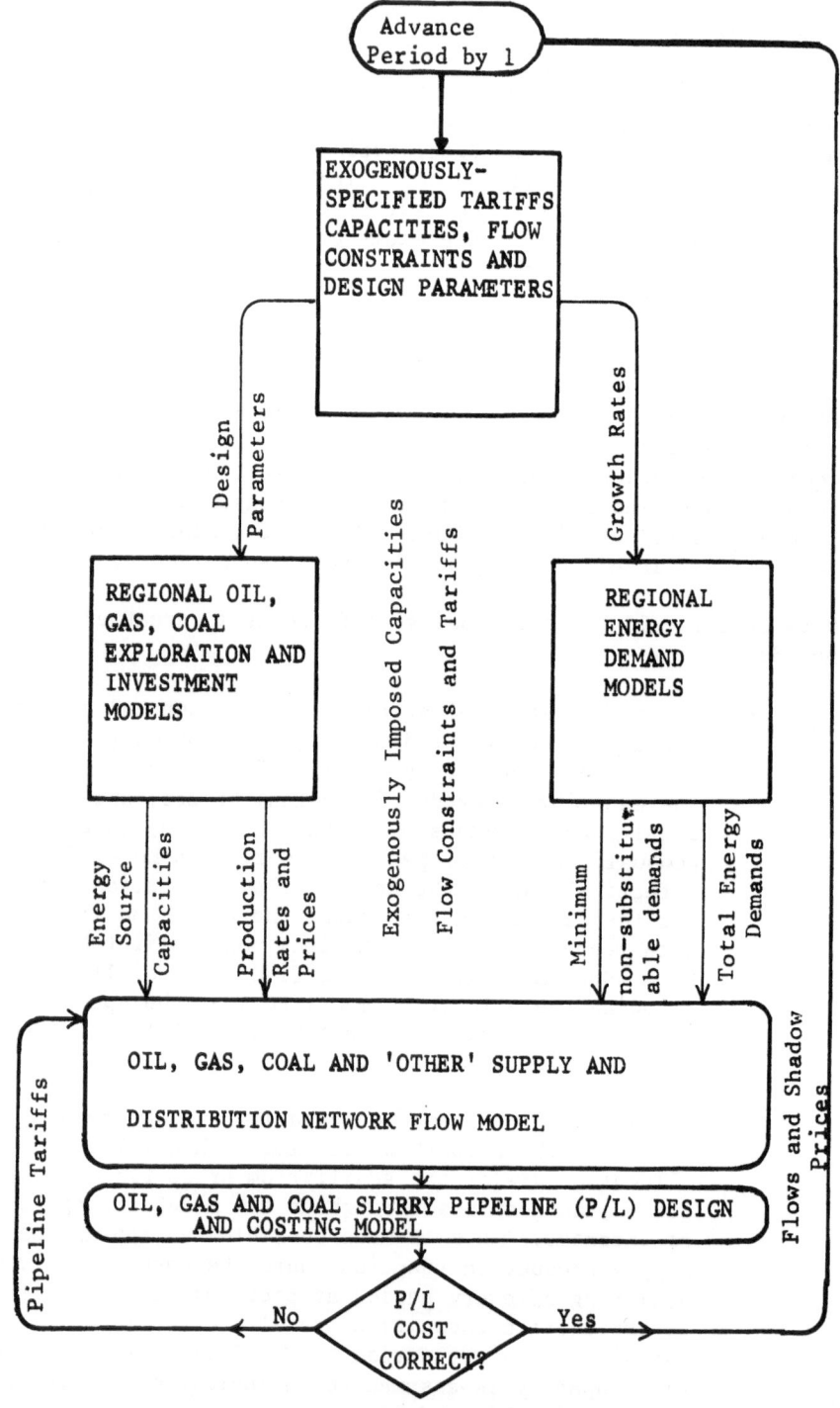

FIGURE 1.a - Flow chart and information flows of model II

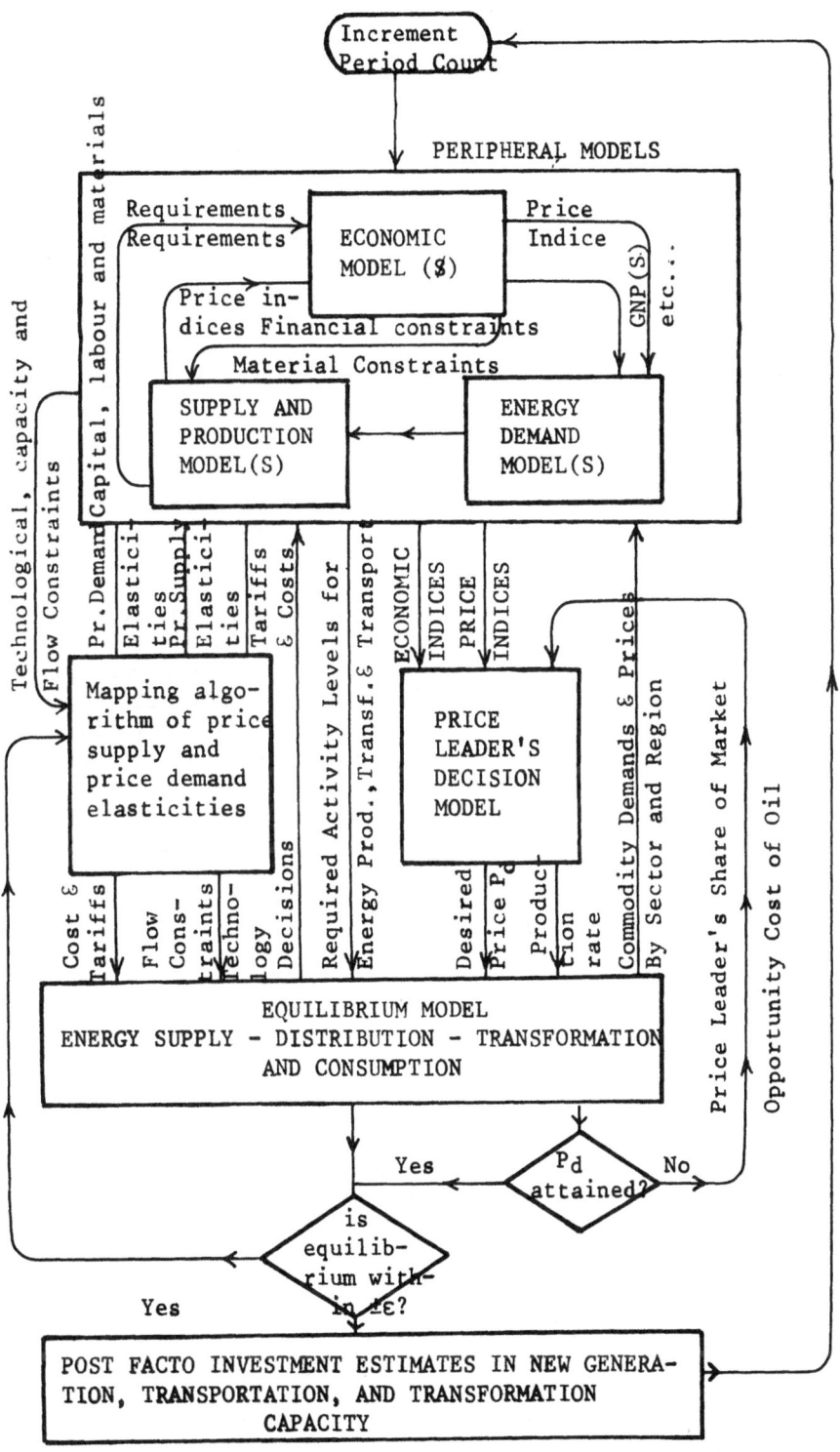

FIGURE 1.b – Flow chart and information flows of model III

- Determination of "equilibrium" investments, prices, and demands in the network flow allocation provides feedback information to the finance, resources, and demand models in order to determine if the capital, labour, material, and equipment requirements consistent with the investments prescribed by the equilibrium network flow solution can be implemented. If, for example, the labour and capital requirements exceed the availability of these resources as determined by the peripheral models, their energy prices must be increased in order to reduce demand for energy and reduce investments in new energy production, transmission, or transformation capacity in the next equilibrium network solution. As illustrated in Fig. 1b for Model III, new price levels are transmitted to the network allocation model as region specific economic rents and import duties which affect demand(s) for energy and related investments.

- Model II does not feature information feedback from the network allocation model to the peripheral models, as shown in Fig.1a. This is justified within the context of this less elaborate model because investments in energy producing capacity are determined outside the network allocation model. More specifically, investments in energy generating capacity, i.e. hydrocarbon reserves and production capacity are estimated at the end of every period; and the resulting changes in energy production capacity become effective for the following period. Any constraints on capital, material or labour could be taken into account in these investment models which are "peripheral" with respect to the supply-distribution network.

2. NETWORK MAPPING OF SUPPLY AND DEMAND

2.1 Network Mapping of Supply

The mapping of commodity price-supply elasticity curves SS' onto a capacitated network is straightforward if the price-supply function is convex. A step function approximation of the SS' curve can readily be represented as a set of arcs having a common origin i and a common destination j as in Fig.2.

A cost minimizing network flow algorithm seeks to activate the least cost (i,j) arc first and, if the upper limit to flow, H_{ij} is reached, the arc $(i,j)_2$ with a unit cost C_2 would be activated, and so forth until a general equilibrium or minimum cost solution is obtained in the supply distribution network.

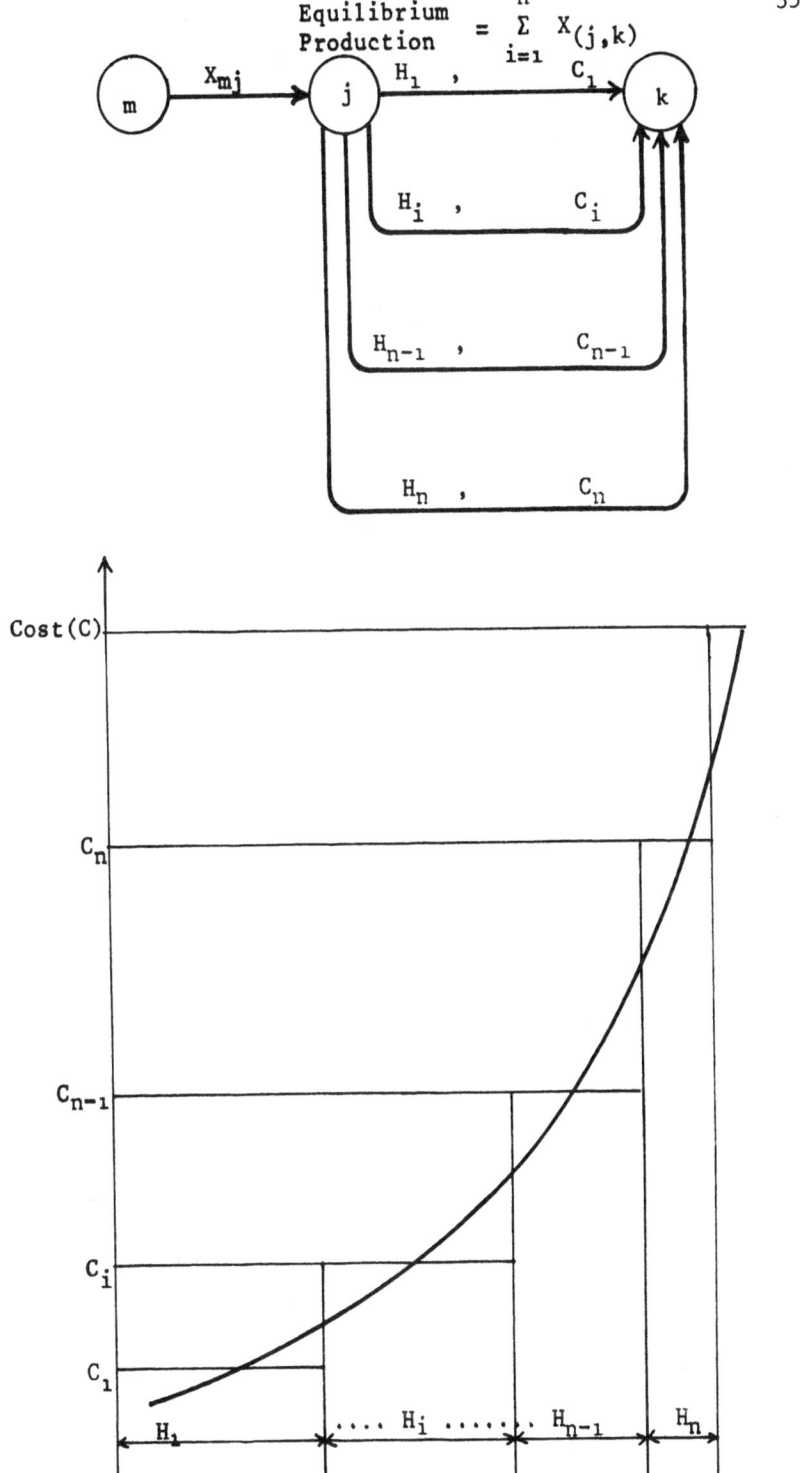

FIGURE 2 - Network mapping of a convex price supply elasticity
function

The price-supply elasticity curve mapping, illustrated in Fig. 2, is not explicitly featured in the supply-distribution network of Model II. Instead, an investment model in new hydrocarbon reserves together with an exploration function to estimate the expected cost of new reserves and a production function determines the production capacities for oil and gas at all production centers represented in the model at the end of every period. These capacities are in turn specified in the supply-distribution network for the optimal allocation of the next period in Model II. Similar exploration and production functions are used in Model III, albeit in a different way, since costs versus capacities are mapped onto the supply-distribution network as illustrated in Fig.2.

Other forms of energy supply are treated in an analogous manner using costing functions that estimate the expected cost of new production capacity for the commodity of interest and a production function that determines unit costs of production as a function of capacity, i.e. a price-supply elasticity curve for the commodity of interest. These curves are discretized for representation in the supply-distribution network, as in Fig.2.

2.2 Market Equilibrium

Before describing the representation of price-demand elasticities in a cost minimizing network [5] and the solution for equilibrium prices, production levels, and consumption rates in the supply-distribution network, it would be useful to recall here the assumptions underlying the purely competitive market equilibrium concept and the rationale behind the notion of supply shortage or surplus and consumer's utility surplus.

Fig. 3 shows the price demand elasticity curve DD' and the price supply elasticity curve SS' for a given commodity C. It is assumed that SS' and DD' represent the respective behaviours of suppliers and consumers of commodity C and that consumers and suppliers operate independently from each other. If the price of the commodity were higher than P_e, the equilibrium price, more commodity would be produced according to SS' than the market could absorb according to DD', thereby creating a commodity surplus, driving the price down which would in turn cause some marginal production to be shut down (i.e. reducing production rates according to SS' and increasing demand according to DD'). This trend would continue until P_e is reached when supply X_e would equal demand and the marginal cost of the last unit produced would be equal to the market clearing price of commodity C.

If P is lower than P_e, then given the supplier and consumer behaviour curves SS' and DD', a commodity shortage would prevail which would drive the price upwards thereby stimulating production and reducing demand until equilibrium is re-established at e.

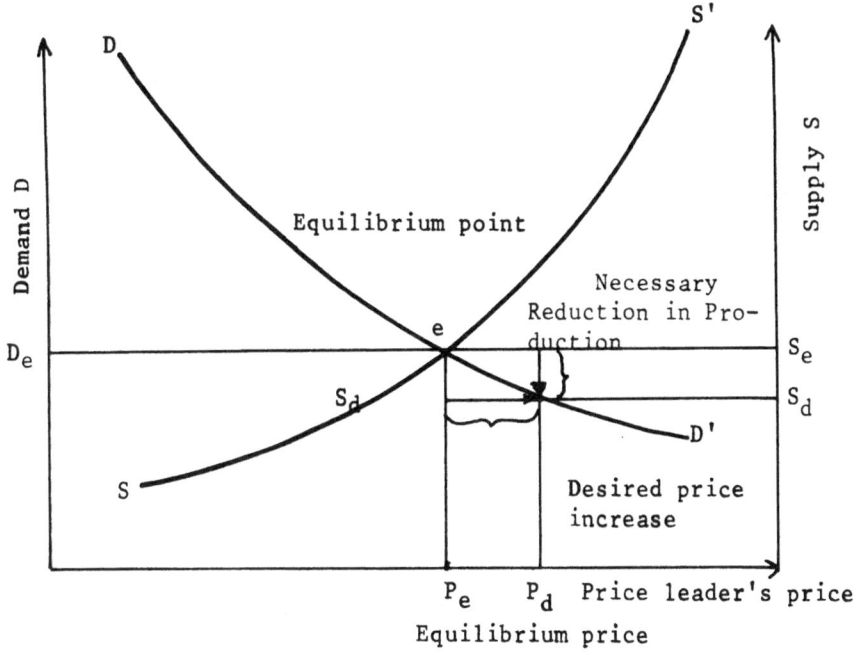

FIGURE 3 - Supply-demand equilibrium and price leader's
control of supply and price

The above discussion of the market equilibrium mechanism leads to
the case of the price leader of group of producers seeking to control price. In order that price leadership be effective, it is necessary that this producer be able and willing to eliminate any commodity surplus corresponding to his desired commodity price by restricting his own production. The higher the desired commodity
price P_d, the larger the (price leader's) production cutback in
order to achieve market equilibrium at P_d, as illustrated in Fig. 3.
The supply curve SS' is changed to SS^d by the price leader who cuts
back production by X in order to achieve or maintain a desired
price level P_d.

2.3   Network Mapping of Demand

The representation of price-demand elasticity curves in a supply-
distribution network was first reported in 1975 (3,4). This method consists in mapping monotonically decreasing demand versus
price curves onto a standard out-of-kilter cost minimizing network
at any desired level of numerical accuracy, by appropriate interval

choice in the step function approximating the curve. While a quadratic network optimization algorithm for equilibrium single commodity trade flows was recently reported in 1978 [10], the mapping technique is considered more appropriate for regional energy modelling because it allows the representation of as many products, sectors, and regions as warranted by the problem at hand, at any desired level of numerical accuracy.

This network mapping procedure is illustrated in Fig.4 in the case of a price-demand elasticity curve DD' featuring a minimum non-substitutable demand, D_{min}. The procedure consists first in setting the reference demand D^* corresponding to a very low (or zero) price, or to the intercept of the demand curve with the ordinate axis. This reference demand D^* is mapped as a LOW constraint node or as an output constraint at node j.

The respective demand differentials or step heights: H_1, H_2,.....H_n of the n step function corresponding to prices P_1, P_2,......,P_n are respectively mapped onto n dummy arcs $(s,j)_i$, originating at the source of the network and terminating at node j. The arcs $(s,j)_1$, $(s,j)_2$,...,$(s,j)_n$ and in general $(s,j)_i$ are respectively assigned a flow capacity H_i and a cost C_i, where $i = 1,2,...,n$. At equilibrium, i.e. if an optimal allocation is obtained, the flows $X_{(s,j)_i}$ in arcs $(s,j)_i$ and the prices P_s and P_j at s and j must satisfy the following duality constraints:

$$X_{(s,j)_i} = 0 \qquad ; \quad Z_{(s,j)_i} > 0$$

$$X_{(s,j)_i} = H_{(s,j)_i} \qquad ; \quad Z_{(s,j)_i} < 0 \qquad (1)$$

$$0 \leq X_{(s,j)_i} \leq H_{(s,j)_i} \qquad ; \quad Z_{(s,j)_i} = 0$$

where the dual variable $Z_{(s,j)_i}$ is defined as:

$$Z_{(s,j)_i} = P_s + C_{(s,j)_i} - P_j. \qquad (2)$$

Given a "market value" $(P_j - P_s)$ at node j relative to a source node value P_s, the dummy arcs $(s,j)_1$, $(s,j)_2$,....,$(s,j)_i$ would be active and saturated with

$$X_{(s,j)_i} = H_{(s,j)_i} \qquad (3)$$

for costs $C_{(s,j)_i} < P_j - P_s$ corresponding to $Z_{(s,j)_i} < 0$.

Dummy arcs are inactive with $X_{(s,j)_i} = 0$ for costs $C_{(s,j)_i} > P_j - P_s$ corresponding to $Z_{(s,j)_i} > 0$. Should an equilibrium dummy arc $(s,j)_e$ exist, i.e. if $C_{(s,j)_e} = P_j - P_s$, corresponding to $Z_{(s,j)_e} = 0$, then

$$0 \le X_{(s,j)_e} \le H_{(s,j)_e}. \tag{4}$$

The sum of flows $X_{(s,j)_i}$ in the saturated or partially active dummy arcs corresponds to the total reduction in real demand at node j, with respect to the reference demand D^* and is equal to

$$D_j^{*'} = \sum_{i=1}^{e} X_{(s,j)_i} \quad ; \quad C_{(s,j)_i} > P_s - P_j. \tag{5}$$

The real demand at node j at equilibrium price $P_j - P_s$ is then equal to:

$$D_j = D_j^* - D_j^{*'}. \tag{6}$$

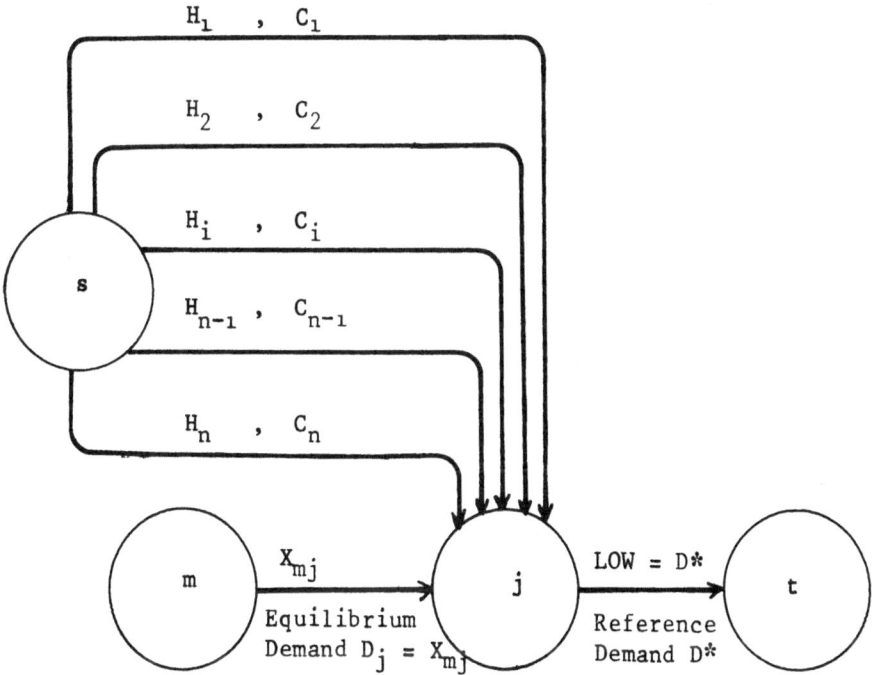

FIGURE 4 - Network mapping of a price-demand elasticity curve

356

Figure 4 continued....

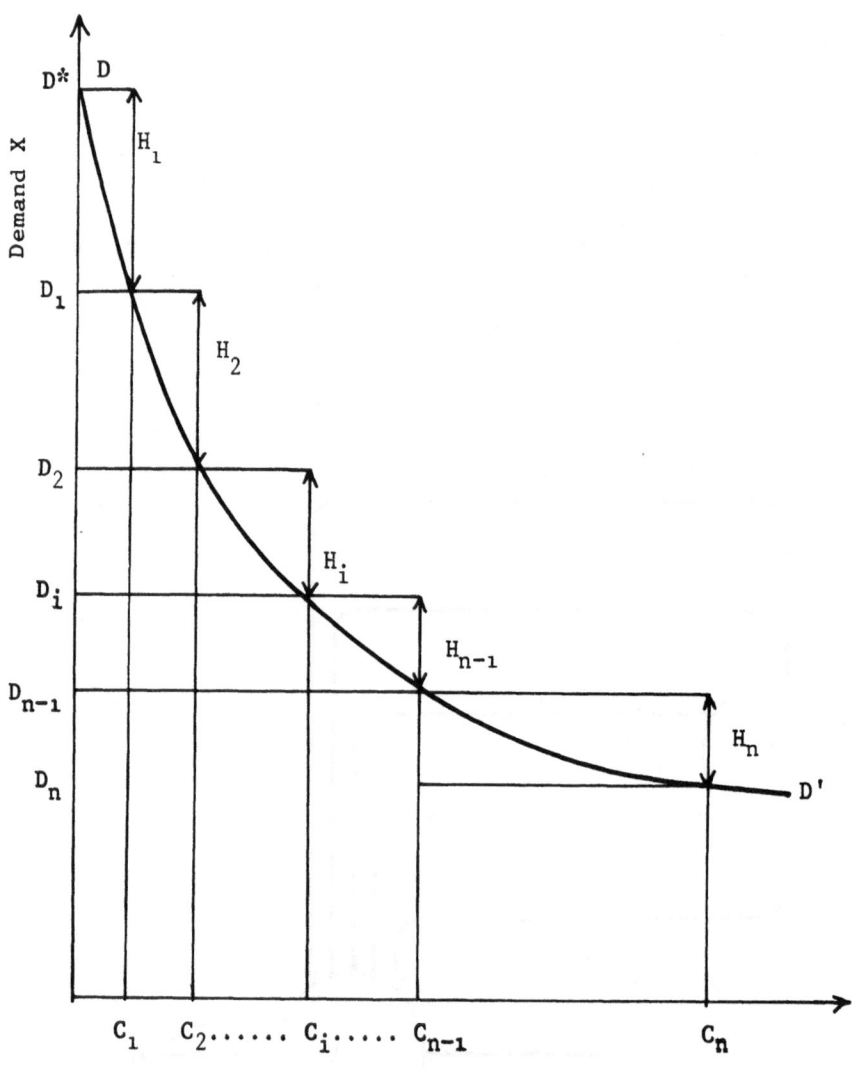

3. ENERGY SUPPLY AND PRICING ANALYSIS WITH GENERALIZED NETWORK MODELS

3.1 The Model II Network

The network representation of a regional energy supply-distribution system in Model II is a standard trans-shipment network summarized in Fig.5. Arcs (i,j), are characterized by upper bounds in the case of oil, gas, and coal production sources and by lower and/or upper bounds to commodity flows for transportation arcs, as well as lower bounds to minimum and nonsubstitutable demands. Note that the single commodity restriction characterizing all standard trans-shipment network algorithms is circumvented as described in [3,4] by expressing all flows in the same energy units, e.g. BTU's or calories and restricting the merging of commodities at demand nodes only. Furthermore, an average energy content is adopted for each commodity, i.e. oil, gas, and coal, while energy losses due to transmission must be neglected or at best indirectly approximated as additional energy demands at certain demand nodes. Energy transformations, e.g. from coal or oil to electric power cannot be represented in Model II.

The costing and capacity setting functions used in Model II and outlined in Fig. 1a are quite detailed as described in [4,7] and feature engineering type design models for pipelines and hydrocarbon exploration and production [5]. These functions determine the unit cost C_{ij} and the upper or lower flow constraints H_{ij}, L_{ij} of every arc (i,j) belonging to the respective arc type. Besides C_{ij}, H_{ij} and L_{ij}, some arcs, notably those representing thermal power generation, are also characterized by a process efficiency E_{ij} equal to the fraction of input energy units which is transformed into electric power units.

3.2 The Model III Network

The supply-distribution network of Model III is inspired by the same regional energy allocation considerations as Model II, but in addition, provision is made for transmission and transformation losses, convex production versus cost functions illustrated in Fig. 2, as well as price-demand elasticity functions illustrated in Fig. 4. In addition, the Model III supply-distribution network features as explicit representation of economic rents which can be used in scenarios featuring a price leader, or interaction with an economic model. Finally, the Model III network provides for discrete investments in energy generation, transportation, and transformation capacity to allow for choice between competing technologies, for example.

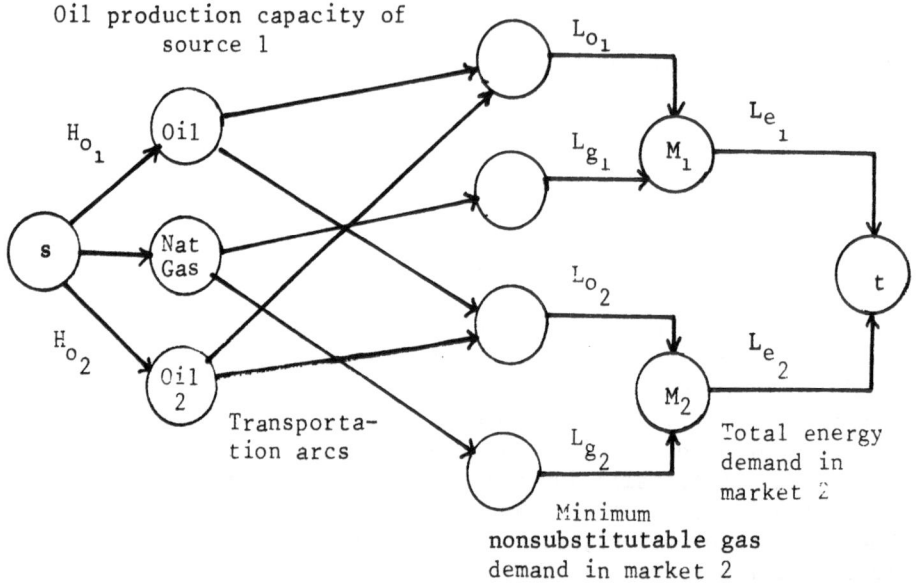

FIGURE 5 - Supply-distribution network of Model II

It may be in order to mention in this evolutionary expose on network based regional energy models that Model IV, which is next in line for development, would be a multi-period optimizing model featuring the same network allocation capability as Model III. Looking further down this evolutionary process a Model V is envisaged which would be a multi-period optimizing version of Model IV, plus the capability of featuring "flow split" processes such as petroleum refining.

3.3 Network Nodes

The supply distribution network N of Models II, III, and eventually Models IV, V feature a number of arcs (i,j), each characterized by an origin node i and a destination node j. Network nodes are characterized by one internal attribute, namely a serial number of arbitrary rank order and one or more external attributes relating every node to the real world problem being modelled, na ely:

- An alphanumeric identification usually refers to a geographical region, e.g. NA, for North America.

- One or more coded attributes specify:

    a. The economic sector, e.g. Industrial, Commercial, Transport, Power Generation, etc....

    b. A functional code, e.g. Demand, Production, Processing, Oil Import, Gas Import, or simply "Trans-shipment".

    c. The characteristic input or output commodity, e.g. the power generation sector may have several commodities as input, but it is characterized by its output commodity, i.e. electric power. Likewise, the oil demand node of any sector is characterized by oil as input commodity. A trans-shipment point would have the same commodity as input and output. Note that all attributes of any node except a flow split node are used for reporting reasons only.

This scheme of external attributes is adequate to describe any node to the user of the model, e.g. NAOIDM 115 specifies that Node number 115 is North America's Industrial Sector Oil Demand node.

3.4   Network Arcs

The arcs of a network represent functions, e.g. production, demand, transportation, transformation or trans-shipment. Moreover, arcs are characterized by one commodity each, e.g. oil, natural gas, coal, electricity, etc.... In addition to the above type attributes, arcs (i,j) are each assigned numerical attributes which may vary in the course of a model "run", namely upper and lower constraints H_{ij} and L_{ij} to flow X_{ij} and a unit cost C_{ij} per unit of flow traversing arc (i,j). In the case of Model III (and eventually Models IV, V) arcs (i,j) have an efficiency or "gain" E_{ij} per unit of flow which is smaller than unity if transmission or transformation losses are incurred proportionately to the flow X_{ij} in arc (i,j). The gain E_{ij} could be larger than unity if unit conversions, e.g. from volumetric to energy units are required and also to model options with fixed costs as described in [5]. In this case arcs must be characterized by an integrality constraint, i.e. that the value of X_{ij} must be an integer, e.g. 0 or 1.

Arc type specification

Arcs are categorized in types corresponding to different updating and reporting treatments as well as different numerical treatment by the network optimization algorithm, or its peripheral functions.

Two attributes are specified in order to determine an arc "type", namely:

    i.   the commodity flowing in the arc e.g. crude oil; and

    ii. the function of the arc e.g. maritime tanker route arc, or production arc, etc....

Commodity types are restricted to oil, gas, coal, and "other" for arcs in Model II; however, the list of allowable commodity types in Model III is open ended, i.e. new commodities can be added by the user to the following list of commodities and their respective mnemonics, by specifying new attribute mnemonics and names in the appropriate attribute and name fields of the arc data record or card image:

    OI   Crude Oil
    NG   Natural Gas
    HC   Hydrocarbons (Oil and Gas)
    SY   Syncrude
    CL   Coal
    EL   Electric Power
    UR   Uranium
    FF   Fossil Fuels

Arc Function Specification

Arc function attributes are restricted to six in Model II, namely, production, pipeline, tanker, distribution, demand, import quota, and dummy arcs. The repertoire of arc functions is open ended in Model III and currently includes the following:

    ER   Economic Rent
    MT   Maritime Tanker
    MC   Maritime Cargo
    ML   Maritime LNG
    OR   Overland Rail
    PL   Pipeline
    TH   Thermal Power Generation
    NU   Nuclear Power Plants (conventional)
    RN   Renewable Power Generation
    EX   Extraction (of depletable resources)
    DM   Demand
    DS   Distribution
    TR   Trans-shipment
    PR   Preparation
    TO   Technological Option

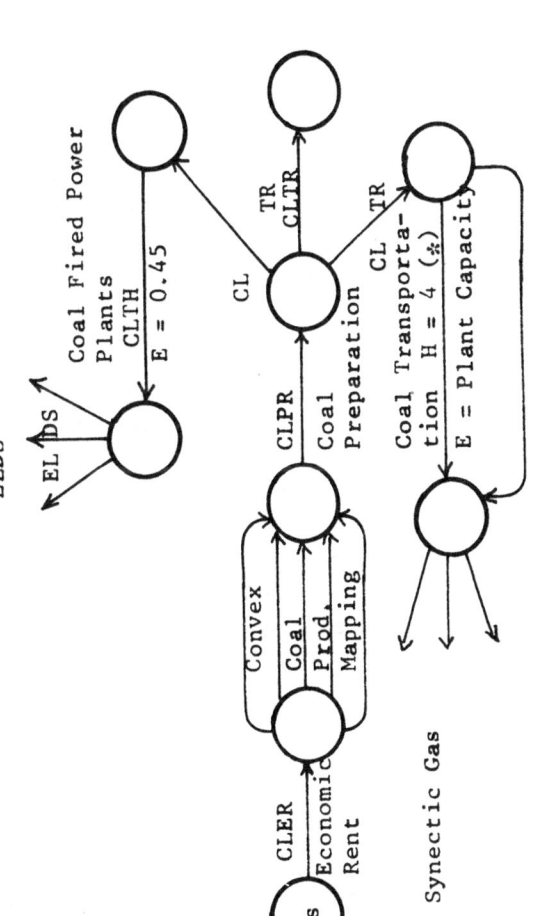

FIGURE 6 - Network representation of coal extraction, preparation transportation and transformation

New functions could be introduced by the user of Model III and assigned to certain arcs, provided that appropriate subprograms be also supplied by the user in order to update the parameters of such arcs.

The following arc functions warrant further explanation:

- The Trans-shipment (TR) function is a general purpose arc which can be used to model a variety of situations, e.g. overseas and/or overland oil import quotas and import tariffs, etc.

- The preparation (PR) function takes into account commodity preparation costs, e.g. coal preparation prior to combustion in a thermal power generation (TH) arc, as illustrated in Fig. 6.

- Efficiencies E_{ij} can be specified in Model III for process arcs (i,j) characterized by the loss of a fraction $(1 - E_{ij})$ of the input commodity X_{ij} into the process. For example, a thermal power generation arc (TH) may be assigned a process efficiency E_{ij} of 0.45 because only 45% of the input energy is transformed into electric power output.

A limitation of Models II and III is that the "splitting" of a flow stream into component commodities is not allowed. For instance, a hydrocarbon (HC) arc requiring a "flow split" into a crude oil (OI) arc and a natural gas (NG) arc in the proportions:

$$0.4X_{jh} + 0.6X_{jm} = X_{ij}$$

as illustrated in Fig.7, cannot be specified. Note that the "flow split" process may feature more than two commodity outputs provided that the sum of split fractions F_{jk} be equal to unity:

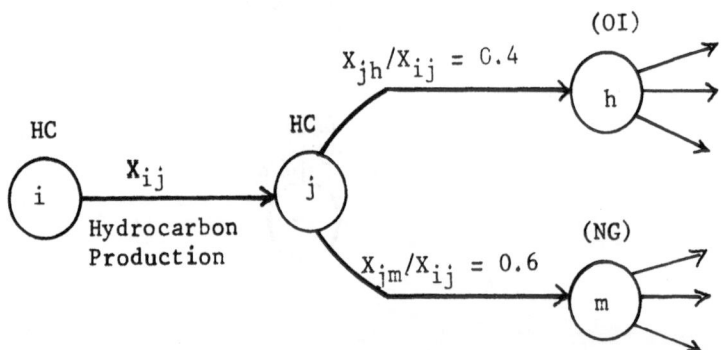

FIGURE 7 - Representation of a flow split of hydrocarbons into oil and natural gas according to a 4/6 gas/oil ratio.

This "flow split" feature is to be incorporated in Model V, pending the development of an appropriate network optimization algorithm capable of taking into account "flow split" constraints.

4. INVESTMENTS IN NEW CAPACITY

Capacity expansions are treated in two different manners in Model II, namely: either within the optimal network flow allocation, or outside this allocation as shown in Fig. 1a. All capacity expansions are determined by the optimal network allocation in Model III, as shown in Fig. 1b. In both models, the investments corresponding to the schedule of capacity expansions are estimated outside the network allocation, except for fixed charge investments treated as integer variables within the supply-distribution network of Model III.

4.1 Internal Capacity Expansion in Model II

Capacity Expansion for pipeline arcs is determined in Model II by relaxing the capacity constraints on the arcs of interest, then solving for the optimal allocation $[X_{ij}]$, then correcting the corresponding arc cost $C_{ij} = f(X_{ij})$ and iterating on arc costs as illustrated in Fig. 1a. The maximum production capacities of oil and gas producing regions are set in Model II at the beginning of the period of interest by appropriate hydrocarbon exploration, investment, and production models, as indicated in Fig. 1a.

4.2 Internal Capacity Expansion in Model III

The capacity expansion treatment in Model II differs from Model III in the following respect: instead of keeping track outside the network of installed capacity H_{ij} of any expandable arc (i,j), then relaxing this capacity, H_{ij} is maintained on arc (i,j), and the allowable expansion capacity H'_{ij} is assigned to expansion arc $(i,j)'$ having the same origin i and the same destination j as arc (i,j). The allowable capacity expansion H'_{ij} in Model III is performed in three different manners, depending on the nature of the cost function that determines the unit cost C'_{ij}.

If C_{ij} is constant and equal to C'_{ij}, the unit cost of "flow" or activity X_{ij} for installed capacity, then no shadow arc $(i,j)'$ is needed, and it suffices to set H_{ij} sufficiently high in order to obtain the optimal capacity expansion of arc (i,j) from the equilibrium solution for X_{ij}.

If C_{ij} is a concave cost function, i.e. if a fixed charge is required to appropriately represent a discrete capacity expansion

of facility (i,j), then arc (i,j)' is treated as an integer variable. The topic of options with fixed charges is discussed in more
detail in section 4.3.

If C_{ij}, hence C'_{ij} is a convex cost function, Model III takes advantage of the convex production versus cost mapping of production
functions illustrated in Fig. 3 to directly solve for the optimal
expansion of energy producing facilities. This method has the
advantage of not requiring iteration on cost, since the convex
cost versus production capacity function is incorporated in the
supply distribution network for production of all depletable energy resources such as oil, gas, uranium, and coal. This is because
the unit cost to find, develop, and produce such resources increases as potential new sources of supply are depleted. Moreover,
renewable resources may also be represented by a convex production
function since the most economical prospects, e.g. hydro, are developed first, while the more expensive prospects are relegated
for future development, should the opportunity cost of such energy
resources be sufficiently high.

Given a convex production versus cost mapping as illustrated in
Fig.2

$$[(j,k)_i]_{i=1}^{i=n} = (j,k)_1, (j,k)_2, \ldots, (j,k)_n \tag{7}$$

characterized by arc costs

$$C_{(k,j)_i} < C_{(k,j)_{i+1}} \tag{8}$$

and given equilibrium node prices P_j and P_k at the common origin
node j and destination node k of the production function
$[(j,k)_i]_{i=1}^{i=n}$, then any arc $(j,k)_i$ would be saturated, i.e.
$X_{(j,k)_i} = H_{(j,k)_i}$ if its dual variable $Z_{(j,k)_i} < 0$, hence if:

$$C_{(j,k)_i} < P_k - P_j. \tag{9}$$

Moreover, an equilibrium arc $(j,k)_e$ may exist such that

$$C_{(j,k)_e} = P_k - P_j \tag{10}$$

in which

$$L_{(j,k)_e} \leq X_{(j,k)_e} \leq H_{(j,k)_e}. \tag{11}$$

The sum of flows $\sum_{i=1}^{e} X_{(j,k)_i}$ in production function $(j,k)_i$ represents the net total installed capacity available at node k in order to satisfy the equilibrium (i.e. optimality) conditions of the network allocation.

At the beginning of every period, the first arc $(j,k)_1$ of every production function is assigned a cost $C_{(j,k)_1}$ equal to the variable or unit production cost for installed capacity at the end of the last period. The capacity $H_{(j,k)_1}$ of this arc is set equal to the net effective installed capacity at the beginning of the period (or the end of the previous period). The successively more expensive arcs of the production function $[(j,k)_i]$ are respectively assigned a cost:

$$C_{(j,k)_i} = K \frac{d(I)}{d(H)_i} + C_{(j,k)_1} \tag{12}$$

where $d(I)/d(H)_i$ = The marginal cost of new capacity at the cumulative capacity H_i defined as:

$$H_i = \sum_{i=1}^{i} H_{(j,k)_i} \tag{13}$$

I = The cumulative investment function characterizing the particulat energy source represented by $[(j,k)_i]$

$C_{(j,k)_1}$ = The variable or unit production cost assigned to arc $(j,k)_1$

K = A cost of service factor defined as that fraction of the investment I corresponding to the unit cost of the energy commodity delivered at node k, not including variable costs (i.e. operating cost and fuel).

This cost of service factor K is defined as follows: Knowing the unit cost $C_{(j,k)_i}$ assigned to the expansion arc $(j,k)_i$. This unit cost of service is defined as:

$$K = A_i + rI \tag{14}$$

where I = The marginal investment per unit of increased capacity at node K.

A_i= The annuity required to retire an investment I per unit of energy generating capacity over the expected life of the facility.

r = The rate or return on the unit rate base, in this case investment I. This rate of return would be sufficient to cover overheads and operating costs per unit of energy output, not including fuel costs, e.g. uranium or fossil fuels required in the energy generation process, since fuel costs are taken into account by the supply-distribution network optimization algorithm which features transmission and transformation efficients E_{ij}.

Calculation of incremental investments

These investments are calculated at the end of every period after an equilibrium network allocation satisfying all market, capital, and physical constraints is achieved.

Calling $H_{jk,t}$ and $H_{jk,t-1}$ the installed capacities at production source k at the end of period t and t-1, the gross capacity added during period t to achieve market equilibrium is:

$$\Delta \overline{H}_{jk,t} = (H_{jk,t} - H_{jk,t-1})/LF \tag{15}$$

where LF = the average load factor characterizing energy source K.

Having determined the incremental increase in gross capacity $\Delta \overline{H}_{jk}$ of every "expandable" activity in the network, the corresponding incremental investments can be directly estimated by integrating the appropriate investment function I($\overline{H}$) between $\overline{H}_{(j,k)_{t-1}}$ and $\overline{H}_{(j,k)_t}$. Typical investment functions for depletable and renewable energy sources are presented in Section 5.

4.3 Investment Options with Fixed Charges

Besides continuous capacity expansion functions such as existing pipeline links or hydrocarbon producing regions, certain investments are of the "go" or "no go" type and are represented in Model III by integer variables. Choices between technological options or certain projects, e.g. the linking of two regions by a new pipeline, fall in this category.

Fig. 8 illustrates three mutually exclusive energy generating options: (jk), (jℓ), (jm) respectively characterized by investments or fixed costs I_{jk}, $I_{jℓ}$, I_{jm} corresponding to new installed capacities H_{jk}, $H_{jℓ}$, H_{jm}. These potential capacities are represented by "gain" coefficients E_{jk}, $E_{jℓ}$ and E_{jm} respectively assigned to (j,k), (j,ℓ) and (j,m).

Note that all variables (i,j), (j,k), (j,ℓ) and (j,m) in Fig. 8 must be declared integer, in this case 0 or 1, which is indicated

by a (*) on the arcs in Fig. 8, the same network notation convention as in [11]. In addition, however, it is possible to specify
in Model III whether the integer variable representing the investment option must be fully utilized if the option is selected, e.g.
a tar sand plant with a guaranteed market, or whether the new installed capacity may be fully or partially utilized, depending on
equilibrium (i.e. market) conditions.

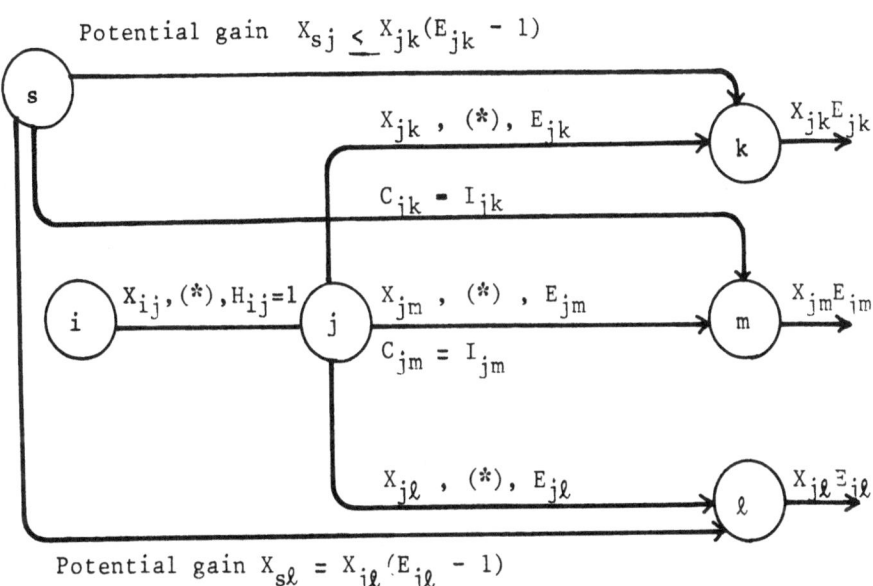

FIGURE 8 - Representation of three mutually exclusive
options in energy generating capacity

Note that the energy generation options illustrated in Fig. 9
imply net energy inputs into the system which originate at s, the
network source. This is the case of a fusion energy project, for
example.

Arcs (j,k) representing fixed charge investment options with positive energy inputs into the system have a dual variable Z_{jk} defined as follows:

$$Z_{jk} = P_j + I_{jk} - X_k P_k + (X_k - 1)P_s \tag{16}$$

where P_s is the node price at the source node s of the network and

X_k is the commodity output at node k. If the selected source output is guaranteed, as in the case of a tar sand plant, then the appropriate dual variable would be obtained by replacing X_k by E_{jk}, the "gain" of option (j,k).

Another category of fixed charge activities, e.g. investments in energy transformation capacity such as coal gasification or coal liquefaction plants which are characterized by a net loss in primary energy can also be represented in Model III. The representation of these variables in the supply–distribution network is illustrated in Fig. 9. The treatment of such variables differs from that of energy generating options illustrated in Fig. 8 by the definition of their dual variable:

$$Z_{jk} = P_j + I_{jk} - X_k P_k \qquad (17)$$

where X_k is the total commodity output at node k. The respective values of Z_{jk} as defined in Eqs. (16) or (17) are used to satisfy the optimality conditions (see Eq. 1) in the generalized primal–dual network algorithm developed for Model III. Replacing X_k by E_{jk} results in the full utilization of (j,k) if this expansion is retained in the equilibrium solution.

The generalized network optimization algorithm featuring the two types of integer variables respectively characterized by dual variables Z_{ij} defined in Eqs. (16) and (17) will be described in a forthcoming paper [8].

In summary, some integer programming capability is provided in Model III, and eventually in subsequent versions of this model (i.e. Model IV and Model V) in order to take into account certain types of decisions characterized by concave cost production functions e.g. a new tar sand plant, or energy transformation functions, e.g. a coal liquefaction plant, with or without a guaranteed market.

5. INTEGRATING NORMATIVE AND DESCRIPTIVE ALGORITHMS WITHIN A NETWORK BASED ENERGY MODEL

Most production functions in Model III are convex, notably those representing depletable resources and also the renewable energy resources such as hydro, where the most economical prospects are developed first. The following is a description of these two categories of functions.

5.1 Depletable Resources

Depletable resources such as oil, gas, coal and uranium oxide are currently represented in Model III by a relationship between cumulative investments I_r and cumulative proven reserves r in which I_r tends towards infinity as cumulative proven reserves r approach R, the ultimate "provable" reserves of the commodity of interest in the region of interest. The first version of such a depletable resource production function is a probabilistic model with replacement, described in [2] which was used for oil in Model I. In this model, the sample space remains constant. The improvements in geophysics brought about by the advent of the electronic computer in the 1960's led to a more appropriate model without replacement, in which the explored portion of the basin of interest is removed from the sample space. This model described in [4] was used in Model II for hydrocarbons.

Since then, a probabilistic learning model without replacement featuring learning and saturation effects was developed [6]. This model is currently being generalized in order to take into account technological breakthroughs in exploration or exploitation of hydrocarbons.

In the meanwhile, a modified version of the probabilistic model described in [4] has been incorporated in Model III and is described as follows:

$$I_r = C_1 (R - r)^{-C_2} \qquad (18)$$

where C_1 and C_2 are positive constants,

> R = ultimate "provable", reserves of the commodity of interest in the region of interest
>
> r = cumulative proven reserves
>
> C_3 = an integration constant equal to:

$$C_3 = -C_1 R^{-C_2} \qquad (19)$$

corresponding to the boundary condition $I_o = r = 0$.

By "proven" and "provable" reserves it is here meant reserves that have been discovered, geologically delineated and developed for production. Accordingly, the marginal cost $d(I_r)/d(r)$ or a new unit of proven reserves referred to in Eq. (12) is distinct from the unit cost of production $C_{(j,k)_1}$.

This marginal cost of new proven reserves r as a function of r is readily obtained by differentiating Eq. (18) with respect to r:

$$d(I)/d(r) = C_1 + C_2/(R - r)^{-(C_2 + 1)}. \tag{20}$$

Given an equilibrium network solution in which the i'th arc $(j,k)_i$ of a convex production mapping of source k is active and the $(i + 1)$'th arc inactive, the equilibrium cost $C_{(j,k)_i}$ on arc $(j,k)_i$ determines the marginal cost of the last increment of reserves proven at source k. This marginal cost can be obtained from Eq. (12), knowing the unit production cost $C_{(j,k)_1}$ and the cost of service factor K, provided that a relation between the utilized capacity H_i and the cumulative proven reserves r can be established.

5.2 The Production Life Index (PLI)

This relation is given by the Production Life Index or PLI, expressed in periods (i.e. time) and defined as:

$$PLI = \frac{r - Q}{H} \tag{21}$$

where Q = cumulative production

$r - Q$ = net remaining proven reserves

H_i = cumulative utilized capacity also equal to

$$\sum_{i=1}^{n} X_{(j,k)_i}.$$

In turn, the $PLI_{k,t}$ for a given producing region k is empirically calibrated as a function of time t as described in [4]; we have:

$$PLI_{k,t} = C_o + e^{C - Ct}. \tag{22}$$

Since the PLI is uniquely a function of time for a given region, it is fixed for period t, which allows the substitution of $(r - Q)/PLI$ to H_i in Eq. (12). Note that H_i represents the cumulative used capacity in the network mapping of source k as defined in Eq. (13). Having solved for the cumulative proven reserves r_t corresponding to the equilibrium network solution at the end of period t and knowing the cumulative reserves r_{t-1} at the end of the previous period, then the investments during period t to prove up incremental reserves equal to

$$\Delta r_t = r_t - r_{t-1}$$

is given by:

$$I_{r_t} - I_{r_{t-1}} = C_1[(R - r_t)^{-C_2} - (R - r_{t-1})^{-C_2}] \tag{23}$$

It is important to note that the above determination of equilibrium production H_i, the corresponding cumulative reserves r_t and investments I_{r_t} hinges on the use of the PLI to relate reserves, cumulative production, and production capacity. The definition of the PLI in Eq. (22) is adequate for medium to long term scenarios as the PLI of a producing region does decline over the long term, as the region "matures", until a floor PLI or lower asymptotic value of the PLI equal to C_0 is reached. North America's aggregate oil PLI for example has been close to its floor PLI of 11 years for the last decade as North America is on the aggregate a "mature" oil producing region.

On the other hand, OPEC countries had an aggregate oil PLI exceeding 39 years in 1977 which is typical of a "young" producing region. This PLI is expected to decline over time as the OPEC producing regions "mature", i.e. approach exhaustion in terms of potential new reserves. Yet, a sudden decline in production corresponding to a quantum increase in the PLI, as was the case in 1979 for Iran, cannot adequately be accounted for unless new coefficients C_0, C_1 and C_3 of Eq. (24) in Model III are exogenously modified for the region of interest. A possible improvement of the PLI equation to be incorporated in later versions of Model III and subsequent models is to take into account the latest PLI as well as the latest net recoverable reserves $(r - Q)$ such as:

$$PLI_t = PLI_{t-1} \; e^{C_0(r_{t-1} - Q_{t-1})} \; e^{C_1 - C_2 t}. \tag{24}$$

5.3   Production Functions for Renewable Energy

The convex cost function for renewable energy adopted in Model III relates cumulative investments I with production capacity H as follows:

$$I = C_1(e^{C_2 \overline{H}} - 1) \tag{25}$$

where C_1 and C_2 are calibration constants for the resource or aggregate group of resources represented by this production function. H is the cumulative installed energy generating capacity of this resource or group of resources.

372

The marginal cost of new energy generating capacity is readily
obtained by differentiating Eq. (25) with respect to H and is
expressed as follows:

$$d(I)/d(\overline{H}) = C_1 C_2 \, e^{C_2 \overline{H}}. \tag{26}$$

The corresponding cost of service per unit of energy produced is
given by Eq. (12), knowing the cost of service factor K and the
variable or operating cost $C_{(j,k)_1}$.

Given an increase in capacity from $\overline{H}_{t-1}$ at the end of the (t-1) th
period to $\overline{H}_t$ at the end of the t th period, the incremental investment is expressed as follows:

$$\Delta I_t = I_t - I_{t-1} = C_1 (e^{C_2 \overline{H}_t} - e^{C_2 \overline{H}_{t-1}}). \tag{27}$$

The new value of the installed capacity at the end of the t th
period is given by:

$$\overline{H}_t = MAX(H_{t-1}, \, X_{ij})/LF \tag{28}$$

where LF = The load factor of energy generating source k

X_{ij} = The activity in the arc incident to node j,

the common origin of the n $(j,k)_i$ arcs making up the convex cost
production function at region or node k. Because of flow conservation around the nodes, we have from Fig. 2:

$$X_{ij} = \sum_{i=1}^{n} X_{(j,k)_i} = H_t. \tag{29}$$

5.4   The Price Leader's Supply Function

Given a price P_d desired by the price leader, where P_d is larger
than the equilibrium price P_e in the market place, the price leader
must reduce his share of the production by ΔS as illustrated in
Fig. 3 in order that a new supply-demand equilibrium be attained
at the desired commodity price P_d. If the production curtailment
ΔS is unacceptable to the price leader, this latter must reduce P_d.
An iterative scheme is built into Model III to achieve supply-
demand equilibrium at price P_d set by the price leader, subject to
a minimum production constraint that may also be imposed by the
leader.

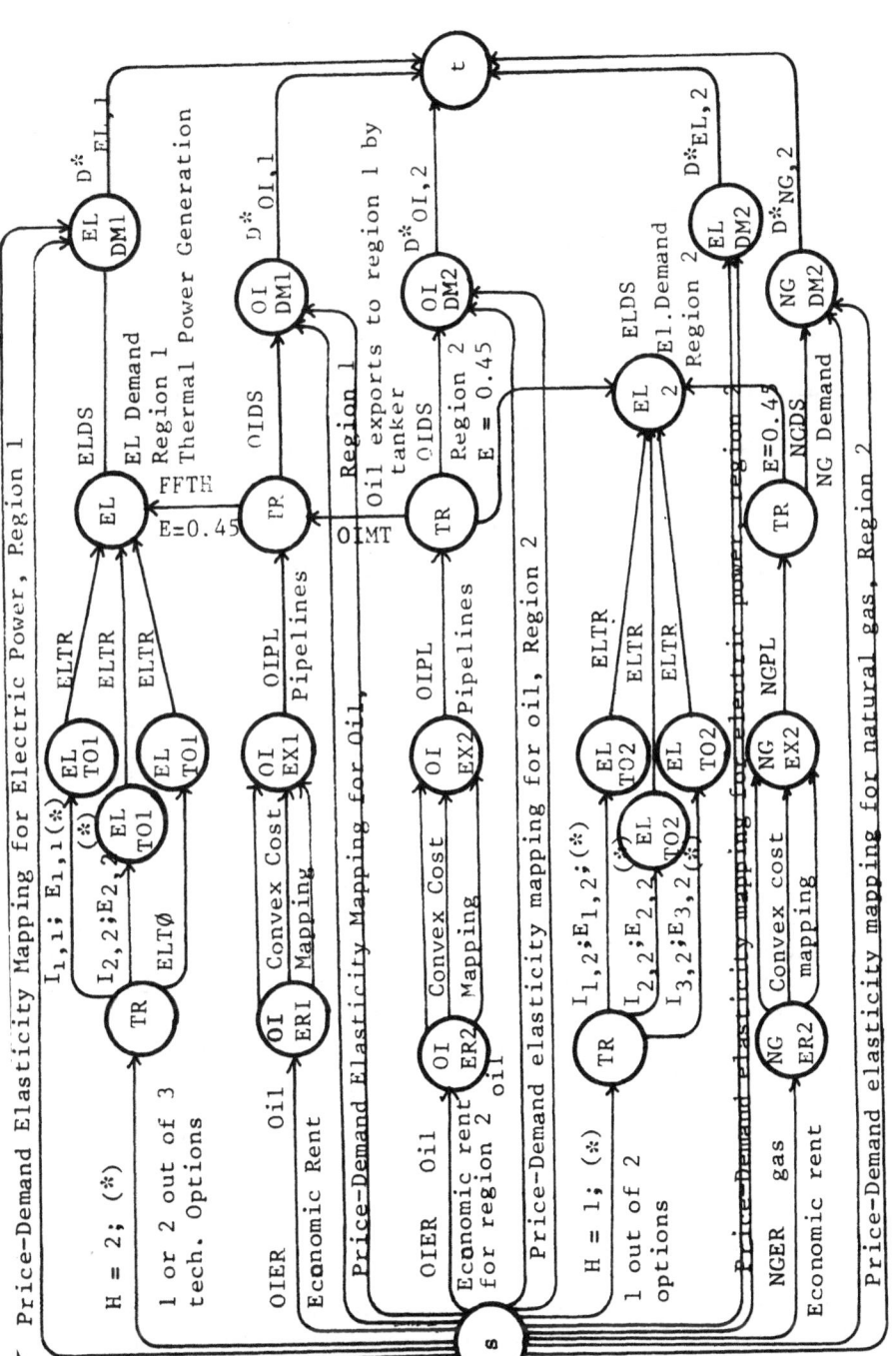

FIGURE 9 - Two region, three commodity supply-demand equilibrium representation in Model III

6.   SUMMARY AND CONCLUSIONS

This expose on increasingly comprehensive network based regional
energy supply and distribution models highlights the congruence
between modelling comprehensiveness with network flow optimization
algorithms and the advancement of the state of the art in two
network related fields: Increasingly more powerful and more
"general" network optimization algorithms, and, new network mapping
constructs to model particular characteristics of supply and distribution systems, e.g. price-demand elasticities.

Implementation of large scale network based models as handy and
efficient planning and decision aids requires progress in yet
another domain, namely in the design of appropriate network generation, initialization, updating, interfacing, and reporting
modules; in short, progress in network modelling infrastructure
appears at this time as the bottleneck or weak link in the successful implementation of large scale network based energy models.

Some of these aspects e.g. the design of an appropriate arc and
node attribute convention or the interfacing of peripheral models
e.g. the highly nonlinear investment models featured in an energy
supply-distribution network were discussed in some detail in this
paper to highlight the importance of infrastructure in network
modelling. Another aspect of a network model infrastructure is
the crucial problem of network initialization which was not discussed in this paper as this cannot be done without a detailed
discussion of every variable represented in the model, a topic
which would be more appropriately treated in the Reference Manual
of Model III.

This paper ends with an illustration in Fig. 9 of a simple, yet
typical energy supply distribution problem using Model III. This
two region problem features two depletable energy commodities produced in every region, plus electric power which is generated by
energy transformation or from renewable energy sources treated as
a technological option. The nomenclature of commodities and arc
functions presented in 3.4 is adopted in Fig. 9. Note the use of
integer variables (*) to represent technological options characterized by investments I and production capacities E.

ACKNOWLEDGEMENTS

In as much as this paper covers a succession of energy models
acknowledgement is due to the various sponsors of this research,
culminating in the development of the current version of Model III,
namely: The Transportation Systems Center of the US Department
of Transportation, The Canada Council for the Social Sciences,
Canada's National Research Council, The Organization of Petroleum
Exporting Countries, and Dow Chemical Inc. (US).

REFERENCES

1. R.E. Brooks, "GASNET: A Mathematical Programming Model for the Allocation of Natural Gas in the United States", Working Paper No. 34-1976, Graduate School of Business Administration, University of Southern California, Los Angeles, 1976.

2. J.G. Debanne, "A Continental Oil Supply and Distribution Model", J.Pet.Tech., Sept. 1971.

3. J.G. Debanne, "A Systems Approach for Oil and Gas Policy Analysis in North America", Proc. 8th. World Pet. Congr., Moscow, Elsevier Publ., 1971.

4. J.G. Debanne, "A Regional Techno-Economic Energy Supply-Distribution Model for North America", Comp. and O.R., Pergamon, Publ., Vol. 2, pp.153-193, Dec. 1975.

5. J.G. Debanne, "Cybernetics and Large Systems Optimization", Cybernetics and Modelling of Large Systems, Doc.3, Assoc., Int. Cyb., Publ., T.I. Oren (Ed.), Namur, Belgium, 1976, pp.57-86.

6. J.G. Debanne, "Probabilistic Learning Theory of Hydrocarbon Reserve Economics", U. of Ottawa Working Paper 77-16, April 1977, submitted to Canada J. of Petr. Tech.

7. J.G. Debanne, The Optimal Design of Natural Gas Pipelines, U. of O. Working Paper No. 77-28, submitted to Can. J. of Petr. Tech.

8. J.G. Debanne, "A Generalized Primal-Dual Network Optimization Algorithm", paper under preparation.

9. D.R. Fulkerson, An 'out-of-kilter' Method for Minimal Cost Flow Problems , SIAM Journ., No. 1, March 1961.

10. R.C. Glassey, A Quadratic Network Optimization Model for Equilibrium Single Commodity Trade Flows , Math. Progr., 14, North Holland Publ., pp. 98-107, 1978.

11. F.H. Glover, J. Klingman, and J. Stutz, Generalized Networks: A Fundamental Computer-Based Planning Tool , Management Sc., Vol. 24, No. 12, pp. 1209-1230, August 1978.

DISCUSSION

The point was made that the piecewise step approximation of the
demand and supply functions (the supply function was a piecewise
step function in practice anyway) did not guarantee convergence
of the algorithm. Debanne admitted that was a shortfall of the
method, but that he had circumvented the problem by introducing
a parameter for the step length, ε, centering around the previous
equilibrium and rerunning the model. If the model failed to converge, ε was reset and the model run again.

A limitation of Debanne's method was that he could not address
the same problems as Deam tackled in his LP   model as regards
mix of refinery products.

The economic specification in the model was also questioned.
The model assumes that the price leader satisfies residual demand.
If supply is very elastic - for example, in the case of the Saudi
Arabians, the supply curve looks like an inverted L, infinitely
elastic to a point, but inelastic thereafter - the convergence
mechanism breaks down. Debanne replied that the supply elasticities which he operated with were not so high as to create problems.

The model included direct price elasticities, but no cross-price
elasticities. It was suggested that cross-price elasticities
should be taken care of in the same way as in the PIES model
where one used an iterative procedure outside the model. Analysis of the PIES model had shown that 2/3 of the total demand
elasticity arose as a result of interfuel competition. Consideration of cross-price elasticities would thus considerably enhance
the credibility of the model results.

Debanne was asked whether he allowed for negative investments
(write-offs) or depreciation in the periods of his model. He
replied that there was no depreciation in a given period, but
that one could make adjustments between periods to take this
effect into account.

It was suggested to Debanne that his model could be extended from
considering the case of comparative equilibrium as at present to
that of a price leader trying to maximize profit. This could be
achieved by replacing the supply curve in Debanne's equilibrium
model with the price leader's marginal revenue curve. Debanne
agreed that this would be a fine extension of his approach.

Debanne was asked whether his model considered lags in the installation of production capacity. He replied that as the model
was specified at present it did not include this. To consider

lags correctly, one needed a multi-period model. One could, however, pay lip service to construction delays by not allowing more supply than was feasible in the supply curve at any one time. The importance of lags was also a function of period size. At the moment, he operated with 3-year periods.

APPLICATIONS OF MULTIPLE OBJECTIVE LINEAR PROGRAMMING TO
ENERGY MODELLING

Erick C. Duesing

School of Management
University of Scranton
Scranton, Pennsylvania 18510, USA

ABSTRACT. The multiple objective linear programming (MOLP) problem
is described together with some of its actual and potential applications to energy modelling. Relationships between the MOLP
problem and fixed coefficient, finitely generated models of production are explored, and the state of the art of MOLP algorithms
is assessed. The multiple objective framework is currently being
used to perform tradeoff analysis between conflicting social objectives. Other potential applications are foreseen in economic
evaluations of proposed technologies and in constructing submodels
of productive entities in general equilibrium models. In order
that these possible applications can be better understood, some of
the mathematical properties of the MOLP model and their economic
interpretations are discussed.

1. INTRODUCTION AND OVERVIEW

A variety of models have been used to represent the technical,
behavioural, and institutional relationships which describe interactions between economic, social, and environmental aspects of the
production and consumption of energy. One might attempt to classi-
fy such models as computer simulation models, econometric models,
and mathematical programming or optimization models. Each of these
modelling techniques has both proponents and critics, and continuing dialogue between modeller and policy analyst, with occasional
input from scientists and decision makers, has increased our understanding of the strengths and limitations of each technique.

In this lecture we will introduce and discuss a generalization of
such fixed coefficient, finitely generated models of production as

are examplified by linear programming, process analysis, and in-put-output analysis. However it will be pointed out that this generalization does not by itself constitute a model; it fails to be "analytically closed" since it does not model the demand side of the market. We believe that this lack of closure is a virtue rather than a shortcoming. We also believe that future energy models will be increasingly eclectic in design, incorporating ele-ments of econometric, simulation, and optimization modelling with-in a single model. In such hybrid models one may gain the advan-tage of adapting the modelling technique best suited to the con-struction of each submodel. We, of course, believe that our model of production offers some unique advantages for building submodels of productive entities in such models.

Our own perspective on the energy modelling scene is not that of a model builder or user, but instead arises from an interest in seeing how our interests in conceptual issues in economic theory and mathematical programming relate to issues in energy modelling. What began as a rather casual inquiry into the design and economic specification of large scale economic models in several fields has evolved to a more active and focused role. While our personal in-terest in constructing economic models of energy supply is limited, we are predisposed more towards the role of "problem solver" than the role of "critic" in reviewing existing models. As such, we believe that theoretical and algorithmic advances on the multiple objective linear programming (MOLP) problem are creating a set of tools of considerable potential usefulness to energy modellers. It is our purpose here to describe the state of the art (as we view it) of the MOLP problem, and to indicate some of the possibilities which we foresee for the applications of MOLP techniques to energy modelling.

The MOLP problem is the mathematical programming counterpart of the linear activity analysis model of production, which was developed and extensively analyzed by Koopmans [1] shortly after Dantzig's formulation of the linear programming (LP) problem [2]. Although the Koopmans model has strongly influenced the modelling of pro-duction in general equilibrium theory and in the theory of economic growth, these theories are not constructive and generally incor-porate assumptions not satisfied by the Koopmans model itself. The analytic emphasis in the economic theory of production has turned to transformation sets which have unique supporting hyper-planes except at the origin, as smooth transformation sets give rise to supply functions and input demand functions rather than correspondences. The desire for analytical convenience has had the unfortunate consequence of isolating economic theorists from developments in the combinatorial aspects of the theory of convex sets. In particular, developments in the theory of linear inequal-ities and polyhedral convex sets made in the field of mathemati-cal programming have had little impact on those parts of economics where one still finds reference to the Koopmans model.

Unlike the LP problem, for which the simplex algorithm was developed within a few months following the initial formulation of the problem, there are still unresolved computational issues surrounding the MOLP problem. It is likely that lack of effective computational procedures caused interest in the Koopmans model to languish. One computational expedient which has been widely suggested is to constrain the values of all objective functions except one, and to maximize the remaining objective. This and other methods for analyzing the economics of production with the Koopmans model by using LP as a computational tool was termed "process analysis" by Manne and Markowitz [3]. It is perhaps a result of these factors that the term "activity analysis" has become, at least within the mathematical programming community, largely a generic term for "linearly constrained, linear objective function" continuous and discrete optimization problems.

The point to be emphasized from the discussion in the preceding two paragraphs is that the Koopmans linear activity analysis model has contributed to a number of developments in economic theory, mathematical modelling, and mathematical programming, yet is not itself used in ways which closely resemble Koopmans' description of his model. Moreover, the tale does not end there, for modern developments on the MOLP problem generally do not reference the Koopmans model. To this economist, the Koopmans model is the canonical interpretation of the MOLP problem, just as the product mix problem is the canonical interpretation of the LP maximization problem.

In section 2 the MOLP problem is defined and some properties which influence the design of algorithms for its solution are discussed. In section 3 several studies are discussed which utilize the multiple objective structure to analyze tradeoffs between conflicting objectives. A generalized linear activity analysis model is discussed in section 4, emphasizing some relationships between the geometric structure and certain economic interpretations. In section 5 we discuss some possible uses of this model in making economic evaluations of alternative technologies. An outline of how the generalized linear activity analysis model might be utilized to design submodels of productive entities within a general equilibrium model is sketched in section 6.

2.   THE MULTIPLE OBJECTIVE LINEAR PROGRAMMING PROBLEM

In this section the MOLP problem is defined and its relationship to the Koopmans linear activity analysis model is discussed. Some notational conventions will first be established.

2.1 Notational Conventions

Upper case Latin letters are used to denote matrices, with the same letter referring to a row of that matrix when subscripted or a column of the matrix when superscripted. Thus if M is a matrix, M_i is its i th row and M^j is its j th column. The elements of M, M_i and M^j would all be denoted by m_{ij}.

Vectors not defined as the rows or columns of some matrix will be denoted by lower case Latin letters, while Greek letters will generally be reserved for scalars. Algebraically all vectors not specifically designated as row vectors will be considered to be column vectors. Moreover we do not distinguish between the vector spaces R^{m+p} and $R^m R^p$. If $x \in R^m$ and $y \in R^p$ we denote by $(x;y) \in R^m R^p$, the vector whose first m components are components of x and whose last p components are the same as the components of y.

Order relationships between vectors x, $y \in R^n$ are defined as follows: $x > y$ means that $x_i \geq y_i$ for each i = 1,...,n; $x \geq y$ means $x \geq y$ and $x \neq y$; and $x > y$ means $x_i > y_i$ holds for i = 1,...,n.

The symbol $e \in R^n$ will be reserved for the vector which has all components one. We depart from our matrix notation in the case of the identity matrix, where wi utilize $E^j \in R^n$ to denote the j th column of the n x n identity matrix.

The letter "T" used as a superscript denotes transposition; $(M^j)^T$ denotes the transpose of the j th column of the matrix M.

2.2 Defining the MOLP Problem

The MOLP problem, like the LP problem which it generalizes, constrains its feasible solutions to a polyhedral convex set. Since there are several objective functions in the MOLP problem, and it is unlikely that all of these will achieve an extremum simultaneously at any one feasible solution, one needs a new concept of optimality for this problem. Instead of looking for a maximum or a minimum, one searches for vectors of objective function values which are "nondominated", "Pareto-optimal", "admissible", or "efficient". Among these synonyms we generally choose the first, although economists may be more comfortable with the second.

Let $S \subset R^n$ be given; we say that $x \in S$ is <u>nondominated in S</u> if $y \in S$ and $y \geq x$ imply x = y. This may be restated as saying that $x \in S$ is nondominated if there is no $y \in S$ such that $y > x$. Should we compare two nondominated points x, $y \in S$ for which $x \neq y$, if $x_i > y_i$ for some $i \in \{1,...,n\}$, then there must exist at least one index j for which $x_j < y_j$.

The multiple objective linear programming problem is defined in terms of the k x n matrix C, the m x n matrix A, and the vector b ε R^m. Define

$$X = \{x \in R^n | Ax = b, x \geq 0\} \quad \text{and} \quad (1)$$

$$Z = \{z \in R^k | z = Cx, x \in X\}. \quad (2)$$

The MOLP problem is that of determining the nondominated points in Z, together with the pre-images of these points in X. Formally, define

$$Z^+ = \{z \in Z | \quad z \text{ is nondominated in } Z\} \quad \text{and} \quad (3)$$

$$X^+ = \{x \in X | Cx \in Z^+\}. \quad (4)$$

Thus to "solve" the MOLP one must determine both Z^+ and X^+.

The set X, which contains all feasible solutions or "plans", is a polyhedral convex set. Since the mapping between X and Z defined by the matrix C is linear, clearly Z is also a polyhedral convex set. Neither Z^+ nor X^+ is convex, although they consist of unions of faces of Z and X respectively.

We digress briefly to remind the reader that every polyhedral convex set S can be described in two ways. The description

$$S = \{x \in R^n | Mx \leq m\} \quad (5)$$

defined by some q x n matrix M and m ε R^q is called the external representation of S. The hyperplane $H_i = \{x \in R^n | M_i x = m_i\}$ supports S if S $H_i \neq \phi$; indeed S H_i is a face of S. A face consisting of a single point is called a vertex of S. If S contains an interior point (i.e., S is not contained in any hyperplane) then some of the hyperplanes H_i contain n affinely independent points of S; such hyperplanes are called tangent hyperplanes, and their intersection with S is called a facet. The set of linear inequalities derived by relaxing the equalities of equations defining the set of tangent hyperplanes is a minimal external representation of S. An internal representation of S is of the form

$$S = \{x \in R^n | x = \sum_{i=1}^{r} \alpha_i y^i + \sum_{j=1}^{s} \beta_j v^j, \sum_{i=1}^{r} \alpha_i = 1, \quad (6)$$

$\alpha_i \geq 0$ for all i, and $\beta_j \geq 0$ for all j}.

Here Y = $\{y^1, y^2, \ldots, y^r\}$ is the set of points and V = $\{v^1, v^2, \ldots, v^s\}$ is the set of directions defining S; V may be the null set, but Y must contain at least one element. The faces of S have a representation very much like (6). If H = $\{x \in R^n | d^T x = \delta\}$ supports S and contains the face $F \subset S$, then $d^T y^i = \delta$ whenever $\alpha_i > 0$ in the internal representation of F, and similiarly $\beta_j > 0$ implies $d^T v^j = 0$. Here we assume that the normal vector $d \in R^n$ is given an "outward" orientation; or that $S \subset \{x \in R^n | d^T x \leq \delta\}$.

Published algorithms for the MOLP problem have focused exclusively on building up an internal representation of the set Z, and finding the subsets of the sets of points and directions generating Z which are sets of generators for faces in Z. The strategy is based on several considerations from the computational theory of linear programming. The pivot selection rules in linear programming are designed to produce sequences of basic feasible solutions. Every vertex of X may be described in terms of at least one basic feasible solution to the defining inequalities (1), so that enumerating basic feasible solutions allows one to enumerate vertices of X. While many of the vertices of X might be expected to be mapped into the interior of Z, it is the case that every vertex of Z is the image of some vertex of X. Thus some way must be found for detecting which vertices of X are mapped to vertices of Z, and moreover which of these vertices of Z belong to Z^+. This is accomplished by various nondominance tests.

It is the case that every nondominated point in Z lies in some hyperplane $\{x \in R^n | d^T x = \delta\}$ with outward normal $d \in R^n$ satisfying $d > 0$. Conversely, if $d \in R^n$ satisfies $d > 0$, then any $z^* \in Z$ which satisfies $d^T z^* > d^T z$ for all $z \in Z$ also satisfies $z^* \in Z$ [4]. These facts, the Kuhn-Tucker theorems, and an assortment of theorems of the alternative can be utilized to derive various tests for nondominance. Fortunately one does not need to enumerate all of the vertices of X to find all of the vertices in Z^+. It can be shown [5] that all vertices in Z^+ are connected by a set of edges in Z^+; the representation (6) of such an edge has $\beta_j = 0$ for all values of j, and for precisely two values of the index i we have $\alpha_i > 0$ possible. It is immediately obvious that if the midpoint of an edge is in Z^+ then both endpoints of the edge must also belong to Z^+. Thus starting from a vertex in Z^+, one can test the midpoint of all edges containing this vertex to see if the midpoints belong to Z^+. All bounded edges passing this test have a vertex at the opposite end which belongs to Z^+. Thus, the procedure can be iterated, testing in turn each of the known vertices in Z^+. Since Z has only finitely many vertices, the procedure must terminate.

Representative algorithms for determining vertices and edges, both bounded and unbounded, which belong to Z^+ have been published by Evans and Steuer [6], Yu and Zeleny [7], and Ecker and Kouada [8]. The algorithmic research of Steuer has been particularly impressive and he has made his computer codes publicly available at nominal costs; the latest version is documented in [9]. More recent research has been aimed at determining faces of Z^+ of all dimensions.

Isermann [10] and Gal [11] have published algorithms for accomplishing such enumerations, and Ecker and his students have actively studied this problem.

3.   TRADEOFF ANALYSIS OF INCOMMENSURABLE OBJECTIVES

Perhaps the most obvious interpretation of the MOLP problem, and certainly the one encountered most frequently in the literature, is that each objective function measures attainment of some index. However, the different indices represent disparate quantities which we cannot compare; the values of the various objective functions are incommensurable or lack a basis of comparison.

For example, we may think of a situation where we are comparing various fuels for heating a boiler. Whatever the constraints (1) (which may reflect, e.g., limitations on amounts of solid, liquid, and gaseous fuel to be used or interdependencies between fuels) we might define one activity vector $(C^j; A^j)$ for each alternative fuel which we are considering. The components of C^j may reflect various criteria by which we evaluate each fuel. For example, c_{1j} may be the negative of the cost of the unit quantity of the j th fuel, c_{2j} the number of BTUs we would get burning a unit quantity of the j th fuel in our boiler, c_{3j} the negative of the amount of sulfur dioxide that burning this fuel would emit from our smokestack, and so on. With our MOLP formulation, we can now analyze the tradeoffs between the various criteria which influence our choice of fuel, considering only nondominated but feasible alternatives. That is, we have defined the indices so that "more is better" by using the negatives of coefficients for things which we would rather do with less of than more of (i.e., cost, SO_2). Each vector in Z^+ summarizes an alternative mix of fuels and the corresponding values of the indices for using this mix; generally we should expect to find as many activities being utilized at positive levels as their are structural constraints in (1), or m. Comparing nondominated feasible alternatives from Z^+, we see what tradeoffs or "substitutions" are imposed upon us by the technology of the available alternatives.

The MOLP techniques have been applied by Zionts and Deshpande [12,13] to analyze U.S. energy policy. In this study, the structural constraints (1) taken from the Brookhaven Energy System Optimization Model (BESOM), a linear programming model of the U.S. energy system [14]. There are six objective functions employed measuring indices of environmental quality, resource use, petroleum imports, capital requirements, and two measures of annual costs. The model was solved interactively, retaining at each iteration the vector (vertex) in Z^+ which the decision maker preferred from among the alternatives already presented. The vertices adjacent to the preferred vertex were then enumerated; either the coordinates of the adjacent vertices or the implicit rates of substitution between objective function values at the chosen and alternative vertices can be presented to the decision maker. If the vertices differ considerably in their coordinates, the decision maker may prefer to move only part of the distance between the two vertices along a nondominated edge, and for this reason the substitution rates may be more useful than the coordinates of the vertices.

There are several possible difficulties with this approach to multi-criterion evaluation. In a combinatorial sense a MOLP problem may be very badly behaved, with the number of vertices in Z^+ growing exponentially in the number of objective functions k. It is not clear a priori how many of these vertices a decision maker might have to examine before finding a vertex from which further improvement is impossible. If the decision maker's preference function is quasi-concave, so that its upper level sets are all convex, then the intersection of Z and any upper level set are convex. It is easy to see in this case that the iterative procedure must terminate. Of course, real decision makers may not have transitive preferences, may forget or exhibit inconsistencies. However, Zionts and Deshpande report rather rapid convergence to a preferred solution on the part of each individual decision maker, although no comparison was made of how "close" the choices are of the various individuals.

It should also be pointed out that Steuer has done a great deal of work on this problem of "too much data" and has developed methods of selecting "representative" vectors from Z^+ that are widely dispersed rather than adjacent. At each iteration, the representative vectors are chosen from alternatives somewhat closer to the most preferred vector of the preceding iteration. This procedure, which makes no a priori assumptions about the decision maker's preference function, has also shown rapid convergence in experimental applications [15].

Another difficulty which is mentioned in Zionts and Deshpande [13] is that decision makers may encounter difficulty in the interpretation of the alternatives which have been presented to them, and thus may encounter difficulty in making meaningful choices. This is an enormously difficult problem, but one that is worth examining in more detail.

Multiobjective optimization techniques are not analytically closed models which determine an optimal solution. If one had a model of the decision maker (i.e., knew his/her preference function defined on the indices corresponding to the objective functions) an optimal solution or solutions could be derived. However, we only know preference functions as they are revealed to us through choices. Moreover, there is a great deal of information relevant to the formation of preferences which is not a part of the record summarized by the preference function. Since the decision maker may have no previous experience making choices among alternatives summarized by our indices, often we cannot estimate the preference function using econometric techniques. However, in the artificial situation of the "laboratory" where the decision maker is presented with novel alternatives, the added information affecting preference formation may be unavailable. In addition, the effects of previous choices do not feed back or cause learning on the part of the decision maker regarding agreement between ex ante and ex post evaluations.

There are a number of alternative analytical techniques which are being developed to assist decision makers choosing among multi-attribute alternatives. Some of these are prescriptive, and attempt to insure consistency and rationality of choice. The philosophy of the multiobjective optimization approach, such as our MOLP techniques, is quite conservative by comparison. It attempts to assist the decision maker by identifying the small subset of non-dominated alternatives from a larger feasible set, and by explicitly quantifying the tradeoffs between the attainment of several criteria. If it is to be faulted by presenting too much information and making the decision maker's task "difficult", then it can be defended on the ground that extraneous information can be discarded. The process of analysis and explication of why one alternative is preferred to another may facilitate both the understanding of one's preferences and the understanding of the model itself.

Anyone who wishes to learn more about our subject should consult the book of Cohon [16], which examines a number of multiobjective optimization models and techniques for their usefulness in planning. It also contains an example of power plant siting which may be of interest to energy analysts. The book is particularly useful for its insight into the formulation and use of multi-objective optimization models. The bibliography is also quite useful and broadly based. However we should caution the prospective reader that the development of analytical tools is rapid in this relatively young field.

4. A GENERALIZED LINEAR ACTIVITY ANALYSIS MODEL OF PRODUCTION

A generalized linear activity analysis model of production is defined by interpreting each of the coefficients in C, A, and b as the quantity of some good or service. We distinguish between those goods and services identified with rows of C, which we call commodities, and those identified with rows of A and b which are called resources. There is asymmetry in our definitions of the technical coefficients, as we interpret C^j as "supply" coefficients and A^j

as "resource demand" coefficients for the j th activity. If $c_{ij} > 0$ the i th commodity is produced by the j th activity, while $c_{ij} < 0$ signals a demand for the i th commodity as an input in the j th activity. The sign conventions are reversed in Aj; $a_{ij} > 0$ indicates the amount of the i th resource used by operation of the j th activity at unit intensity, while $a_{ij} < 0$ implies that more of the resource becomes available if the activity is operated at a positive level. Generally $b_i > 0$ holds; b_i may be interpreted as the amount of the i th resource which is available for production.

A fairly detailed interpretation and many theoretical results for this model are contained in [5]. A few highlights can be presented here, as it will be useful to refer to them in discussing energy modelling issues. Z is called the transformation set in this interpretation, and the commodity vector $z \in Z$ is sometimes called a transformation or a supply/demand vector. Any commodity vector $z \in Z^+$ is said to be technically efficient. Solving the MOLP problem by identifying Z^+ is essentially equivalent to deriving a production function description of the technology described by the activity vectors.

4.1   Feasible Transformations

X, the set of feasible production plans, and Z, the set of feasible transformations, are defined by equations (1) and (2) respectively. X has no economic meaning; elements of $x \in X$ are instructions regarding at what level the various activities are to be operated, and thus have significance only within the productive entity modelled by the generalized linear activity analysis model. All of our interpretations of coefficients in C and A are focused on columns. Mathematically, the components x_j of $x \in X$ should be thought of as scalar multiples operating on columns of C and A. The importance of this point of view will now be demonstrated.

Define the technology in the model by

$$T = \{(z;y) \in R^k \times R^m | z = Cx, \quad y = Ax, \quad x \geq 0\}. \tag{7}$$

T is a convex polyhedral cone, which we assume to be pointed, so that if $(z^0;y^0) \in T$ then $(-z^0;-y^0) \in T$ is impossible. We also define

$$R(b) = \{(z;y) \in R^k \times R^m | y = b\}. \tag{8}$$

R(b) is an affine set in R^{k+m}; that is, a translation of a linear subspace. Here we think of translating the linear subspace consisting of vectors of the form (z;0) for arbitrary $z \in R^k$ by the vector $(0;b) \in R^k \times R^m$. Now Z as defined by (2) can be found by first finding T ∩ R(b), and then discarding the last m components of any vector in T ∩R(b).

It is worthwhile to see what this prescription says in the special case of linear programming. In this case k = 1, and so the affine set R(b) is a line through (0;b) which we can think of as being "vertical" if the first coordinate axis points "up". If T ∩ R(b)≠φ, so that the vertical line intersects the technology cone, we have $Z^+ = \{(z^*;b)\}$, which is the endpoint of T ∩ R(b), or else R(b) ⊂ T if the problem is unbounded. Thus, in the "nice" case of a bounded solution the value of the optimal solution is the first coordinate of the point in Z.

We have now described our geometric picture of the problem of feasibility of transformations. Let us recall from section 2.2 that we asserted that every convex polyhedral set has two descriptions. Now equation (7) is the internal description of the technology cone, so there must also be an external description of this cone. The external description of T can be found in terms of an internal description of the cone polar to T, which is defined as

$$T^* = \{(u;v) \; \epsilon \; R^k \; x \; R^m | (u;v)^T(z;y) \leq 0 \text{ for all } (z;y) \; \epsilon \; T\}. \quad (9)$$

If T is of full dimension, which will always be the case in our model, the fact that it is also pointed leads to the conclusion that T^* is also pointed and of full dimension. Now a pointed cone contains no lines, has the origin as its only extreme point, and is the convex hull of its extreme rays which are halflines emanating from the origin [17]. We can pick any point other than the origin along each extreme ray and generate all of the points on the extreme ray as nonnegative scalar multiples of the chosen point. However, it is also the case that the coefficients of the points which generate the extreme rays of T^*, the internal description of T^*, are the coefficients of the homogeneous linear inequalities which provide the external description of T.

Now T^* is finitely generated, so there exists an integer q which is the number of extreme rays of T^*. We can find a q x k matrix F and a q x m matrix G such that

$$T = \{(z;y) \; \epsilon \; R^k \; x \; R^p | [F,-G](z;y) \leq 0\}. \quad (10)$$

From the discussion following equation (8) we now conclude that

$$Z = \{z \; \epsilon \; R^k | Fz \leq Gb\}. \quad (11)$$

This provides a method for determining Z once we learn how to find the external description of T from data for the internal description; a problem solved in complete generality in [5]. Without further analysis this method is not very efficient as a large proportion of the constraints $F_i z \leq G_i b$ may not be binding for any $z \in Z$.

The question of nondominance is nearly trivial in this geometric interpretation. Koopmans [1] showed that the nondominated faces of Z are precisely the same as the nondominated faces of

$$Z + N = \{z + x | z \in Z, \ x \in R^k, \text{ and } x \leq 0\}. \tag{12}$$

The nondominated faces of Z can be identified if the set of activity vectors contains $(C^j; A^j) = -E^j \in R^{k+m}$ for $j = 1, \ldots, k$; that is, the first k activity vectors are the negatives of the first k columns of the identity matrix of order k+m. It is easily shown that F is nonnegative if we include these <u>commodity disposal activities</u> defined in this way. If $\{z \in Z | F_i z = G_i b\}$ is not empty it is a proper face of Z, and this face is nondominated if and only if $F_i > 0$.

4.2 Price-quantity Relationships

Questions of feasibility and nondominance of transformations are posed entirely in terms of quantity vectors. Commodity prices will now be introduced into the analysis. We will suppose that the economic entity being modelled operates in competitive markets for all commodities; if this is not satisfied, the price vectors discussed below must be reinterpreted as vectors of marginal revenue products for each commodity.

The set of transformations $\{\bar{z} \in Z | p^T \bar{z} > p^T z \text{ for all } z \in Z\}$ defined for some $p \in R^k$ for which $p > 0$ is a face Z that belongs to Z^+. Interpreting p as a vector of commodity prices, we see that every transformation in the face becomes a "revenue maximizing" transformation at these commodity prices. The fact that we associate a face (which will contain more than one point unless it is a vertex) with a given commodity price vector means that the association between commodity prices and quantity vectors will in general be a correspondence or point-to-set mapping. We would like to be able to give an explicit representation of that correspondence.

Each face in Z^+ is the intersection of facets of Z. Each facet of Z corresponds to just one index i for which the equation $F_i z = G_i b$ holds for all points z in the facet, which establishes a one-to-one correspondence between the facets of Z and the <u>essential</u> inequalities in (4.5). Each essential inequality is part of a minimal description of Z. If $f \subset Z^+$ is a face, let I(f) index the facets of Z whose intersection is f. Then if $f = \{\bar{z} \in Z^+ | p^T \bar{z} \geq p^T z$ for all $z \in Z\}$ for some $p \in R^k$, it can be shown that p^T is a nonnegative linear combination of the rows of F indexed by I(f). As a consequence, we can associate with every face in Z^+ a cone of commodity price vectors which is generated by the set of outward normal vectors to hyperplanes containing the facets whose intersection is the face. Unless we have unbounded faces in Z^+, which is unlikely when m>0, this set of price cones will cover the entire nonnegative orthant of R^k. We call the collection of price cones the commodity price space decomposition for the technology T.

What is particularly remarkable about this result is that it defines a mapping which can be extended to associate shadow prices for resources with the commodity price vectors. We want to find a supply/demand correspondence from the space of commodity price vectors, generally the nonnegative orthant of R^k, to the revenue maximizing transformations in Z^+. We know that each commodity price vector is mapped to a face in Z^+; the price cone associated with this face shows all of the price vectors for which this mapping is "locally constant"; i.e., has the same set of images. For any price vector p in the price cone, the same set of nonnegative scalar multiples which are used to express p^T as a linear combination of rows of F can be utilized to express a vector $w^T \epsilon R^m$ defined by the same linear combination of rows of G. The vector w so defined is a shadow price vector for the resource vector b. Each of its components can be shown to be the value of the marginal product of the corresponding resource, and if z* is revenue maximizing at the prices p, then $p^T z^* = w^T b$. Thus we can impute factor incomes for any commodity price vector once we know which rows of F generate the price space decomposition.

One can use the coefficients of the matrices F and G to define various marginal rates of substitution. This allows one to study the effects of perturbations of the resource vector b, and to derive some interesting theoretical propositions.

It may be worthwhile to reexamine the above results from a geometric perspective. It is clear that the rows of the matrices F and G are interpreted as price vectors, and since the transposes of these matrices generate T* it follows that we must interpret T* as the cone of prices which is polar or dual to the quantity vectors in T. We have interpreted T in terms of quantity vectors (z;y) and T* in terms of price vectors (u;v). We take as given or exogenously determined the resource quantity vector y=b and the commodity price vector u=p. From this information the "unknown" quantity vectors zϵZ and shadow price vectors v=w are recovered.

What we are doing is building a general equilibrium model of supply, although if $z_i < 0$ the economic entity will be a net demander of the i th commodity rather than a supplier. In the general equilibrium model there are 2(m+k) variables of economic interest; k+m are prices and k+m are quantities. Knowing the technology T

cuts down the manifold of economic variables in the general equilibrium model from dimension 2(k+m) to dimension k+m; nothing else
is consistent with competitive economic equilibrium.

5. ECONOMIC EVALUATION OF TECHNOLOGICAL ALTERNATIVES

The usefulness of linear programming models for analyzing a number
of issues relating to energy supply has come to be widely appreciated. Econometric models are limited by the fact that only a
small subset of the space of relative prices of commodities has
been "explored by Nature," and little confidence can be placed
on regression equations when extrapolating beyond the range of
observations. Often, one is interested in comparing existing
technologies with newly proposed alternatives which are not yet
employed; the lack of data renders econometric estimation impossible. At other times, one may wish to study the effects of
government intervention in energy markets, environmental regulation, or other similiar "discontinuities" in the energy system.
There may be no natural way to model these phenomena in the econometric models.

By way of comparison, linear programming models are quite flexible in design and can readily incorporate technological information. The same bag of tricks which linear programmers have
developed can readily be applied to the MOLP formulation. In
particular, many of the data sources which have been developed
for LP models of energy supply may be adaptable to the generalized
linear activity analysis model with little change. The theoretical and practical issues regarding the estimation of technical
coefficients are the same in both models. What is different is
that LP models are essentially partial equilibrium models in which
the data can be perturbed along 1-dimensional manifolds, and the
MOLP models are general equilibrium models for which many multi-
parametric perturbation techniques appear to be feasible. We do
not <u>know</u> what this additional modelling flexibility portends for
energy modelling it may take "many cooks making many stews" to get
a reasonably accurate picture.

5.1 Reformulating Linear Programming Problems

The most immediate use for MOLP techniques may arise in performing sensitivity analysis in linear programming models. Frequently, there are only a small number of structural constraints

for which the constant term will be perturbed during sensitivity analysis. Even so, the results which one gets by perturbing the right hand side vector may become difficult to interpret if several components are changed by sizable amounts. With MOLP techniques available one would completely relax all of the structural inequalities for which large perturbations will be made in the constant term. Each of these structural inequalities and the original objective function can be treated as objective functions in the MOLP reformulation. The coefficients in the objective functions of the MOLP problem that were derived from structural constraints in the LP problem must correspond to the relational operator "greater than or equal to," so the sign of the coefficients must be changed if the structural constraint was originally of the "less than or equal to" variety. Similiarly, if the original LP objective function was a minimization problem, the signs of all coefficients must be reversed.

The precise use which one might make of this MOLP reformulation of an LP problem will, of course, depend on the purposes to which sensitivity analysis is put in the original analysis. When the transformation set is defined by equations (11), it is very easy to define such things as cost curves or demand functions for variable inputs. Generally one generates families of such curves or functions by specifying the values of all but two components of the vector z and treating these two components as variables. On the other hand, one may have preselected prices for certain goods or services which ought to equal the shadow prices of certain coefficients of the right-hand-side vector. Using the price space decomposition one can determine precisely which right-hand-side vectors are consistent with the shadow prices of resources being equal to the prespecified prices, and what the value of the objective function is for each such right-hand side.

5.2   Technology Evaluations

Often one is faced with the problem of evaluating technological alternatives which are untried commercially but which are sufficiently well understood technically so that one can specify the inputs and outputs of some proposed process. It may be the case that the technology is being imposed on reluctant adopters, such as may occur with environmental controls, or entrepreneurs may be searching for substitute products which promise a financial advantage. In such cases one may estimate activity vectors for the generalized linear activity analysis model and study the economic consequences of the proposed technologies. The utility of such studies is generally increased if one already possesses a "library" of activity vectors for existing technologies and various proposed additional technologies.

We will not specify exactly how such models are defined, except to note that all technical coefficients in a given model must be defined in terms of flows or in terms of stocks; here "or" is used in the exclusive sense. The inescapable fact that production cannot occur without capital congealed as stocks of fixed factors of production creates dilemmas no matter how one proceeds. Some analytical difficulties are also created by the fact that certain models incorporate structural constraints which must be equality constraints without slack or surplus variables. Such a constraint makes it impossible that the cone T be of full dimension, complicating the analysis of prices.

The generalized linear activity analysis model of production is particularly well suited to determine answers to questions regarding both technical efficiency and economic efficiency of technological alternatives. An activity vector is technically efficient if there is some set of prices at which the activity becomes economically efficient. The question of technical efficiency depends on the situation regarding the coefficients of the resource vector b. In the broad sense, an activity is technically efficient if it generates an extreme ray of T; we continue to assume that T is pointed. In a narrower sense, the activity vector is technically efficient for the resource vector b if the activity is used at a positive level for one or more vectors in Z^+. Since the components of b will generally represent capacity limitations, one is generally forced into making a study of short run and long run adjustments in a complete technology evaluation.

The sorts of things which can be done with the generalized linear activity analysis model in technology evaluations frequently involve analysis of relative prices. For example, the price space decomposition is useful in several ways. For a given resource vector b, one can determine all of the relative commodity prices at which a given activity vector is economically efficient. This gives some indication of the likelihood of employing any particular activity vector. However, one must also model the demand side of the market if an accurate evaluation is desired. If the demand side is estimated statistically, it may be possible to incorporate uncertainty into the model. Clearly, if one knows both the price space decomposition and a probability distribution for commodity prices it is possible to assign probabilities to vectors in Z^+ and thus to activity vectors.

One thing which we would like to know but have not succeeded in analyzing is the question of how much the technical coefficients of a technically inefficient activity must be "improved" to make it technically efficient. It is fairly easy to determine an answer for efficiency in the broad sense, but it appears to be much more difficult when structural constraints are present in the model.

When technology is imposed, such as in the case of environmental controls, there are a variety of studies which can be made to determine the impace of alternative methods of control. In such cases the price space decomposition retains its usefulness, and in addition the related identification of shadow prices for each commodity price vector becomes even more helpful. This is due to the fact that the resource vector often shows capacity limitations, and the shadow prices then indicate a distribution of income among resources. The "harm" imposed on various resource owners by the imposed technological alternatives can be assessed for whatever commodity price vectors appear to be relevant. There is a limited discussion in [5] regarding the application of the production model to studies of the economic consequences of alternative environmental policies.

6. TOWARD GENERAL EQUILIBRIUM MODELLING

We believe that the greatest potential usefulness of our generalized linear activity analysis model of production will occur in the design of general equilibrium models, where our model will represent submodels of particular production entities (e.g., firms, industries, regions, national economies,etc.).There are several aspects of our model which make it particularly attractive. Both prices and quantities are variable in the model in the sense that the supply/demand correspondence can be "inverted" by mapping each quantity vector in Z^+ to the price cone containing all price vectors which make this quantity vector economically efficient. Thus,one can specify either a set of prices,or a set of quantities or in fact prices of some commodities and quantities of other commodities and then find the complementary sets of prices or quantities. The association of shadow price vectors with commodity price vectors allows one to incorporate the incomes of resource owners into the model. One result of these considerations is that the linkages between the production submodel and the rest of the general equilibrium model consist entirely of variables which have economic meaning.

Another aspect of this approach to modelling is that the "messages" between submodels can be kept relatively small. For example, the description of the production submodel may have five, ten, or twenty times as many structural constraints as objective functions and an even larger ratio of activities to objective functions. However coordination between submodels will generally involve vectors or linear inequalities that have as many coefficients as there are objective functions in the MOLP problem representing the submodel.

We have begun a preliminary investigation into some of the issues arising in the design and computation of general equilibrium models. This study recognizes the desirability of an eclectic approach to modelling, and considers the role of econometric and simulation modelling in such models. The availability of efficient computer software for computing the full range of results necessary to implement our generalized linear activity analysis model is and is likely to remain for at least another year the greatest bottleneck to implementation of these results.

ACKNOWLEDGEMENTS

The author wishes to express his thanks to Professor İbrahim Kavrakoğlu and the sponsors of the Advanced Study Institute for the opportunity to attend the ASI and to present this lecture. A travel grant from the RCA Faculty Development Fund, University of Scranton, is also gratefully acknowledged.

REFERENCES

1. T.C. Koopmans, Analysis of Production as an Efficient Combination of Activities, in: Activity Analysis of Production and Allocation, Wiley, New York, 1951.

2. G.B. Dantzig, Linear Programming and Extensions, Princeton University Press, Princeton, 1963.

3. A.S. Manne and H.M. Markowitz, Studies in Process Analysis, Wiley, New York, 1963.

4. J. Philip, Algorithms for the Vector Maximization Problem, Mathematical Programming II, 207-229, 1972.

5. E.C. Duesing, Polyhedral Convex Sets and the Economic Analysis of Production, unpublished Ph.D. dissertation, University of North Carolina, Chapel Hill, 1978.

6. J.P. Evans and R.E. Steuer, A Revised Simplex Algorithm for Linear Multiple Objective Programs, Mathematical Programming V, 54-72, 1973.

7. P.L. Yu and M. Zeleny, The Set of all Nondominated Solutions in Linear Cases and a Multicriteria Simplex Method, Journal of Mathematical Analysis and Applications, XLIX, 430-468, 1975.

8. J.G. Ecker and I.A. Kouada, Finding All Efficient Extreme Points for Multiple Objective Linear Programs, Mathematical Programming, XIV, 249-261, 1978.

9. R.E. Steuer,"Operating Manual for the ADBASE/FILTER Computer Package for Solving Multiple Objective Linear Programming Problems", working paper No. BA 7, College of Business and Economics, University of Kentucky, Lexington, April 1978.

10. H. Isermann, The Enumeration of the Set of All Efficient Solutions for a Linear Multiple Objective Program, Operational Research Quarterly, XXVIII, 711-725, 1977.

11. T. Gal, A General Method for Determining the Set of All Efficient Solutions to a Linear Vector Maximum Problem, European Journal of Operations Research, I, 307-322, 1977.

12. S. Zionts and D. Deshpande, A Time Sharing Computer Programming Application of a Multiple Criteria Decision Method to Energy Planning--A Progress Report, in: Multiple Criteria Problem Solving, ed. by S. Zionts, Springer-Verlag, Berlin Heidelberg, New York, 1978.

13. S. Zionts and D. Deshpande,"Energy Planning Using a Multiple Criteria Decision Method,"Working Papers Series No. 398, School of Management, State University of New York at Buffalo, March 1979.

14. E.A. Cherniavsky,"Brookhaven Energy System Optimization Model", Brookhaven National Laboratory Publication 19569, 1974.

15. R.D. Steuer and F.W. Harris, Intra-Set Point Generation and Filtering in Decision and Criterion Space, Computers and Operations Research, forthcoming.

16. J.L. Cohon, Multiobjective Programming and Planning, Academic Press, New York, 1978.

17. R.T. Rockafellar, Convex Analysis, Princeton University Press, Princeton, 1970.

DISCUSSION

In response to a question, Duesing explained that the difficulty was that the linear programming problems solved at the efficient edges were most probably degenerate. The LP solutions for the nondominated solutions themselves were also degenerate. This caused two main problems:

i. Cycling
ii. Pivot rejections

and these increased the necessary book keeping and calculations
as the number of objective functions increased.

According to Duesing, there were three alternative techniques for
multi-objective problems:

    a.   goal programming models,
    b.   process analysis models, and
    c.   Lagrange multipliers method.

Goal programming minimizes the absolute difference between un-
attainable goals, it can find some of the nondominated edges,
but these may not be the desired tradeoffs. In process analysis
models, one approach is to combine the objectives by means of
shadow prices; but the determination of appropriate shadow prices
is a major problem. Another approach is to do the optimization
according to one objective and add constraints representing the
other objectives. These two approaches generate solutions only
for a given set of shadow prices. In the Lagrange multipliers
method, the objectives are collected by means of the multipliers,
but the numerical techniques are not yet well-developed and convergence is still a problem. Presently, studies were being made
on multi-objective integer, dynamic, nonlinear, and discrete
control models. The basic problem was that in nonlinear programming the feasible set is not convex and the adjacent nondominated edges had not be connected. (See for example, "Survey on
Multi-objective Optimization, by Joland, Decision Sciences, 1978.)
There are already some working papers in the power generating
systems involving four objectives (MOLP and Applications, by Core,
Academic Press).

As far as implementation of results were concerned, the MOLP generated many nondominated solutions and therefore presented too
large a set of alternatives for the decision maker to be practical.

AN OPEN DYNAMIC LINEAR ACTIVITY ANALYSIS MODEL OF PRODUCTION

Erick C. Duesing

School of Management
University of Scranton
Scranton, Pennsylvania 18510, USA

ABSTRACT. A finite horizon multi-period model of production is described which utilizes fixed technical coefficients but allows for both joint products and any arbitrary finite number of fundamental activity vectors. After a brief description of closed models of accumulation, it is shown how vectors of "imports" and "exports" across the boundaries of the production system can be modelled. The resulting open dynamic model of production is a multiple objective linear programming problem whose constraints exhibit staircase structure. The potential uses of such models for designing energy planning models are discussed.

1. INTRODUCTION

Decision makers responsible for the planning of research, development, and capital investment expenditures in the energy sector face many difficult choices between perplexing alternatives. At present energy consumption in industrialized nations is largely based on the utilization of fossil fuels, for which economic reserves represent from several decades to several centuries supply at present rates of use. It is clear that long term prospects for the continued prosperity of industrial society requires that a transition be made from dependence on fossil fuels to the utilization of new sources of energy. There is no shortage of potential technological alternatives; many proposals have been advanced for exploiting new sources of energy and for developing new techniques for conversion of resources to useful energy.

In addition to a comparison of relative costs, the economic analysis of these alternatives involves the consideration of several aspects of time. Not only are there dynamic aspects of demand for useful energy and of supply of energy resources, but there are also lags between the initial commitment of resources to research, development, demonstration and construction of facilities and the first appearance of usable output from such activities. These considerations suggest that an energy planning model becomes more useful to decision makers if it is designed to include several features. It should be possible to incorporate a wide spectrum of technological alternatives in the model, including alternatives proposed by engineers but untested in practice. The model should be specified as a dynamic model to trace the time paths of energy demand and resource depletion and to incorporate lags in economic activities.

The question of economic evaluation of alternative technologies in a static setting is discussed in our previous lecture [1]. The analytical technique proposed there involves utilizing computational technique for the multiple objective linear programming (MOLP) problem to study the linear activity analysis model of production. The activity vectors in the linear activity analysis model usually have technical coefficients which represent quantities of goods and services. The derivation of correspondences between prices and quantities is discussed in the fourth section of [1].

The attractiveness of the linear activity analysis model for economic evaluation arises from several considerations. The technical coefficients defining each activity are physical quantities, as are the coefficients in the resource vector b. Consequently this data can be specified by scientists and engineers who need not concern themselves with economic variables. Given a particular collection of activity vectors and a resource vector, the entire set of prices at which any activity vector is utilized and the corresponding imputations of resource prices can be determined. This imposes stringent but rather transparent consistency conditions on possible equilibrium configurations in the model which contains the linear activity analysis production subsystem. In particular, the internal consistency of a set of parameter values which constitute an analytical scenario can be checked for such models. The fact that MOLP techniques are utilized for computation makes it possible to include as objective functions some measures of the attainment of noneconomic goals.

Our objective in this lecture is to show how the multiple objective linear programming/linear activity analysis framework can be extended to construct a dynamic model of production. This model is not analytically closed, as no choice mechanism is specified to select an optimal production plan from the feasible alternatives found by the model. As will be seen, closure of the model requires specification of a preference function to choose a preferred "export" vector of goods in each time period. A valuation function defined over terminal stocks of goods is also required for closure. Lacking these preference and valuation functions, what remains is a dynamic general equilibrium model of supply.

As described by the title of this lecture, our model may be characterized as being an open, dynamic, linear activity analysis model. By open we mean that two things can occur. Part of the resources used for production in each time period can be "imports" of goods from an unspecified source outside of the productive entity being modeled. Similiarly, part of the output which is produced in any time period may be "exported" from the productive entity. The possibility of transport of goods and services in either direction across the boundaries of the production subsystem is what makes our model open.

The model is dynamic in the limited sense of being a multi-period, finite horizon model. It represents production as occurring in a finite number of time intervals of equal length. While these intervals may be days, weeks, or years, we will refer to any particular time interval as a "period". The linkage between subsequent periods is due to the fact that some of the output in a given period may be used as an input in the following period. Durable goods may be stored, and the possibility of accumulation or using up stocks of goods allows one to model stocks of both produced capital goods as well as non-produced natural resources. Lag structures can also be specified in such a model.

By describing our model as a linear activity analysis model, we mean that production is carried out by a finite set of processes which transform inputs into outputs in fixed proportions. Each such process is called an activity. It is assumed that all processes can be scaled up or down arbitrarily without affecting input and output proportions, and that processes do not interfere with one another. Thus an arbitrary representation of the process by a vector with fixed numerical coefficients is possible; such a vector is called an activity vector. The scaling and noninterference properties can be summarized by stating that positive scalar multiples and sums of activity vectors are also activity vectors.

As should be clear from our description, the model can account for those features which were described in the second paragraph of this section as being particularly relevant to energy planning. We will return to the question of the appropriateness of the model for energy modeling after describing the model and some of the aspects of pricing associated with economic efficiency.

2.   DEFINING THE MODEL

It will be assumed that there are m goods and that production occurs in the T time intervals numbered 1,...,T. The vectors of exports, imports, and production in the t-th time period are denoted by x^t, y^t, and z^t respectively. We assume x^t, y^t, $z^t \in R^m$, and that the i th component of each vector refers to the same good. All goods in the model are to be regarded as "stocks" which have the dimension of a physical quantity (e.g., tons) services are measured in terms of a unit measuring the source of the service and the length of the time period in the model (e.g., person-years of labour.)

The letter A will denote a m x n matrix of input coefficients, and C will denote a m x n matrix of output coefficients. That is, a_{ij} represents the amount of the i th good required as an input for unit operation of the j th productive activity. Similarly c_{ij} is the output of the i th good when the j th productive activity is operated at a unit level. If desired, the matrices A and C can vary over time, in which case they can be represented as A(t) and C(t) respectively. Although technical change requires that input and output coefficients will change over time, we sacrifice realism for typographic convenience.

2.1   The Closed Model of Production

The first model of production which will be considered is a closed model. In this model the goods which are produced in period t - 1 are used as an input for production in period t. Symbolically, the intertemporal dynamics of the closed model are given by

$$z^t = Cu^t, \quad Au^t = z^{t-1}, \quad u^t \geq 0 \qquad (2.1)$$

for $t = 1,\ldots,T$, which presupposes a fixed initial stock of goods z^0. The production plan u^t has as many components as there are fundamental activities, which are the columns of the matrices C and A.

If we substitute for z^{t-1} in (2.1) by using the first relation, we get

$$Au^t - Cu^{t-1} = 0 \qquad \text{for} \quad t = 2,\ldots,T \qquad (2.2)$$

together with the boundary conditions

$$Au^1 = z^0, \quad Cu^t = z^T \qquad (2.3)$$

and the nonnegativity conditions $u^t \geq 0$. In this model we have a MOLP problem with m objective functions Cu^T ($= z^T$). The constraints are the equations (2.2), the initial condition $Au^1 = z^0$, and the nonnegativity conditions $u^t \geq 0$, for $t = 1,\ldots,T$.

This closed model of production is analyzed in detail in the author's dissertation (2). Let Z(T) denote the set of all feasible

values of $z^T = Cu^T$, and let $Z(T)^+$ denote the nondominated points
in $Z(T)$. Then $z^T \in Z(T)^+$ implies that there exists a price vector
$p \in R^m$ with all components positive such that z^T lies in a face of
$Z(T)$ contained in a hyperplane with outward normal p. Let u^{-t} be
a feasible solution to the model which satisfies $z^T = Cu^{-T} \in Z(T)^+$.
Then $z^t = Cu^{-t}$ for $t = 1,\ldots,T$ defines an optimal production sequence. Moreover there exists a set of vectors $q^t \in R^m$ for $t = 1,..,$
T which are the shadow price vectors in the model. These shadow
price vectors are precisely the price vectors which must prevail
to clear competitive markets in each of the T periods at the quantities z^t.

Economists generally refer to our closed model of production as a
model of accumulation or a von Neumann growth model. Afriat [3]
has made an extensive study of the growth properties in such models.
When A and C are both nonnegative, the model is an example of a
polyhedral monotone process of concave type studied by Rockafellar
[4]. Fujita [5] has also studied multi-period versions of Rockafellar's monotone processes and applied them to various economic
problems. It should be noted that none of these studies considers
computational issues, whereas the theory in [2] is constructive.

2.2   The Open Model of Production

To persons with somewhat more practical inclinations than are displayed by economic theorists and mathematicians, the closed model
of production may seem rather pointless. After all, what is the
purpose of producing goods if the only use of these goods is to
produce still more goods? This defect in the specification of the
closed model will be removed in two steps. It will first be shown
how one gets goods "out" of the model (e.g., for consumption or
trade), after which it is shown how to get productive resources
(e.g.,labour services) "into" the model. Hopefully this will add
an air of realism to the modelling process.

The model of production is envisioned as a submodel of a more complex model. For our purposes, there is considerable flexibility in
how the analytical closure of the model is specified. For this
reason we have chosen the term "exports" to describe the vector of
goods x^t removed from the production sector in period t. The word
"export" should be regarded as synonymous with "removal from the
production subsystem", and does not imply anything about the ultimate disposition of these goods. What will generally be of concern
is the effect on future production of satisfying the demand for
goods x^t which arises outside the production subsystem.

The timing convention which will be used in our model assumes that
the goods produced in period t - 1 will be allocated between exports
and input into production in period t. If the matrix A includes

disposal activities which utilize one of the m goods as input but produce no output, then the intertemporal balance equation reflecting our timing convention is given by the equation

$$x^t + Au^t = Cu^{t-1} \; .$$

(2.4)

Here we must constrain x^t to satisfy $x^t \geq 0$; it is assumed that the nonnegativity conditions on the u^t continue to hold.

Equation (2.4) reflects the fact that the output of production in period t-1 is allocated between exports and input to production in period t. To get the export vector outside the production subsystem the MOLP problem is formulated which has the following objective functions:

$$Iv^1 \; (= x^1), \; Iv^2 \; (= x^2), \ldots, Iv^T(= x^T), \; Cu^T(= z^T)$$

(2.5)

The constraints on the MOLP problem include

$$Iv^t + Au^t - Cu^{t-1} = 0 \quad \text{for} \quad t = 2, \ldots, T;$$

$$Iv^1 + Au^1 = z^0; \quad \text{and} \quad v^t \geq 0, \; u^t \geq 0$$

(2.6)

for t = 1,...,T.

Notice that we distinguish, perhaps unnecessarily, between the export vector x^t and the vector of activity levels v^t for the "transport activities" which remove the exported goods from the production subsystem. The matrix I in (2.5) and (2.6) is the m x m identity matrix.

It will now be supposed that the stock of goods that is produced in period t-1 and allocated between exports and inputs in time t can be augmented by "importing" a vector of goods across the production subsystem boundaries. The issues of timing and of dependence must be considered along with the question of how the input vector will be represented.

The timing convention that will be utilized here assumes that the import vector y^t represents goods available in the t th period. That is, (2.4) can be extended to

$$x^t + Au^t = Cu^{t-1} + y^t$$

(2.7)

rather than the alternative

$$x^t + Au^t = Cu^{t-1} + y^{t-1}.$$

Our preference for (2.7) rather than (2.7*) reflects the fact that the economically most significant import into the production subsystem in many economic models will be the import of labour services.

Moreover, to the extent that the import vector represents goods
actually imported from a foreign country, it should be noted that
the definition of the export vector x^t already incorporates a lag.
From the standpoint of the productive entity being modeled, the
relevant issue is not when the imported goods are produced, but
rather when they become available within the production subsystem.
To the extent that x^t and y^t represent goods being traded abroad,
both the imports and exports can be presumed to have been produced
and in transit during period t-1.

It will be noted that our use of (2.7) rather than (2.7*) allows a
more appealing initial condition when t = 1.

It will be assumed that the import vector y^t is selected from the
set

$$\{y \in R^m | Dy \leq d\}. \tag{2.8}$$

If desired, the defining inequalities $Dy \leq d$ may be made to depend
on time by writing $D(t)y \leq d(t)$. It is assumed that D has m columns,
but the number of rows is relevant only to the extent that it affects
computational efficiency and consistency of definitions of D and d ,
or $D(t)$ and $d(t)$.

An argument can be made that the import vector y^t should not be
autonomous, as is implied by (2.8), but should depend on present
or lagged values of the activity levels u^s and v^s ($s \leq t$). For
example, if a component of y^t represents labour services, the amount
of labour offered depends both on the disutility of labour and the
utility received from consumption. This might be represented by
writing the defining inequalities in (2.8) as $Dy^t + Ev^t \leq d$. The
objections which can be raised to this reformulation are that

    i.   not all exports x^t are consumed by labourers, and
    ii.   consumption demand depends on the distribution of
         income, and the income variable in the model will
         be a function of shadow prices.

This argument can be extended to include the issue of whether or
not (2.8) is the proper method of representing choices of import
vectors. Recall from our previous lecture [1] the definitions of
the internal and external representations of a convex polyhedral
set. Equation (2.8) is an example of an external representation
of y^t that expresses y^t as the sum of a convex combination of points
plus a nonnegatively weighted sum of directions. If this representation of y^t is available it can obviously be substituted directly
into (2.7).

In actual models of some complexity it is likely to be the case
that y^t will be specified exogeneously to the production subsystem

and thus can be substituted directly in (2.7). The formulation which is "right" or "best" depends on the context of the modeling effort. The preceding discussion should be viewed as a preliminary survey of several possible alternatives.

The open model of production can now be completely defined. It is a MOLP problem whose $mT + m$ objective functions are given by (2.5). The constraints of the model include

$$Iv^t + Au^t - Cu^{t-1} - Iy^t = 0 \quad \text{and} \quad Dy^t \leq d \tag{2.9.a}$$

$$\text{for } t = 2,\ldots,T;$$

$$Iv^1 + Au^1 - Iy^1 = z^0 \quad \text{and} \quad Dy^1 \leq d; \quad \text{and} \tag{2.9.b}$$

$$u^t \geq 0, \ v^t \geq 0, \quad \text{and } y^t \geq 0 \text{ for } t = 1,\ldots,T. \tag{2.9.c}$$

A three period example is given below in Table 1 in detached co-efficient format.

TABLE 1. A 3-period Open Model of Production

Variables:

v^3	u^3	y^3	v^2	u^2	y^2	v^1	u^1	y^1

Objective Functions:

v^3	u^3	y^3	v^2	u^2	y^2	v^1	u^1	y^1	
									$(= z^3)$
I	C								$(= x^3)$
			I						$(= x^2)$
						I			$(= x^1)$

Constraints:

v^3	u^3	y^3	v^2	u^2	y^2	v^1	u^1	y^1	
I	A	-I		-C					$= 0$
	D								$\leq d$
			I	A	-I		-C		$= 0$
				D					$\leq d$
						I	A	-I	$= z$
							D		$\leq d$

With all variables v^t, u^t, $y^t \geq 0$

3.   PRICING IN THE OPEN MODEL OF PRODUCTION

The values of the objective functions (2.5) will be denoted by the vector $(x^1; x^2; \ldots; x^T; z^T) \in R^m \times \ldots R^m = R^{mT+m}$. The set of all feasible objective function values may be denoted by $Z(T)$, and its nondominated boundary by $Z(T)^+$. $Z(T)$ is a convex polyhedral set, and using the results in [2] we know (at least in principle) how to find both internal and external representations of $Z(T)$. Every point in $Z(T)^+$ is contained in at least one face of $Z(T)$ which can be supported by a hyperplane with a strictly positive normal. $Z(T)^+$ can always be defined so as to be of dimension $mT + m$ by including "disposal activities" in every objective function row.

There are two sets of questions regarding prices which are worth exploring for the open model of production. The topic of efficiency prices is discussed first, after which pricing will be discussed in regard to the analytical closure of our model.

3.1   Efficiency Prices

It will now be assumed that the point $(x^1; \ldots; x^T; z^T) \in Z(T)^+$ is contained in the supporting hyperplane whose outward normal is $(p^1; \ldots, p^T; q^T) \in R^m \times \ldots \times R^m = R^{mTm}$, and $(p^1; \ldots, p^T; q^T) > 0$. Regarding $(p^1; \ldots, p^T; q^T)$ as a commodity price vector, the discussion in [1] would lead us to expect that there must be a shadow price vector associated with these commodity prices. The shadow price vector will be denoted as $(r^1; s^1; r^2; s^2; \ldots; r^T; s^T)$, where $r^t \in R^m$ is associated with the constraint $Iv^t + Au^t - Cu^{t-1} - Iy^t = 0$ and s^t is associated with $Dy^t \leq d$.

The original exposition of the linear activity analysis model by Koopmans [6] contains a result as the "weak complementary slackness principle". This principle states that

     i.   if an activity is used, it "prices out" at the commodity price-shadow price pair, and

     ii.   any activity which would cause a loss at these prices will not be used.

Essentially this result states that shadow prices are imputed in such a way that each resource can be paid the value of its marginal product in all activities actually used; activities unprofitable at these opportunity costs for resources are not used. We will now examine what results this principle implies when it is applied to our open model of production.

One, $v_i^t > 0$ implies that $p_i^t = r_i^t$, while $p_i^t < r_i^t$ implies $v_i^t = 0$. Now $v_i^t > 0$ if and only if some of the i th good is exported in period t. If the good is exported, $p_i^t = r_i^t$, which means that the value of this good is equal in its two uses — as an export and as an input into production. If the highest price an exporter would pay is less than the value of the marginal product of the i th good when it is all used for production, then this good will not be exported.

Two, $u_i^t > 0$ implies that $(r^{t+1})^T C^i = (r^t)^T A^i$, while $(r^{t+1})^T C^i > (r^t)^T A^i$ implies $u_i^t = 0$. In interpreting this result, recall that output produced in the t th period is used in period $t + 1$. $(r^{t+1})^T C^i$ is the value of the outputs resulting from unit level operation of the i th activity, valued at shadow prices prevailing in time $t + 1$ when these goods will be used. The term $(r^t)^T A^i$ is the value of the goods used as inputs for unit level operation of the i th activity in the t th period. Thus activities which are employed $(u_i^t > 0)$ must "price out" with $(r^{t+1})^T C^i = (r^t)^T A^i$. Activities for which input costs exceed the next-period value of outputs are never employed.

If all goods are exported, so that $v^t > 0$, then $p^t = r^t$ according to the first result, and r^t in the second result may be replaced by p^t.

Three, we have $y_i^t > 0$ implying $(s^t)^T D^i = r_i^t$, while $(s^t)^T D^i > r_i^t$ implies $y_i^t = 0$. The shadow price r_i^t is the value of the marginal product of the i th good used as an input for production. The components of s^t measure, in terms of value per unit, the benefit of unit increases in the components of d.

3.2 Prices in a General Equilibrium Model

There are a number of ways which the open model of production can be analytically closed. One possibility might be to find econometric estimates of demand for all goods by various sectors of the economy (i.e., households, government, and foreign), build models of income distribution, government, financial, and foreign sectors, and simulate the evolution of the economy. A less ambitious approach is to ignore the government and foreign sectors, set up preference functions for goods consumed in each time period and for final stocks, and solve the resulting intertemporal optimization problem. The latter approach will be discussed here very briefly.

The intertemporal optimization problem for a finite planning horizon is a linearly constrained nonlinear programming problem. We may suppose that the preference function is given by $U(x^1; x^2; ...; x^T)$ and the valuation function is $V(z^T)$. The constraints on the problem

$$\text{maximize } U(x^1; x^2; ...; x^T) + V(z^T) \tag{3.1}$$

are that

$$(x_i^1 x^2; \ldots; x^T; z^T) \; \varepsilon \; Z(T) \cap R^{mT+m} \; ; \qquad (3.2)$$

that is, we only consider nonnegative exports and final stocks which belong to $Z(T)$. The constraints (3.2) are linear inequalities since $Z(T)$ is a convex polyhedral set.

Rather than keep track of a number of superscripts, we will use a simple example to illustrate the point we wish to consider. To maximize the function $f(x):R^n \to R$ subject to $g(x) < 0$, where $g:R^n \to R^m$ is defined by $g(x) = Bx - b$ for the $m \times n$ matrix B and $b \; \varepsilon \; R^m$, one sets up the Lagrangian function $L(s,y) = f(x) - y^t(0 \; Bx - b)$ and applies the Kuhn-Tucker conditions. If the maximum is attained at $(\bar{x},\bar{y})$, then $\nabla f(\bar{x}) < \bar{y}^T B$, $\bar{y}^T(B\bar{x} - b) = 0$, and $[\nabla f(\bar{x}) - \bar{y}^T(B\bar{x} - b)]$ $\bar{x} = 0$, where $\nabla f(\bar{x})$ is the gradient of the function $f(x)$ evaluated at $x = \bar{x}$. Geometrically these conditions state that $\nabla f(\bar{x})$ is a nonnegatively weighted sum of the outward normals, of constraints which hold with equality at $x = \bar{x}$. These normals are precisely the rows of B, say B_i, for which $B_i\bar{x} = b_i$ is satisfied. Equivalently, one can assert that $\nabla f(\bar{x})$ is the outward normal of a hyperplane which supports the set $\{x \; \varepsilon \; R^n | Bx \le b\}$ at $\bar{x}$.

Pulling together the threads of the discussion of this and the preceding sections and section 4.2 of [1], we see that the gradient of the objective function $U(x^1; x^2; \ldots; x^T) + V(z^T)$ plays precisely the role in the analysis of the intertemporal maximizing model as the price vector $(p^1; p^2; \ldots; p^T; q^T)$ played in section 3.1 above. Thus, one can substitute the appropriate partial derivatives of $U(\;) + V(\;)$ for the components of p^t and q^T in the weak complementary slackness results. The outcome is a set of conditions for intertemporal efficiency in a finite horizon general equilibrium model.

4. POSTSCRIPTS

We know of no model which is designed according to the specifications described here; our work exists solely as a conceptual exercise. Of the energy models known to us, our intertemporal optimizing model most closely resembles the Stanford PILOT Energy/Economic model developed by Dantzig and coworkers [7]. Indeed, it was of interest to see how one could overcome the objectionable features of the objective functions imposed by the choice of linear programming as an analytical tool that led us to our intertemporal optimizing model. The PILOT model uses linear programming methods for the energy sectors of the model and couples these with an input-output model for other goods. The effect appears to be the same as representing production by an input-output matrix which has variable coefficients wherever energy is involved and fixed coefficients elsewhere.

There are several advantages and disadvantages of the approach we propose in comparison to a model design such as is employed in PILOT. The advantages are that production can be modelled more flexibly and that there is complete freedom regarding how the objective function is specified. When modelling production, our technique allows joint products in each activity and as many activity vectors as one cares to specify. Thus, all sorts of substitution possibilities can be modelled. Moreover, there is no a priori specification of prices, and thus both producers and consumers can freely alter demand in face of changing prices.

On the other hand, there is uncertainty regarding how efficiently computation can be carried out in a model such as the ones we propose. Moreover, there are unresolved theoretical problems regarding the relationships between the utility function $U(x^1;x^2;...;x^T)$ and the valuation function $V(z^T)$. The second objection applies equally to all intertemporal models, not just our own. The first problem arises directly from our choice of MOLP as an analytical technique. We hope to have some preliminary estimates on the amount of computation involved within the next two years.

In summary, the modelling techniques proposed here would appear to fulfill the desiderata required for energy planning models. They can incorporate a great deal of detail on alternative technologies, and these need to be described only in terms of quantities of inputs and outputs, not in terms of economic variables. The models can readily handle a variety of lag structures through alternative timing conventions. Stocks of natural resources can be included in the initial stocks z^0, with the unused stocks transferred to the following period. Technological change can readily be incorporated in the model, and the MOLP formulation makes the incorporation of environmental and other social indices into the model an easy task. Thus, it appears to be a conceptual approach to modelling which is well worth further development.

REFERENCES

1.   E.C. Duesing, Applications of Multiple Objective Linear Programming to Energy Modeling, this volume.

2.   E.C. Duesing, Polyhedral Convex Sets and the Economic Analysis of Production, unpublished Ph.D. dissertation, University of North Carolina, Chapel Hill, 1978.

3.   S.N. Afriat, Production Duality and the von Neumann Theory of Growth and Interest, Volume XI of Mathematical Systems in Economics, Verlag Anton Hain, Meisenheim am Glan, 1974.

4. R.T. Rockafellar, Monotone Processes of Convex and Concave Type, Memsius of the American Mathematical Society No. 77, American Mathematical Society, Providence, Rhode Island, 1967.

5. M. Fujita, Spatial Development Planning, North-Holland, Amsterdam, 1978.

6. T.C. Koopmans, Analysis of Production as an Efficient Combination of Activities, in: Activity Analysis of Production and Allocation, Wiley, New York, 1951.

7. G.B. Dantzig, et. al.," Stanford Pilot Energy/Economic Model," EA-626, Volume 1, Electric Power Research Institute, Palo Alto, California, 1978.

DISCUSSIONS

In the discussion that followed, Duesing explained that in the model whatever was produced in one period was used in the following period, either as export or as input for production, which was a rather big assumption. No feedback was imposed on the model, but there were some studies on this subject and on closed models of production.

It was found difficult to include utilities into this model. First of all, it was difficult to model the utility in a linear equilibrium model like this. Secondly, utility functions are non-linear and this is difficult to handle in multi-objective algorithms. But if one knows the utility function, he may use the activity vector directly by searching only the nondominated set within the transformation set.

In response to another question, Duesing stated that returns to scale problems could also be modelled within a multi-objective open dynamic linear activity analysis model.

PART VI

COOPERATION IN MODEL IMPLEMENTATION

Models in the French Energy Planning Process

The Energy Modelling Forum: Past, Present and Future

MODELS IN THE FRENCH ENERGY PLANNING PROCESS

D. Finon

Institut Economique et Juridique de l'Energie
Université de Grenoble
France

ABSTRACT. A model is always a simplified representation of reality taken from a certain point of view which reflects the status and the interest of the modeller (or of the one who is in charge of it). This way one will find purely cognitive models, in other words, models meant to improve a knowledge of reality (physical, social, cultural...), and decision models, the use of which would help a person or an institution to make the best possible decision. In the sense of a similar decision field, this last one will vary along with the status, the functions, the temporal horizon, the space of reference... of the determining instance. In dealing with energy matters, for example, one usually distinguishes between

1.   the corporate models which help in choosing a commercial strategy or a long term investment strategy in a given market (coal, oil, natural gas, electricity...) and

2.   the sectorial models eventually put into practice by public authorities in order to have a better understanding of the behaviour and the manoeuvrability of the global energy system.

1.   THE INSTITUTIONAL STRUCTURE OF THE FRENCH ENERGY SYSTEM

For historical reasons, the entire French decision system in the energy field is extremely closed, with no intervention of Parliament or different social groups. All decisions are centralized in the high level administration (Ministry of Industry, Ministry of Finances, Energy Commission of the Planning Board,...) linked very strongly with the General Directorates of the Public Companies which dominate the French energy sector. More precisely, this sector is composed of:

- subsidiaries of multinational firms which control about 50% of French petroleum market

- public or mixed enterprises, either competing (CFP and Elf-Aquitaine in the oil branch/ or monopolistic (EDF, CDF, GDF, CEA,...).

The first ones show a changeable autonomy according to the structure and the strategy of the firm upon which they depend. At any rate, they never decide by themselves since the stakes are of some importance (large investments in refining or transportation, for example).

The second ones show large power which is, however, far from being complete because their public status makes them subordinate to the state authority. This subordination varies, however, according to the extent to which the enterprise shows a monopoly or not (the guardianship of the state is less constraining for firms like CFP and ELF-Aquitaine which compete with branches of the multi-national oil enterprises) and can face or not face its investment charges (EDF which had become more than 50% self-financing before 1974, had acquired a much larger autonomy than CDF or GDF). No matter what their exact impact in the final decision - especially concerning new investments - all these firms resort to models in order to diminish the uncertainty which discourages the evolution of their market, prices of their newly imported materials, and technologies to which they resort.

Relative positions of the various organizations

	Importation Extraction	Commercialisation	Ministerial Administration		
Coal	Charbonnages de France (CDF)		Direction du Gaz, de l'Électricité, du Charbon (DGEC)	Direction de l'Énergie et des Mines	Ministre de l'Industrie
Gas	Elf-Aquitaine Gaz de France	Gaz de France (GDF)			
Electricity	Électricité de France (EDF) Self-Producers				
Petroleum	Elf-Aquitaine et CFP	et Multinational oil companies	Direction des carburants (DICA)		
Nuclear Energy	Groupe CEA et PUK	Groupe CEA			
			Commission de l'Énergie	Commissariat Général au Plan	

CFP: Compagnie Française des Pétroles

CEA: Commissariat à l'Energie Atomique

Since the foreign oil firms have entered France at the beginning
of the century, they have tried to influence a legislation which
has not always been favourable to them, and, in order to do this,
have installed their aerials as close as possible to the state
power. Later on, since the great wave of nationalization of 1945,
rationalization was synonymous with uniformization - standardization - centralization... as a reaction against the crumbling away
of the mechanism of production (especially electricity and gas),
which was the result of a slightly dynamic and slightly concentrated capitalism. Between all these heads, the osmosis is considerably easier and speeded up by another aspect of the French
centralization, namely the uniform fabrication in the large schools
of engineers also concentrated in the Paris region. Through flows
of civil servants towards the general directorates of public and
private firms, the group of decision-makers is very homogeneous.

Another aspect of this situation is the monopoly of expertise in
the energy field by the departments of economic studies, of the
different energy companies, particularly of Electricite de France.
No real expertise capabilities exist at the level of the public
administration (Ministry of Industry, Energy Commission of the
French Planning Board...). The weakness of infrastructure of
public economic studies on energy makes studies of a truly global
character impossible at this level whereas certain public companies
have the capacity to do so.

On the other hand, the parliament has no group of experts attached
to it to assess the choices of high technical content. In this
context, independent firms of consultants and a fortiori university
research groups have great difficulties to exercise their compe-
tence in the modelling field for the aid of energy decision-making.

2.   THE CORPORATE MODELS

2.1 The Oil Branch Models

Each oil corporation has elaborate refining models which optimize
the management and the running of a given refinery, taking into
account its technical characteristics, different oil qualities,
the supplies, and the production programme imposed by the central
office of the company; taking into account its market in the viewpoint of the refinery. In order to do this, the expenses are
minimized under constraints of a given production while taking into account specifications of oil products, with the help of linear
programming techniques. Some corporations also have models of
transport and distribution which minimize the transport expenses,
while taking into account the availability of different oil products,
the siting of refineries, of district storage places, and of main
areas of consumption. These models are most often regional.

Much more general and global models do exist which try to optimize the strategy of oil companies by planning their investments of exploration, of refining, of distribution and by optimizing their strategy on the markets of the different products (fuel oil, gasoline...).

2.2 The Gas Branch Models

In addition to the models which optimize the management of a gas pipeline, taking into account the possible extension of different regional markets and the availability of gas (national resources, contracts of import), and in addition to models of reservoir management with underground storage in order to regulate the peaks of demand, few models have been developed in France by Gaz de France concerning gas. One must also specify that different methodologies taking the place of informal models have been utilized either to analyze the competitiveness of gas, or to help in choosing the investments.

With a first methodology, the outlets of gas are studied case by case by considering the different areas where gas is usable and/or used;* one determines a price of equivalence of gas from the price of the competing fuel, while taking into account the profit of utilization and the costs of equipment of the usable gas installations.

With a second methodology [1], "Gaz de France" studies the profit of investment projects; with the help of a test, it determines profitable operations (in other words those of which the profit rate is superior to the discount rate imposed by public power), which attribute the best financially estimated result. It is indeed necessary to cut down on less profitable projects for the amount of investments allowed yearly by the Ministry of Finances is limited.

2.3 The Electricity Branch Models

This branch has been a delicate object particularly from the part of the modelers: the first models of linear programming used in France were developed this way in 1954 at the EDF for the choice of electrical investments. And since this date, the researchers of this enterprise are at the point of progress and utilize new techniques of calculation (non-linear programming, dynamic programming, optimal control, etc....). Let us first point out the existence of very specific models such as the optimization of the

---

*These areas are just as specific and precise as the tubular boilers, different types of heating, baking of burnt earth...

cycle of the nuclear fuel,* the optimization of the network of electricity transport, the maximizing of the security of this network, etc... [2]. But it would be a good idea to dwell upon more general models and in particular on the models of demand forecast and the models of choosing electricity investments.

The forecasting models of electricity demand used by EDF [3] are relatively simple and are based on extrapolation of the past starting from statistical relations of simple or multiple regression type. These relations (generally logarithmic) correlate at the global level (or at the level of highly aggregate sectors such as the residential and tertiary sector, or the industrial sector) the quantities of electrical energy (C) with time:

$$\log C_t = a + bt$$

or with the economic operation represented by an operational economic index of the national economic activity such as the GNP, or the Industrial Added Value

$$\log C_t = a + b \log GNP_t.$$

The forecasters of the EDF have indeed estimated that these models obtain the best results and that all the efforts to tie the electricity consumption in with other variables (such as the relative price of capital, of labour or of competing fuels in the industry, or the income and/or the amount of households in the residential and tertiary sector) prove to be unsatisfactory. One must also point out that this econometric approach assumes that the consumption of electricity was inelastic to the price, and that the outlets of electricity developed themselves in a relatively autonomous manner in the specific uses of electricity. The commercial strategy of EDF of expanding its market in the domestic space heating and the increase in price of oil products, however, cancel all partition between the markets of different forms of energy (electricity versus fuel). The econometric forecasts are presently confronted with the commercial objectives of the firm and completed by a prospective carried out in terms of scenarios. One could never do without methods of extrapolation, but beyond a horizon of more than five to ten years, the forecasts obtained must be used with much caution. In other respects, an econometric model is used to forecast the different load curves output per hour according to the hour, the day, the week and the month from the extrapolation of different characteristics coefficients [4].

---

*Cf. Simulation model of the nuclear fuel cycle, Charpentier-Naudet-Paillot, Commission of Atomic Energy, France, 1973, and Model SEPTEN, Service of Nuclear studies, Direction of the equipment of the EDF, France 1972.

Let us now consider the models for electricity investment choices.
An important bibliography exists on this subject [5]. These models
minimize the discounted cost of electricity production over a long
period of time; the production objectives are determined by the
forecasts of the global electricity demand brought into effect with
the help of econometric models which have just been mentioned, and
by a representation of this demand with the help of weekly load
curves. The different types of equipment for electricity production are explicitly considered and are characterized by their
capacity and their availability during the different hours of the
yearly load curve. The risks of failure associated with climate
(hydroelectricity, yearly peak consumption) are taken into account
with the help of probabilities.

The present model [6] uses the theory of optimal control. The
objective function is composed of three terms (investment, operating cost, cost of failure).* The controle variables are the equipment capacities created year by year, and the constraints express
an obligation of satisfaction of future demand of power with the
forced (or limited) development of certain types of equipment
(breeder reactors, for instance). The algorithm runs twofold:
firstly, the control variables are determined,and secondly, the
optimal management of the given power equipment is defined. The
programme allows the optimal equipment plan at the national level to
be obtained as well as the duration of economic life of equipment,
the probability of failure, the marginal costs of production of a
kWh (according to the hour, day and month) and the dual value of
equipment. The marginal costs which have also thus been determined,
serve to establish the electricity tariffs. In order to do this,
one adds to them the marginal transport costs and costs of distribution calculated in other respects and a "toll" which permits the
EDF to reach the budget equilibrium and even to possess a certain
self-financing capacity.

These models of investment choice are particularly complex since
the system of French electricity production is a mixed hydraulic-
termal system, which makes necessary a rather detailed representation of the management of the different hydraulic equipment
(run of the river, lake, pumped storage...).

Among the successive EDF models of investments, only one model
(the one built in 1965) was disaggregated in 5 regions connected
with the other ones by variables of interregional exchanges. The
objective was not to determine those ones, but to try to trace a
primary scheme of optimal location of the production equipments.
But the model has been considered as too complex. One another
point, these models of investment only integrate the private costs,
and by no means social costs as far as the degradation of the

---

* The cost of power-failure is a non-linear function increasing in
relation to the duration and the amplitude of the power failure.

environment or the national security of supply is concerned.

3. THE SECTORIAL MODELS

As has been mentioned, the public expertise in the energy field is mastered by certain public companies. And, at the sectorial level, no global energy model has been built by public administration or by these companies to help the energy choices.

Previously, a method of energy planning - kind of informal model which determined the energy supply structure at lowest cost with respect to objectives of security of supply - was used in the framework of elaboration of the 4th Economic Plan (1961-1965) and of the 5th (1966-1970). But the evolution of the French economy since 1965, and particularly of the French energy sector, was synonymous with a "deplanning" movement i.e. of a diminution of the role of the French Plan. At the level of the energy system, the very fast substitution of oil for national coal meant that the public power no longer had actual control over the sector. The planning process is only a confrontation of the objectives of the different producers, and the content of the Reports of the Energy Commission of the Planning Board for the 6th Plan (1971-1975) and the 7th Plan (1976-1980) is very light and deceiving.

In this context, the global energy models (developed in either the supply field or the demand field) has no use at all. Let us, however, mention the existence of a simulation model for the financing of the energy sector[*] by the Direction de la Prévision (Forecasting Department) in the Ministry of Finance (this Direction plays an important role in the definition of short and medium-term energy policies). This model permits the assessment of the consequences of modifications of energy policies (tariffs, taxes, investment, regulation...) on the financing, employment and annual needs of investment and on the budgets of the public companies.[**]

The other efforts of modelling the global energy system have been made in two directions by our University Group of Research, the Institute of Energy of Grenoble: a model of optimization of the energy sector EFOM (developed between 1971 and 1974 independent of public power or the energy companies) and a model of simulation of the long term energy demand MEDEE developed between 1973 and 1976 with ties to EDF, CEA and Elf-Aquitaine.

---

[*] Model FINER - built by D. Blain (1972).

[**]This model has been tested for instance on the policy of recession of the national extraction of coal.

The first model (which is detailed in another paper) [7] has never aroused interest from the French Public Administration and the French energy companies.

Curiously, Electricité de France (EDF) announced five years ago that it had started work on a global energy model to establish the energy structure which would be the best for itself and consequently for the collectivity (sic). But firstly, this model has never seen the light of day;* secondly, it is curious, to say the least, that it was not the ministry responsible for the French energy sector that initiated this study; lastly, the EFOM model of Grenoble has always been the object of a polite indifference from all the official actors. This could be explained by the French decision-making process, but also by the monopoly of EDF on the programming models.

Institute of Energy of Grenoble was condemned to work on the optimization model in a certain academic way without relation to decision-makers. Our main goals were to study the behaviour of the system subject to endogenous or exogenous events (what happens if?) and the limits of manoeuvrability (what can we do?), and not to define the optimum of the energy system and so the framework of the energy policies. In 1975, The Commission of the European Economic Communities (EEC) was interested in EFOM and asked us to develop a detailed model usable for each of the nine member-countries. And, by the detour of Brussels, EFOM is presently used by certain French public organisms for their own exploration of the future. The normative aspect of optimization seems to frighten certain decision-makers a lot who also distrust results from large, complex models which are difficult to comprehend and to operate and which need great amounts of data of unequal quality. In fact, they fear that they will find their part of power reduced.

On the other hand, the model MEDEE has been much better adopted by energy companies and more recently by public administration. Aware of the limits of econometric methods for long term energy demand forecasting (strong reliance to the past, high level of aggregation...), our Institute has attempted to develop more flexible approaches using simulation models and scenario techniques. The model [8] MEDEE is based on a detailed and integrated analysis of the energy demand determinants of a country, and on a scenario description of the economic and social development of the country.

---

* Only certain parts of energy system (electricity branch, district heating...) have been formalized.

The philosophy of the analysis consists in:

1. identifying end-use categories or "modules" in which the energy demand can be grasped in an homogenous and aggregate way;

2. identifying in each module the social need or the economic activity which creates the energy demand, the factors determining the level of useful energy implied by this need or this activity and finally those which determine the level of final energy demand necessary to supply this useful energy;

3. identifying the factors and relationships which explain the evolutions of these factors, and then, the evolution of the energy demand of this module.

The scenario description consists in selecting among the energy demand determinants identified in the analysis those which describe in an exhaustive way the social and economic patterns of development of the country under consideration. Then these variables are organized in a hierarchical structure which will be a guarantee of consistency for the scenarios. Finally, assumptions are made about the evolution of these variables in a plausible and consistent manner, following the order resulting from the hierarchical structure, and using a simplified DELPHI method linked to a cross-impact technique.

The partitioning of the socio-economic system has been made in three submodels (urban, industrial and transport); each one is disaggregated in a set of "modules" corresponding to the various end-uses of energy (space heating, waterheating, intra- and intercity transportation, cooking, steam uses, kiln uses...) and to the various economic agents. For instance, we could briefly characterize the residential space heating energy need module by the following scheme:

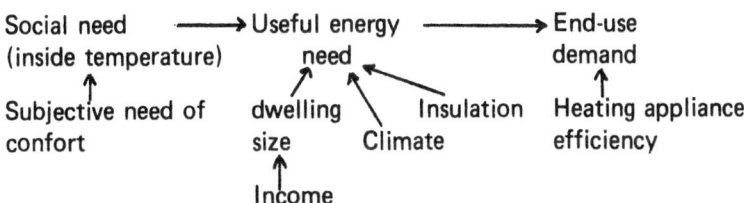

The global model contains about 500 recurrent equations which simulate the evolution of energy demand determinants on a five-year step basis, and to calculate, each five years, the resulting energy demand.

At the official level, the econometric methods had largely been used until 1974. But after the crisis, some energy companies, with long decision horizons, were convinced of the necessity of the use of such a model. Its prospective aspect seems to be less constringent than the normative aspect of optimization for a decision-maker. Moreover, the basic concepts are easily understandable by him. And, presently, the Planning Board is about to use MEDEE to explore the future of energy demand (horizon 2000) in the framework of the preparatory works of the 8th plan.*

REFERENCES

1. Cf. Toromanoff, The choice of investments at the Gaz de France, Revue Française de l'Energie, No. 260, February 1974.

2. J.C. Dodu, "Probable model in the study of the alimentation security of a transport network", EDF, 1973.

3. D. Finon, Forecasting the consumption of Energy and Electricity: the methods used in France, in Proceedings of French American Conference on Energy Systems Forecasting and Planning, I.E.S., University of Wisconsin, Madison, October 1974.

4. Y. Pioger, Forecasting Power Consumption and Models for Constructing Load Curves, In Proceedings of the Conference in Madison, pp.49-83.

5. P. Masse and R. Gibrat, Application of Linear Programming to Investments in the Electric Power Industry, Management Science, No. 3, 1957; and,

   F. Bessiere, Methods of Choosing Production Equipment at Electricité de France, European Economic Review, Winter 1969.

6. C.F.D. Levi, D, Saumon, "The New National Model of Investment", Internal Memo of the EDF, May 1973.

7. D. Finon, About the Use of an Energy Optimization Model, in Mathematical Modelling of Energy Systems, This Volume.

8. B. Chateau, B. Lapillonne, Long-term Energy Demand Forecasting: A New Approach, Energy Policy, June 1978, pp. 140-156.

---

*MEDEE has also been developed for the EEC; and linkages between MEDEE and EFOM are being built.

DISCUSSION

The conference expressed their sympathy with Finon after his description of the French institutional framework for treating energy
matters.

The focus on useful energy was felt to be a good idea, but the
problem was that everything became exogenous. Very little was
known about the historical development of "useful energy". It
was also felt that such a bottom-up analysis had difficulty in
taking account of technologies as yet unknown - for example, new
types of appliances in private households. Furthermore, in the
real world prices, technologies, and life-styles are all interde-
pendent.

Finon replied that recent studies, especially in Denmark had shown
that it was possible to increase the efficiency of household appliances significantly. The interesting question was the speed of
market penetration. These factors could be built into a scenario.
It was shown in the theory of econometrics that to disaggregate
added bias to a model.

A very real problem with all models was how to get the decision-
maker to use them. A drawback with MEDEE was that there were so
many exogenous variables that the model was easy to misuse. Much
of the data required for the scenarios in MEDEE were not available
and the modeller should be honest when presenting this to the
decision-maker. MEDEE had no checks on the consistency of the exogenous input.

The greatest benefit from a model was probably its use as an educational tool. It was often very edifying to examine the results
of coupling the whole energy supply system. A holistic view would
show, for example, that to produce electricity in a thermal power
station with an efficiency of 30% and then to use the electricity
in a heatpump with a coefficient of performance of 2.5 gave no
net gain to the system.

In summary, it could be said that both econometric and engineering-
estimate models have their drawbacks and it is difficult to say
which is more useful.

THE ENERGY MODELLING FORUM: PAST, PRESENT AND FUTURE

J.L. Sweeney and J.P. Weyant

Energy Modelling Forum*
Terman Engineering Center
Stanford University
Stanford, California, 94305**

ABSTRACT. The Energy Modelling Forum seeks, through a series of ad hoc working groups, to improve the use and usefulness of energy models in the study of important energy policy and planning issues by serving as a communication bridge between modelers and model users. This paper summarizes the EMF studies, its history and evolution, its basic goals, organizational structure, conduct of studies, and issues fundamental to the future evolution of the Forum.

1. INTRODUCTION

In recent years, especially since the oil embargo of 1973, there has been widespread development of energy models in the executive and legislative branches of government, universities, industry, research institutes, consulting companies, and commercial establishments. Unfortunately, our ability to utilize the models effectively for energy policy-making and planning has not kept pace with this development. The gap between modelers and potential users of models is large and pervasive. Heightened concern about energy problems coupled with the proliferation of analytical tools for addressing these problems has created both the need and the

---

* The Energy Modelling Forum is sponsored by the Electric Power Research Institute.
** The authors would like to thank George B. Dantzig, Wendelin Dintersmith, Martin Greenberger, William Hogan, Douglas Logan, and Alan S. Manne for their insights and comments during the writing of this paper. Any remaining errors, of course, remain the responsibility of the authors.

opportunity for bridging the gap. Finding ways to improve communication between model developers and model users has become an active area of investigation and innovation [6].

The Energy Modelling Forum (EMF) has been one response to this situation. The EMF seeks to improve the use and usefulness of energy models in the study of important energy issues. Sponsored by the Electric Power Research Institute (EPRI) and administered by the Stanford Institute for Energy Studies, the EMF, with the Departments of Engineering-Economic Systems and Operations Research, operate through a series of ad hoc working groups consisting of roughly equal numbers of energy modelers and potential energy model users. Each working group focuses on an issue or set of closely related issues important to energy policy making or planning to which existing energy models can be applied. The group designs, implements, interprets, and communicates a set of tests designed to illuminate the basic structure and behaviour of the models. The issues addressed by the group thus provide a forum to compare and contrast the various models, identifying their capabilities and limitations. At the same time, the issue focus assures that the policy relevant implications of the various models are developed and communicated.

This paper on the EMF is organized in six additional sections:

1. a summary of the five EMF studies initiated to date;

2. a description of the Energy Modelling Forum as it exists today (June, 1979), its basic goals, organizational structure, and the conduct of its studies;

3. the history and evolution of the EMF over the two and one-half years of its existence;

4. several new activities closely related to - and partly motivated by - the EMF, which are performing different and complementary functions;

5. issues fundamental to the future evolution of the Forum; and

6. a summary and conclusion.

2. EMF STUDIES: A SUMMARY

The products of the Forum* - comparative studies of significant

---

* More complete discussions appear in [7] and in the EMF reports and working papers referenced below.

energy issues - include two completed studies, "Energy and the Economy" and "Coal in Transition: 1980-2000", two more studies well under way, "Electric Load Forecasting: Probing the Issues with Models" and "Aggregate Energy Demand Elasticities", and a fifth study just beginning, "U.S. Oil and Gas Supply".

EMF 1: Energy and the economy

The Forum project was initiated in 1976 with a study designed to demonstrate the research concept. The working group compared six models of the link between energy and the economy to isolate the key factors determining the effect of energy system changes on the long run growth of the U.S. economy. The results demonstrate the importance of the value share of energy in the economy, the flexibility in substituting other inputs for energy use, and the link between productivity and capital formation in explaining the behaviour of the models [4].

EMF 2: Coal in transition (1980-2000)

A second EMF working group, organized in July 1977, compared 10 different models in the analysis of coal production, distribution, and utilization. The report documents the greater importance of coal demand issues relative to supply issues and describes various insights into the level and composition of future coal output gleaned from the model's results. Emphasis is placed upon the sensitivity of patterns of future coal use to changes in regional economic conditions and standards on allowable emissions (2).

EMF 3: Electric load forecasting: Probing the issues with models

The third working group examined the use of ten current models in forecasting electric loads. The experiments identified and illuminated prominent load forecasting issues and improved the understanding of the models' capabilities and limitations. An issue identified - but not resolved - was the degree to which combined historical data from many utility regions could be used to estimate relevant parameters in a demand model for a single utility region [3].

EMF 4: Aggregate energy demand elasticities

A fourth working group conducted a specialized test of aggregate price elasticity of demand implicit in the participating energy models. Eighteen models were run under nine scenarios testing the models' responses to variations in the prices of oil and gas,

coal, and other energy sources. Interpretations of model runs and
conclusions are still under heated debate [1].

EMF 5: U.S. oil and gas supply

A fifth working group held its first meeting in January, 1979. The
group intends to examine the effects on domestically produced oil
and gas of alternative world prices for oil, domestic prices for
natural gas, oil price controls, alternative federal leasing rates,
price controls, surprises in price trajectories, changes in the
tax structure, and alternative assumptions about the geological
resource base.

3.   THE ENERGY MODELLING FORUM TODAY

3.1   Goals and Design Principles

The basic goal of the Energy Modelling Forum - to improve the use
and usefulness of energy models in the study of important energy
issues - entails a number of subgoals, some competing, some complementary.

The first set of goals relates to comparison of models to improve
understanding of their limitations and capabilities;

- To identify and compare critical elements of existing
  energy models and to illuminate their major strengths
  and weaknesses.

- To cast light on key modelling issues so as to afford
  a greater understanding of alternative modelling approaches; and

- To provide guidance for the improvement, linkage, and
  extension of energy models and to establish priorities
  for new modelling research.

The second set of goals is related to the development of better
information relevant to energy policy making and planning, and
available through the use of energy models:

- To use major energy models to sharpen insights, improve
  understanding, and explore the implications of selected
  energy decisions and scenarios; and

- To broadly disseminate information about possible energy futures and the impacts of various energy actions on those futures.

A set of design principles guide Energy Modelling Forum activities in pursuit of these goals [11]:

- User Orientation. The EMF should work to improve the use and usefulness of energy models, approaching the studies from the user perspective and maintaining an active user involvement.

- Model Comparison. The EMF studies should compare the capabilities and limitations of many models, and these comparisons should be descriptive rather than normative. This is a unique contribution that the EMF can make, and it avoids the difficult problem of model validation.

- Issue Focus. For the general model user, abstract model comparison should be conducted in the context of the application to an important energy issue. This will provide a direction and a discipline for the model tests.

- Broad Participation. The communication objectives of the EMF are best served if there is a wide participation in the selection of study topics, the formation of the working groups, and the dissemination of the study results.

- Decentralized Analysis. Existing energy models are often complex and require skillful application by the model developer. Despite the inherent advantages of third-party analysis, the EMF must rely on model tests as reported by the individual research group.

In general, the studies have conformed well to the design principles and guidelines. All but, perhaps, EMF 4 have maintained an active user involvement, including users from government agencies, private sector corporations, research institutes, and universities. Each of the studies has included model comparisons, using between six and 18 models, with 10 being the median number addressed in any one study. Each had a strong issue focus although the immediacy of the issues varied significantly among the studies. The two final design principles, broad participation and decentralized analysis, have been fully met in each of the studies.

3.2   Organizational Structure

The current organizational structure of the EMF is illustrated in
Fig. 1. The Senior Advisory Panel, the working groups, the EPRI
staff, and the EMF staff interact with one another and with the
broad community of energy modelers and potential model users.

The heart of the EMF consists of the ad hoc working groups of about
35 members each. The working group chairman and the issue to be
studied are selected before the formation of the working group.
Each working group, composed of volunteer participants, with a
balanced representation of model users and model developers, is
organized around a specific energy issue to ensure both the proper
representation of relevant models and participant interest in the
policy or planning issues addressed. The chairman selects members
with the goal of obtaining a working group which is diversified
geographically, institutionally, and philosophically. With observers and others who closely follow the working groups, and rotation
of the working groups as new issues are selected, over 250 people
have been involved in the five studies to date, and more than 100
might be active in any year.

At each stage in this process, the EMF is assisted by the Senior
Advisory Panel. This group, chaired by Charles Hitch of Resources
for the Future, Inc., is composed of senior energy decision makers
(see Appendix for a list of the membership) who represent the
ideal target audience for the EMF studies. The Panel helps maintain the necessary broad participation and user orientation to
assure the value and immediate relevance of the working group
topics. The Panel meets annually, and provides necessary advice
throughout the year. Its functions are primarily fourfold:

- to suggest appropriate study topics and to critique
  prominent study proposals so as to provide a sense of
  priority;

- to suggest and possibly assist in recruiting appropriate working group chairmen and members;

- to critique the working group's final report in draft
  form both for substance and presentation; and

- to help disseminate the results of the studies.

The overall planning, coordination of daily operations, and administration of the Energy Modelling Forum are handled by the EMF
staff, supervised by an Executive Director. Located at Stanford
University, the EMF staff is affiliated with the Stanford Institute

for Energy Studies, and Departments of Engineering-Economic Systems
and Operations Research. The staff (see Appendix for a listing of
staff members) provides support for the Senior Advisory Panel in
the development and selection of issues for future topics, recruits
the working group chairman, assists the working group chairman in
organizing a study, participates both as members of the working
group and staff to this group, and publishes the final working
group reports.

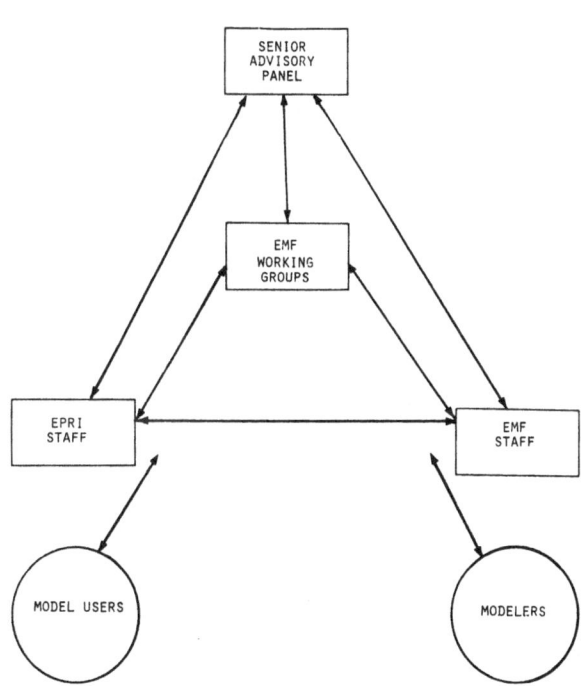

FIGURE 1 - Energy Modelling Forum interactions

The communication function of the EMF is enhanced by close ties
maintained among the various participants in the Forum. The Senior
Advisory Panel includes EPRI representation. Meetings of the Senior
Advisory Panel are normally attended by the Executive Director of
the EMF, at least one working group chairman, and by several EPRI
staff members. Energy Modelling Forum working groups include EMF
staff members as well as EPRI staff members particularly knowledge-
able in the area being addressed. Close coordination between the

EMF staff and the EPRI staff is maintained throughout all phases
of the Forum. Some of the working group members or chairmen were
initially proposed by Senior Advisory Panel members. The community of energy modelers and potential energy model users interact
broadly with the entire process, by participating directly in working groups as members or observers by membership on the Senior
Advisory Panel, or by a one to three year position with the EMF
staff. This community also suggests appropriate study topics,
models to be considered, issues to be addressed, and maintains informal communication with EMF groups.

3.3   The EMF Process

A typical study cycle begins with a broad call to modelers and
potential model users to assist in identifying potential study
areas, moves through phases of working group organization, intense
modelling activities, and result interpretation, and the writing
and publication of the report. The complete process may take as
long as a year and a half, involving typically three to four working group meetings, spaced about three months apart. Publication
is normally followed by an indefinite period of publicizing the
study and development of applications for the work. Table 1 illustrates the various phases of studies and indicates which groups
typically are actively involved in various phases of a study.

The process of selecting a study topic involves a wide range of
participants. Initial identification of potential topics is accomplished by collecting ideas offered by many people. These
suggestions are distilled to a dozen or so major potential study
areas to be considered by the Senior Advisory Panel. The Panel
in discussing the issues provides a sense of priority from a user
perspective as well as providing suggestions for specific issues
within the general areas. Additional preliminary exploration and
issue identification of high priority topic areas follow. This
activity is coordinated by the EMF staff, but involves participation of the community by energy modelers and model users.

This process results in the selection of a topic and a chairman to
direct the working group. Virtually concurrently, the chairman
in coordination with the EMF staff selects the specific project
to be undertaken. This simultaneity assures that the chairman is
not only directing a study on a topic within his area of expertise
and interest, but also one which seems feasible in light of the
existing energy models and the limitations on the current state
of the art.

A working group is recruited primarily by the chairman and the EMF
staff, with the working group chairman guiding the selection process.
This phase, along with the first step of selecting the working group

TABLE 1 - Participation in phases of EMF studies

Study Phases	Senior Advisory Panel	EPRI	Working Group Chairman	Working Group	EMF Staff	Others
Identification of Potential Area	X	X			X	X
Consideration by Senior Advisory Panel	X	X			X	X
Preliminary Exploration and Issue Definition	X	X			X	X
General Topic Area Selection	X				X	
Working Group Chairman Recruiting	X	X			X	X
Specific Topic Selection			X		X	
Working Group Recruiting	X	X	X		X	X
Model Selection			X	X	X	
Issues Identification			X	X		
Scenario Specification			X	X	X	
Selection of Output Variables			X	X	X	
Running of Models				X		
Displays of Results				X	X	
Model Comparisons			X	X	X	
Critique and Interpretation of Runs			X	X	X	
Write Executive Summary			X	X		
Write Summary Report			X	X	X	
Write Appendixes			X	X	X	
Critique Report	X		X	X	X	
Publication of Report		X			X	
Publicizing Study and Applications	X	X	X	X	X	X

435

chairman, is critical to the study's success. The value of the
process and the final report is dependent upon a strong, know-
ledgeable, diverse working group whose members are familiar with
nuances of policy issues and policy models and devoted to improving
applications of models to policy and planning issues.

Once the working group is organized and holds its first meeting,
complete responsibility for the conduct of the study is vested
in the group.

The working group selects the models to be run. Normally, there
is agreement that each existing current model represented by a
working group member can participate. However, the working group
may identify and recruit additional modelers.

During the first meeting, group members identify the most important
energy policy or planning issues to be addressed and those that
cannot be addressed. The participation of modelers and model
users is critical at this point. Capabilities and limitations of
the various models, along with priorities for further model development, soon become apparent. In general, the goal here is to
focus upon the most important issues and to test the capabilities
of the various models to address these issues. This is to be
contrasted with what can easily happen in practice: the issues
to be dealt with are determined by the particular strengths of the
individual models, with secondary consideration given to the importance of the issue for policy and planning purposes.

Study issues normally relate to informational questions significant
to energy policy and planning. For example, the second study asks
to what extent alternative environmental restrictions influence
both the rate of the transition back to coal and the regional distribution of the growth. The fifth study asks how the U.S. supply
of oil or natural gas will be influenced by domestic price controls.

Alternatively, the study question may be a modelling or forecasting
issue. The third study asks how significant changes in the price
of electricity will influence consumption. Differences in the
answers among models seem to depend upon the geographic scope of
the data base used with the models. Thus, the issue becomes: to
what extent could combined historical data from many regions be
used to improve the estimation of parameters for a model applicable
to a single utility service area?

Once the issues are identified, scenarios are generated to capture
the essential features of the issue being considered. The scenario
specification includes a set of standardized input assumptions,
some of which are changed systematically among the scenarios to test
the models and to provide information potentially useful for policy
and planning purposes. In general, each modeler is asked to run
each of the scenarios with the standardizing assumptions.

Differences in the models become apparent. The input data for
one model may be the output of another. Some input variables
may not be included in a model, even though they are believed by
working group members to be particularly significant in influencing
the projections. Some scenarios simply cannot be addressed with
a particular model except in the most ad hoc way, requiring extensive off-line manipulation by the analyst. Documentation of these
differences is important when comparing the results of the various
models and the models themselves.

The definition of scenarios is accompanied by a selection of output variables to be reported by each modeler. These are normally
selected by the working group in order to provide projections of
those variables that are most meaningful for policy and planning
purposes and to give the working group members an opportunity to
observe the inner workings of a model at several critical points.
For example, in the fifth study, output variables to be reported
will include measures of drilling activity, quantities of oil and
gas discovered, reserves of oil and gas, and production rates for
a selected set of years. Whether or not individual models can
calculate these intermediate variables is often important in communicating capabilities, limitations, and dependability. Comparisons of model behaviour at several observation points provide the
raw material for group discussions and report writing.

Each model is run by its key developer or a close colleague. Third-
party model operation is not routine. There would be advantages
in third-party model operation, but these are outweighed by the
practical advantages of the modelers running their own systems.
The developer may best understand the limits of applicability of
the model. Moreover, the modeler knows which sets of equations
or data within the model must be modified to examine specific scenarios. Having worked extensively with a given system(s), he can
make runs without undertaking the enormous learning cost required
for third-party analysis.

Displays of model outputs are designed so as to facilitate interpretation and comparison. Graphic comparisons are prepared by the
EMF staff for use by the working group. The displays of the outputs, if done creatively, can help in interpreting comparative
behaviour of the individual models.

A large portion of working group time is devoted to critique and
interpretation of the runs. Differences among models provide an
opportunity and a motivation for explaining why the different results occur. Discrepancies in results may point to fundamental
differences in model structure, model parameters, basic data utilized, or perceptions about the direct implications of scenario
assumptions. Divergence in model results normally leads to crea-
tive tension among the modelers with each trying to understand why

his model differs from the others. One motivation is to improve the model if appropriate. Another is to show why one model's answers may be more dependable than those developed by some other modeler. This dialogue is a strong motivating force which leads to important understanding about the fundamental model differences relevant for policy or planning purposes, the areas of uncertainty in knowledge about the world, and the significant areas of research potential. The process can result in the revision of a model during the study to account for implementation problems not initially perceived by the model's developer.

These working group discussions enhance the model users' insights about the policy issues and suggest distinctions among policy options that may not be apparent on the surface. For example, many such policy options may increase the cost of producing a given amount of energy, e.g., restrictions on types of coal that can be used by electricity generation or restrictions against the use of nuclear energy. Others, however, may simply influence the price of energy without influencing its cost, e.g., an energy tax that is redistributed through the U.S. Treasury. The first class will have a far greater impact on economic growth than will the second class of options [13]. What may seem to be a subtle distinction in policy options may have a profound impact on the effects of the various options on economic growth.

The working group report is characteristically written and published in two separate components: a relatively short summary, approximately 30 pages long, directed at a broad audience and a much longer series of supporting documents aimed at a smaller, more technical audience.

The summary explains the major commonalities and differences in the models, provides answers (to the extent possible) to the issues raised, identifies limitations of the analysis, and presents recommendations developed by the group. In writing this report, communication is emphasized; the report is intended to be jargon-free and accessible to non-technical readers. A two to three page executive summary encapsulates the key conclusions of the report.

The series of supporting documents varies significantly from study to study. Generally included, however, are descriptions of the individual models, comparisons of the models, a simplified framework for both comparing the models and communicating an intuitive understanding of the results, a detailed description of the scenarios, detailed results from each model with comparative graphics, and a set of technical papers discussing more deeply any modelling and analysis issues that may have surfaced during the study.

The working group report provides one communication vehicle for disseminating the results of individual studies. Other mechanisms also are used. The Senior Advisory Panel is briefed on the report and members have played significant roles in communicating the results. Working group observers bring insights back to their respective organizations and help to disseminate the study's findings more broadly. Working group members typically make seminar or conference presentations based upon the study. EPRI, as well as the EMF, publishes the report and facilitates its distribution. Thus, many people help publicize the study. The study belongs fundamentally to the working group and the communication of results relies heavily on study group members actively publicizing the results.

Although the EMF reports, individual participant presentations, and other vehicles are used to communicate the results of the study, much of its benefit is not easily transferred to nonparticipants. A major EMF focus is on how people can use models more effectively, but effective use is a skill, learned like any other skill. While an EMF study can help modelers and model users alike in the difficult, artistic process of utilizing models for addressing real, complex issues, the skills and insights gained often cannot be fully transferred beyond the participants, except at extremely high cost. Hence, formal communication of results beyond the working group is an important product of EMF studies, but clearly not the only product.

4.   EMF HISTORY AND EVOLUTION

Several forces provided the impetus behind the creation of the Energy Modelling Forum and its location at Stanford University.

Professor Martin Greenberger of Johns Hopkins University was intensly interested in the relationship between modelling and the policy process [5]. He was concerned that models, on the whole, had not been used as effectively as possible for policy making and planning, partly because of the insufficient linkage between the model development and the policy process. In early 1976, while on leave from Johns Hopkins University as manager of the Systems Program of the Electric Power Research Institute (EPRI), Greenberger developed the idea of establishing an institution designed to improve the use of energy models in the policy process.

At Stanford University, the Institute for Energy Studies (SUIES) had developed a very productive programme in energy research. This research, however, was focused fundamentally in technology areas with relatively little attention to energy policy or analysis considerations. Partially to change this focus, Professor Thomas Connolly initiated an energy policy seminar which brought together policy makers and planners with analysts. The interchanges proved productive. Some seminar members, including Professor George

Dantzig, felt the introduction of formal models useful for policy purposes could improve the seminar activities. Dantzig initiated the PILOT energy modelling project to meet this need, and, in a SUIES seminar, advocated the development of an institution to bring together energy modelers and potential users of energy models.

Chauncey Starr, then President of EPRI, and Martin Greenberger responded favourably. Dantzig and Greenberger pooled their ideas and, with the encouragement of Starr and others at EPRI, jointly developed the concept and initial architecture for the Energy Modelling Forum. A competitive request for proposal was issued by EPRI in the spring of 1976.

During this same time period, William Hogan, then Deputy Assistant Administrator for Data and Analysis at the Federal Energy Administration (FEA), was discussing the possibility of joining the Stanford University Departments of Operations Research and Engineering-Economic Systems. Believing that the use of models for policy and planning could be greatly improved, Hogan was attracted by - and ultimately accepted - Stanford's invitation for him to take the lead role in responding to EPRI's request for proposal. Stanford proposed that the EMF be established at Stanford with Hogan as Executive Director.

In July 1976, a workshop was held at Stanford University under the auspices of EPRI and the Institute for Energy Studies to explore interest in a forum and to examine ways to create and structure such a project. Greenberger initially intended the workshop as a means of launching the Forum project, but the host institution for the EMF had not yet been chosen, and the workshop was conducted independently. The approximately 100 people attending the workshop provided a broad representation of model developers and users.

The idea of a forum was received favourably by participants in the workshop and many variants were suggested. Proposals ranging from a clearinghouse for model access to a certification agency for model evaluation and validation were considered. A central theme confirmed by the discussion was that a strong user orientation and broad participation were crucial.

The Stanford Institute for Energy Studies, selected by EPRI as headquarters for implementation of the EMF project, initiated the project in September 1976 as a six-month experimental effort to test the viability of the concept. William Hogan, by then a Stanford faculty member, became the first Executive Director of the EMF and provided leadership in that position for two years.

For its first study, the EMF examined the use of models to study the nature and strength of the feedback from the energy sector to the aggregate economy. An important topic in its own right, this subject was of particular interest because of the closely related work of the Modelling Resources Group of the Committee on Nuclear and Alternative Energy Systems (MRG-CONAES) [9], which did not fully consider this feedback.

Following the principles developed at the summer workshop, a group of 30 interested model users and developers was organized to conduct the study. Because of the experimental nature of the project, Hogan served as working group Chairman as well as EMF Executive Director.

The six models explicitly represented the energy-economic linkage. Each model was for the full U.S. economy, and each was judged appropriate for long run issues but not for short run issues. As in the MRG, common scenarios were constructed by standardizing many input variables. The working group then sought to explain the common results or the causes of model differences. This comparison process was facilitated by the high degree of commonality among the various models.

The key comparative results of the study were estimates of the aggregate elasticity of substitution[*] implicit in the participating models. This parameter was shown in the report to be one of the key determinants of strength of the link from energy to the economy.

The first EMF study contributed importantly to the current structure of the EMF process. In particular, the study involved considerable participation of the model using community, which resulted in careful attention to specifying limitations of the models in studying the energy-economy issue. Additionally, the study had a strong issue focus. Although some questioned the direct applicability of the study's results to the evaluation of the energy policy options available to the federal government, it served to educate many policy makers about the magnitude of the relevant tradeoffs.

Despite its positive contributions, the first EMF study suffered from several problems that had plagued previous model comparison studies. The group was often torn between the sometimes conflicting goals of policy analysis and model comparison. The study suffered from a lack of visibility; and it had a distinctly academic flavour.

---

[*] For small changes in energy prices, the aggregate elasticity of substitution closely approximates the aggregate price elasticity of energy demand.

The issue focus of the study was intended to aid in the model comparison, but this focus seemed to imply that policy recommendations could be drawn from the study's conclusions. A subgroup of study participants, however, contended that they would have conducted a policy study differently from a model comparison study, and therefore argued against the development of policy recommendations.

Despite the active participation of the user community in the study, the study suffered from a lack of visibility at the highest levels of government and industry. Part of the problem was that a professor as working group chairman and six models developed primarily in universities gave the study an academic aura.

The perceived shortcomings of the first study figured in the decision by Greenberger and Hogan to institute a Senior Advisory Panel of high level energy decision makers to review the Forum studies and suggest new topics for comparative analysis. Charles Hitch, President of Resources for the Future, agreed to chair this panel.

The second EMF study, "Coal in Transition: 1980-2000", differed in several respects from the first. The Executive Director did not also serve as working group Chairman. Rather, Dr. David Sternlight, Chief Economist at the Atlantic Richfield Company (ARCO), was chairman of the working group. His industrial affiliation, user perspective, and previous energy policy analysis and modelling experiences proved to be invaluable. This choice allowed the complementary talents of the working group Chairman and the EMF Executive Director to be jointly applied in the leadership of the study.

The models in the second study differed in scope from those in the first. Three types of models were employed: energy sector models with significant coal detail; models of coal supply, transport, and demand; and a resource planning model. It was now more difficult to standardize assumptions: e.g., the energy sector models took exogenous projections of aggregate energy demand as inputs, whereas the coal and facilities planning models required exogenous demand projections for electricity consumption and for non-utility coal demand. Model comparisons were conducted not only in the parallel mode as in the previous study, but also in a complementary mode with different information developed by different models. In the complementary mode, the results of one model, for example, a detailed energy sector model, could be scrutinized by use of another, more disaggregated model - for example, a resource planning model.

The conflict between the model comparison and policy analysis goals that surfaced in the first study was again apparent in "Coal in Transition". Once more, a subgroup of modelers resisted the notion that the study would make policy recommendations.

The third study group was designed to help electric utilities deal with the new complexities and uncertainties of electric load forecasting. Bernard Cherry, Vice President of Corporate Planning at the General Public Utilities Service Corporation accepted the position of working group chairman, thus repeating the second study's successful practice of having a working group chairman from the relevant industry. Again the chairman's problem orientation provided critical guidance and discipline for the study.

The study contrasted with the previous two because it involved models with differing geographical coverages. Most of the models were used for a particular utility's load forecasting, considering only that utility's region. This was a critical issue in the study because regional differences made it undesirable, if not impossible, to standardize the inputs to the models. The standardization problem was circumvented by allowing each modeler to specify a "best information" base case. Input parameters variations between scenarios were specified in percentage terms and scenario results were examined in terms of percentage differences in values of output variables.

The variance in scope in the models provided observations of differences in model behaviour. In particular, for given percentage changes in electricity prices and in competitive fuel prices, applications of the new nationally oriented models in this study showed a larger proportional impact on electricity consumption than did the utility region models. As the differences were addressed, the striking dissimilarities led to debate among the modelers. The observation led one group to conclude that combined data from many utility service regions could be used to more accurately estimate parameters in the demand models than would be possible by using data from only a single utility region. Another group felt that combining this data would reduce the quality of the estimates because of the great differences among regions. Although consensus was never achieved, the interchanges led some participants to consider the benefits of changing utility forecasting practice by estimating parameters on combined data, and it led other participants to further critical examination of the econometric foundations for pooling data from many regions.

Many participants in the third EMF working group initially questioned the potential value of the study and worried about the possibility of misuse of the results. By the time the study was completed, the participants were calling for continuation and expansion of this type of activity. Strong support by the working group members helped the Electric Power Research Institute launch its Utility Modelling Forum (UMF), a comparative analysis project focusing specifically on the problems of the electric power industry, in the autumn of 1978.

During the summer of 1978, Hogan resigned his position at Stanford University to join the Harvard faculty and was succeeded as executive Director by Professor James Sweeney, an Engineering-Economic Systems faculty member. John Weyant, an original EMF staff member, was appointed Associate Director to replace Shail Parikh, who had resigned from that position some time previously. Weyant's expanded role in EMF, together with a period of overlap between the outgoing and incoming Executive Director provided some measure of continuity in the operation of the Forum.

The fourth study "Aggregate Energy Demand Elasticities", is somewhat different from the earlier studies. Motivated partly by the EMF 1 conclusion that the aggregate elasticity of substitution is a critical determinant of the link between energy and the economy, the Senior Advisory Panel recommended that the EMF perform further experiments to improve the precision and level of confidence in the elasticity estimates. They felt, however, that the experiment would be too technical and not tied closely enough to specific policy issues to warrant formation of an EMF working group. Thus during early 1978, the EMF staff, with the aid of many outside experts, designed an experiment to estimate the aggregate demand elasticity implicit in energy models.

By late 1978, 18 models had executed the experiment. Interest in the study's results had escalated to the point where face-to-face meeting of the study participants and interested observers was deemed desirable. Therefore, a working group of approximately 40 people, predominantly model builders, was formed. Hogan, by then at Harvard University, became working group chairman, thus providing further continuity in the transition to a new Executive Director. One meeting of the working group has been held to date, and a second meeting will probably be necessary.

The scope of the models of this fourth working group varied greatly. There were self-standing U.S. models and U.S. models as components of international energy demand models. Some were highly aggregated, others disaggregated by fuels. Some models included the energy sector embedded within the entire economy. Some examined the aggregate of all energy consuming sectors, while others were disaggregated by sector or represented only a single energy consuming sector.

While it has not yet been completed, the fourth EMF study has raised the question of whether a technical comparison study can be conducted apart from the model using community. However, we believe that to be valuable the technical comparison must be focused on one or two relationships widely believed to be significant. The ultimate utility of the study will depend upon our ability to widely communicate the results effectively to the model-using community.

At the time of this writing, the fifth EMF study, "U.S. Oil and Gas Supply", is just getting under way with Ben Ball as working group chairman. As was true in most of the previous studies, the chairman, currently an Adjunct Professor of Management and Engineering at the MIT Energy Laboratory and formerly a Vice President of Gulf Oil Corporation, has fundamentally a user perspective.

It is too early to predict the progress of the study. One difference, however, from past studies is in the effort placed on model examinations and comparisons. Energy Modelling Forum staff members have drafted and made available to working group members a comparison of major oil and gas supply models. It is hoped that deeper model assessments and comparisons will be possible.

Table 2 summarizes the progress of the various studies as of February 1, 1979.

TABLE 2 - Phases of EMF Studies[a]

Phases	EMF 1	EMF 2	EMF 3	EMF 4	EMF 5	Energy/ Envir.?	World Oil?
Identification of Potential Area	C	C	C	C	C	C	C
Consideration by Senior Advisory Panel	C	C	C	C	C	C	C
Preliminary Exploration and Issue Definition	C	C	C	C	C	●	●
General Topic Area Selection	C	C	C	C	C		
Working Group Chairman Recruiting	C	C	C	C	C		
Specific Topic Selection	C	C	C	C	C		
Working Group Recruiting	C	C	C	C	C		
Model Selection	C	C	C	C	●		
Issues Identification	C	C	C	C	●		
Scenario Specification	C	C	C	C	●		
Selection of Output Variables	C	C	C	C	●		
Running of Models	C	C	C	●			
Graphical Displays of Results	C	C	C	●			
Model Comparisons	C	C	C	●	●		
Critique and Interpretation of Runs	C	C	C	●			
Write Executive Summary	C	C	C	●			
Write Summary Report	C	C	●	●			
Write Appendixes	C	C	●	●	●		
Critique Report	C	C	●				
Publication of Report	C	C					
Publicizing Study and Applications	●	●					

C - Completed as of February 30, 1979
● - Ongoing as of February 30, 1979

[a]EMF studies: EMF 1: Energy and the Economy
EMF 2: Coal in Transition: 1980-2000
EMF 3: Electric Load Forecasting: Probing the Issues with Models
EMF 4: Aggregate Demand Elasticities
EMF 5: Oil and Gas Supply Functions

Two other study topics identified by the Senior Advisory Panel as high priority areas are listed in Table 2: "Energy and the Environment" and "World Oil Supply, Demand and Prices". Although, at this time we do not know whether either of these two subjects will be chosen for future studies, the EMF staff is currently in the process of preliminary exploration and issue definition. This process will help us decide whether such studies will be feasible and valuable, as well as educate the EMF staff and facilitate the working group deliberations.

5. ISSUES FOR THE FUTURE

Unresolved issues to be addressed include:

- the appropriate tradeoff between model comparisons and policy analysis,

- the extent to which the EMF should conduct model assessments and evaluations,

- the appropriate role of the EMF in an academic institution such as Stanford University, and

- the extent to which study participants should be compensated for their time and computer expenses.

A tension keenly felt in the first two studies and anticipated in the fifth is the appropriate tradeoff between model comparisons and policy analysis. A typical working group includes a mix of people, some primarily concerned with modelling and some primarily with using information for policy and planning purposes. Indeed, a major goal of the project is improved communication between these two groups. While these two activities are complementary in many respects, in other respects they conflict. The model comparisons considerably improve the policy analyses by refining the quality and the reliability of the information developed by using models. The policy analysis enhances the relevance of the model comparisons by structuring the comparisons to focus attention on similarities and differences most relevant for policy issues.

The first area of conflict between these two goals comes in the choice of scenarios. Because of limited time and resources, only a small number of scenarios can be structured, implemented, and interpreted by the group. Since some scenarios are most useful for model comparisons while others would be most useful for policy analyses, the selection of scenarios represents implicitly or explicitly a choice between model comparisons and policy analyses. The model users would have little interest in a study devoid of policy implications. But the model developers are anxious to examine the comparative advantages of their systems. While efforts have been made to choose scenarios to satisfy both purposes, the tension has been strongly communicated.

The tension between goals also is felt in subsequent interpretations of the results of the model runs, with conflicts on the allocation of group time for model comparisons versus policy analysis. It has been felt generally that users would see little value in detailed, jargon-ridden debates on the equations and data embedded in the models. On the other hand, without such debates, in-depth model comparisons may be impossible. One proposed solution is to extend one or more working group meetings by a day to allow the modelers (and anyone else desiring to participate) an opportunity for in-depth model-oriented discussions. Another would be for the EMF staff to draft more complete model comparison papers and to allow some of the debate to proceed by mail and telephone, directed toward improving the EMF draft.

In developing the final report, the tension over goals also is apparent. The various goals imply different themes and formats for the report. The solution to date has been to write a summary report which focuses on the policy analysis and on the capabilities and limitations of the models as a class. Individual model comparisons then appear in a longer report. While this compromise gives weight to both goals, it tends to downplay the model comparison objective.

The issue of policy analysis versus model comparison may never be resolved, but this may be a sign of health. The tension between the goals provides opportunities for working group members to concentrate their own efforts primarily on aspects of the study most relevant to them and thereby helps to improve the overall study quality.

A closely related issue concerns the depth to which EMF should conduct model assessments and evaluations as opposed to simply model comparisons. The EMF recently has been criticized, particularly by the academic community, for its lack of critical review of the participating models.

The Forum has focused attention on the "ventilation" of models, the simple examination and explanation of their behaviour [18]. This step, of course, logically must precede evaluation or assessment. This comparative study of the behaviour of a number of models allows consequent work to identify differences stemming from data differences, structural differences, differences in explicit assumptions, and often most importantly, differences in implicit assumptions or world view held by the developers. Although evaluation per se has not been conducted by the Forum, the differences identified through the comparisons provide an improved basis for individuals to make their own evaluations of the models.

There are several reasons why the Forum up to this point has not conducted in-depth evaluations. First is the question of the extent to which objective comparisons are possible. Of course, some aspects of assessment can be conducted objectively. One could examine whether the computer code was written as the developer intended or could attempt to replicate the underlying econometrics. Activities of this sort are in fact being conducted by the MIT Model Assessment Laboratory [10]. Some assessments, however, cannot yet be objective with the current state of the art but are based upon subjective peer review judgements. Econometric evidence, for example, can be viewed differently by various professionals. Even more difficult is sorting through and assessing the implicit assumptions and the world view incorporated in the mode. Evaluating which implicit assumptions and world view are more nearly correct cannot be done objectively. Individuals, however, can make their own judgements on a subjective basis, if the behaviour of the models is clearly communicated. This individual, subjective process is facilitated by the Forum studies.

Furthermore, early experience at the Model Assessment Laboratory has demonstrated that a credible in-depth, hands-on, third-party review of a single model can require resources comparable to a full EMF study. A comparative in-depth assessment of, perhaps, 10 models could require in order of magnitude more resources. Faced with a limited budget, even if desirable, such a process is impossible for the Energy Modelling Forum.

Another difficulty with model evaluation is that different models may be particularly useful for different purposes. The type of model useful for forecasting the consumption of gasoline in the presence of new car average efficiency standards may be different from the type desirable for forecasting the market share of station wagons. Different models may be more appropriate for various alternative purposes. For example, for forecasting gasoline prices two months hence, a simple time trend extrapolative model may be far superior to one including a detailed representation of the economic and engineering relationships. Conversely, for evaluating the impacts of oil decontrol on gasoline prices, the time trend extrapolative approach would be useless, while a structural modelling approach could be quite effective (12). Realizing that different models are appropriate for different purposes, the EMF has attempted simply to delineate the capabilities and limitations of various models without undertaking the more difficult task of model evaluation.

Finally, potential working group members must be convinced of the value of the process to them as individuals. Each volunteers time and many contribute computer costs. If the expected rewards for participating are primarily public criticisms of their models, especially criticisms built on weak foundations, then the voluntary participation could be reduced notably. This may still become an issue. However, if the assessments are objective, and if they recognize strengths along with weaknesses, more telling comparisons probably are possible without discouraging the participation of the modelers. In this way, the Forum may include some elements resembling professional peer review, but without the academic apparatus of manuscript referring.

The current movement is toward deeper model comparisons. At the same time, it is crucial that the issue focus not be lost. For the fifth study, the EMF staff devoted to model comparisons has been significantly expanded. Staff members currently are examining methodology differences among oil and gas supply models, to the extent possible without hands-on experience. This examination was started even before the working group convened, with the expectation that the group will encourage and participate in this process. This expectation is being realized. The extent to which the EMF evolves towards deeper model comparisons and evaluations is an open issue. Its resolution depends upon such factors as the future EMF budget, the preferences of working group members, and the willingness of working group members to expose their models to critical review.

Closely related is whether or not third party model operation should be introduced into the EMF. In the current mode of operation, all model runs are made by the model developer, not the EMF staff. Benefits of changing this mode of operation would be associated with the opportunity for a more scientific, objective, and complete examination of individual models. The costs would be several, and high. First, would be the requirement for staff members to learn software that is model-specific. Second, there would be a weaker linkage between the evaluation process and the modelers, possibly resulting in a lesser interchange of information. Finally, model assessment is far more costly than the current EMF procedure. Therefore, it is expected that little if any independent third party operation will be introduced.

The role of the EMF in an academic institution such as Stanford University, where the educational progress of students is a key concern raises another issue. In the past, Stanford students have participated in the EMF, but to a relatively limited extent. Currently, eight Stanford graduate students are participating directly in ongoing EMF studies or in preliminary issue identification and exploration for possible future studies. Many of these students are looking into methodological issues or are comparing methods, supporting data, or econometric techniques underlying participating models. This activity contributes simultaneously to the academic goals of Stanford University and toward deeper, more telling model comparisons and evaluations. Thus, responsiveness to the educational goals seems at the same time to allow responsiveness to the call for increased critical evaluation, while maintaining the issue focus.

The final unresolved issue involves costs. Participants in an EMF study volunteer their time and generally computer expenses. Although most feel a sense of professional obligation and perceive a learning experience in participation, which justifies the donation of their time, the expense of running the sometimes quite costly models is not so easily absorbed. However, this policy may have, in the past, and probably will, in the future, exclude some of the models.

Many private and public sector organizations benefit extensively from the studies, but bear none of the costs. Plans, therefore, are under way to raise money to cover the out-of-pocket expenses of study participants whose organizations cannot bear these costs. What kinds of costs should be covered, to what extent, and by whom are unresolved at present.

6.   SUMMARY AND CONCLUSION

The Energy Modelling Forum, organized almost three years ago as an experiment to improve communication between energy decision makers and energy modelers, has been succeeding although many issues remain unresolved. Future success depends upon continuing cooperation from the broad community of energy modelers, planners, and policy makers. Constructive critiques and suggestions and constant efforts to improve interaction will be important factors in fulfilling the objectives promised in the original design of the Energy Modelling Forum.

REFERENCES

1.   Energy Modelling Forum, "Aggregate Energy Elasticity Estimates", Working Paper EMF 4.4, Stanford University, Stanford, California, September 1978.

2.   Energy Modelling Forum," Coal in Transition: 1980-2000," EMF 2, Stanford University, Stanford, California, July 1978.

3.   Energy Modelling Forum, "Electric Load Forecasting", Working Paper EMF 3.11, Stanford University, Stanford, California, August 25, 1978.

Energy Modelling Forum, "Energy and the Economy," EMF 1, Stanford University, Stanford, California, September 1977.

M. Greenberger, M.A. Crenson and B.L. Crissey, Models in the Policy Process: Public Decision Making in the Computer Era, Russell Sage Foundation, New York, 1976.

M. Greenberger, Closing the Circuit between Modelers and Decision Makers , EPRI Journal, No. 8, Electric Power Research Institute, Palo Alto, California, pp.6-13, October 1977.

W.W. Hogan, "The Energy Modelling Forum: A Communication Bridge", paper presented at the 8th Triennial IFORS Conference, Toronto, Canada, June 20, 1978.

W.W. Hogan, "Energy Models: Building Understanding for Better Use", paper presented at the Second Lawrence Symposium on Systems and Decision Sciences, Berkeley, California, October 3-4, 1978.

T.C. Koopmans, et al., "Energy Modelling for an Uncertain Future", the report of the Modelling Resources Group, Synthesis Panel of the Committee on Nuclear and Alternative Energy Systems, National Research Council, Washington, D.C., 1978.

Model Assessment Laboratory, "MIT Model Assessment Laboratory: First Year Report", Massachusetts Institute of Technology, Cambridge, Massachusetts, 1978.

Stanford Institute for Energy Studies, "Stanford-EPRI Workshop for Considering a Forum for the Analysis of Energy Options Through the Use of Models", Electric Power Research Institute, Special Report EPRI EA-414-SR, May 1977.

J.L. Sweeney, and M. Flaherty, Methodologies for Petroleum Product Price Forecasting: A Review , in Topics in Energy, Data Resources, Inc., September 1978.

J.L. Sweeney, "Energy and Economic Growth: A Conceptual Framework", EMF Occasional Paper 3.0, Stanford University, Stanford, California, November 1978. Also appearing in Symposium Papers: Energy Modelling and Net Energy Analysis (Colorado Springs, Colorado, August 21-25, 1978), Institute of Gas Technology, Chicago, 1978; and in Perlmutter, Arnold; Kadiroğlu, O.K.; L. Scott (eds.); Proceedings of the International Scientific Forum on an Acceptable World Energy Future, Ballinger Publishing Company, Cambridge, Mass., 1979 (forthcoming).

APPENDIX

Senior Advisory Panel

Mr. Charles J. Hitch, Chairman	President, Resources for the Future
Dr. Philip Abelson	Editor, _Science_
Dr. Harvey Brooks	Professor, Harvard University
Mr. David Cohen	President, Common Cause
Mr. Gordon R. Corey	Vice Chairman, Commonwealth Edison
Dr. Floyd L. Culler, Jr.	President, Electric Power Research Institute
Mr. Charles Di Bona	President, American Petroleum Institute
Mr. Herman M. Dieckamp	President, General Public Utilities Service Corporation
The Honorable John D. Dingell	Member, United States House of Representatives
The Honorable Joseph L. Fisher	Member, United States House of Representatives
The Honorable William P. Hobby	Lieutenant Governor of Texas
Mr. Jack K. Horton	Chairman, Southern California Edison Company
Mr. W.F. Kieschnick, Jr.	Vice Chairman, Atlantic Richfield Company
Dr. Henry R. Linden	President, Gas Research Institute
Mr. Guy W. Hichols	President, New England Electric System
Mr. John F. O'Leary	Deputy Secretary, United States Department of Energy

DISCUSSION

There was some discussion on the EMF study on electricity load forecasting. Sweeney was asked whether EMF had studied how these forecast models were actually used by the utility companies in the planning of their investment programmes. There were many sources of uncertainty in the models and the basic question was how this was taken care of in the next stages of the planning process. Sweeney replied that uncertainty in the input data (i.e. different economic growth rates, price changes in competing fuels, effects of temperature) were normally taken care of by running a set of different scenarios. Uncertainty in the output as a result of model structure was, however, not normally examined. This was what the EMF study had set out to do.

As regards the bias which the utility companies might have as to over- or under-dimensioning the supply system in regard to their forecasts, a study had been done at the Electric Power Research Institute on the costs of over- or under-dimensioning. Very little work had been put into determining the penalty function (i.e. the cost of not being able to deliver enough power) and this was a great opportunity for future research.

Ontario was quoted as an example of a region which had overestimated the number of nuclear stations required and had placed contracts accordingly. Much of the problem stemmed from not solving for supply and demand of electricity simultaneously. Sweeney replied that none of the models in the EMF study had solved supply and demand simultaneously, but the study had concluded that this would be a natural extension of the models. The models on coal supply, which another EMF study had analysed, included this feedback. Models of oil supply did not normally include this mechanism because of the importance of imports.

It was pointed out that a natural reaction to reducing the risks involved in failing to meet supply was to develop new technology. Such a development was the installation of fuel-cells at secondary transformers to meet peak load. Power pooling was also a way of reducing the risks. This did not help solve the problem though, if all the power companies underestimated demand.

It was pointed out that one of the general insights from the EMF studies appeared to be that engineering models operated with lower estimates of energy elasticities than did econometric models. One of the drawbacks with an econometric model was that it did not identify the engineering (real life) variable which was a function of price. The exogenous inputs to an engineering model implied some sort of price elasticity.

Sweeney replied that in principle the questioner was right: the exogenous imputs to an engineering model did imply a price elasticity. In practice, these estimates were rarely changed when energy price changed so that in practice it was difficult to incorporate price effects in engineering models. Many engineering models were used to assess the potential usefulness of alternative energy technologies. Normally this was done by assuming a constant demand development and so such models would tend to overestimate the usefulness of the alternative technology. On the other hand, it was difficult to take such factors as structural effects, efficiency standards and new technologies into account in econometric models.

Sweeney contended that much of the debate which was going on, especially in the U.S.A., on the comparative usefulness of different classes of models, was misdirected. The most probable development was the merging of engineering and econometric models. This point was supplimented by Saltzman who said that parallel to the econometric model for electricity demand in New York State which he had described at the seminar, another group had developed an engineering model. They planned to come together later in the year to see if it were possible to merge the models in some way. He could see that an engineering model could be enhanced by obtaining better estimates of its exogenous economic inputs from an econometric model, but had difficulty in seeing how engineering information could be incorporated in the econometric model.

Sweeney said that another general insight from the EMF work was that modellers find it hard to come up with conclusions about the real world; it was easier to talk about the results of the models. It was not sufficient to write reports and journal articles, but was necessary to involve the user. The user in this case was often a group of professional staff between the modeller and the decision maker. In response to a question on recommendations for institutional changes to promote this process, Sweeney's immediate reaction was greater support for the EMF to extend its studies.

PART VII

CRUDE OIL PRICING MECHANISM

Panel Discussion

PANEL DISCUSSION: CRUDE OIL PRICES

Panelists

Charles Blitzer	*World Bank, Wash., D.C., U.S.A.*
Joseph Debanne	*University of Ottawa, Ottawa, Canada*
James Sweeney	*Energy Modelling Forum, Calif., U.S.A.*
Robert Deam	*Queen Mary College, London, U.K.*
Leif Ervik	*Ch*. *Michelsen Institute, Fantoft, Norway*

The panelists were asked to express their opinions on the underlying mechanisms that determined the actual price of crude oil,
and the extent to which these mechanisms could be captured within
a formal model.

After an initial round, the panelists were asked to respond to
the views expressed by the other panelists. The Panel was followed
by a discussion session where the audience also participated.

INITIAL REMARKS

Charles Blitzer

I would like to begin by saying that there is no way that anyone
can presume to make strictly accurate forecasts. Prices are settled according to political facts, and according to realities.
To a large extent, fundamental forces at work are more or less
economic. Thus, economists can put some dimensions to the long
run trajectories.

As far as the world oil market is concerned, there are certain
limitations on the supply side. Oil production capacity is not
infinitely expandable, but has a more or less well-defined ceiling.
Oil producing countries are essentially profit maximizers; maybe
not in the short term, but in the long run, at least. In my opinion, in oil models, essential features can be captured quite

simply; it is <u>not</u> necessary to go into details like quality, substitutions of various forms of energy, but in the long run we must
embed the oil model into an <u>energy</u> model. Because, as the long
run prices change, oil will be substituted by other sources of
energy, like coal, uranium, etc.

As the major element of any model, the behaviour of the oil cartel
is an influencing factor. The cartel can either set a price trajectory, or a quantity trajectory, but not both. The ultimate
prices and quantities are settled, on the other hand, by how much
people expect the market for energy to grow. In other words, their
expectations about growth in GNP at the global level, as well as
some expected view of price elasticity and ultimately the growth
in demand will determine prices. And I think once you carefully
model the investment in terms of supply you captured about 90% of
what is important in enabling us to determine what the derived
demand for OPEC oil supply is. When this is done, then I think
simulation models will play a more important role than optimization models in that the cartel itself is highly diffuse in its
own objectives, and some countries are better able to absorb revenues than others; some countries have larger reserves than others; and consequently somewhat different objectives and discount
rates. These have to be taken account of in a very explicit way
which does not readily lend itself in my mind to easy optimization
procedures. On the other hand, by hypothesizing alternative simulations, each of which might be identified with the optimal strategy of particular members of the cartel, and looking at strategies in between, I think a great deal of insight can be gained in
terms of what are realistic expectations of the way that the cartel
might come out in terms of its own internal strategies. I think
that optimization and in particular analytic optimization models
are useful in giving us a frame of reference from which we can
judge the behaviour of what could be called the "optimal cartel",
one in which there are full possibilities for internal repayments,
prepayments, and financial interactions. But the real world is
not working that way, and hence, I think in focusing on numeric
expectations, you want to focus on simulation models. The subject
that will lend itself more readily to such simulation will be
expectations of investment behaviour in alternative supplies.

Joseph Debanne

Before expressing an opinion about the future price of oil, it may
be in order to state at the outset that there is very little relation between the cost and the price of oil. In most other industries one usually can estimate fairly accurately the future
cost of a product,but one gambles as to market conditions for the
product, i.e. its price and the quantities that could be sold.
In the petroleum industry, however, market conditions are usually

known with little uncertainty, if any, but one gambles as to the future cost of new oil. Once a well is drilled or an oil field developed, the major costs of the oil producer are already incurred; i.e. are sunk costs, hence are irrelevant in the decision of the oil producer to produce and sell oil at the going price or to shut down his well or oil field. This is because production costs are usually very small compared with the price of oil. Consequently, as long as the going price of oil exceeds production costs, the producer will produce and sell his oil at the going price, i.e. at the dictated price, in order to maximize his gain or minimize his loss. This "going price" has therefore no short term relation to the unit cost to find, develop and produce the barrel of oil, and can just as easily be set at $ 3.0 a barrel or $ 30.0 a barrel. The OPEC countries know these basic fundamentals about the pricing of oil and are determined to keep control over the dictated price of oil as long as OPEC will be in a position to clear the oil market. OPEC control over the world price of oil ensures, therefore, that the days of $ 1.60 a barrel oil will never come back again, and that the price of oil will be maintained at the level desired by OPEC. In short, an economic imperative bonds the OPEC group and this bond is reinforced by the spectre of financial ruin if OPEC loses control over the price of oil.

Another important cohesive bond in the OPEC group is ideological and has two facets: that of underdevelopment, and that of Islam. All OPEC countries are underdeveloped and feel up to this day exploited and taken advantage of by the technologically advanced OPEC countries. Rightly or wrongly, OPEC countries feel that they are currently being short changed in the barter of their precious depletable oil resources against depreciating dollars that have failed till now to buy them development and a reasonable hope to prosper economically after oil reserves run out. OPEC's attitude towards OECD can be summarized in one sentence: "If you expect us to meet your oil consumption needs, we expect you to meet our development needs" and it stands to reason that a united OPEC should be able to obtain better terms in the oil versus development barter by remaining united.

The second facet of the ideological bond uniting OPEC is a fluke of nature, or rather, of history. It happens that 95 percent of OPEC oil is situated in predominantly Moslem countries. Since Islam is both a religion as well as a civil code, the bond among Moslem countries is much stronger than among other religious groupings thereby reinforcing the cohesion of the predominantly Moslem OPEC bloc. Moreover, the Moslem world, hence the mainstream of OPEC countries, which have been left on the sidelines of world power plays since the haydays of the Ottoman Empire, perceive in OPEC an opportunity to acquire once more a place under the sun in terms of economic power and influence. In other words, oil power has become the major strategic and tactical asset of the Moslem

world since OPEC wrested away control over oil price and oil supply from the multinationals. These countries are not about to allow the dissipation of this power via the dismemberment of OPEC.
This is because oil power is perceived as the surest guarantee
that the underdeveloped Moslem countries within and outside OPEC
will be given a fighting chance to develop modern economies...
OPEC is therefore perceived as the providential instrument that
may bring about another golden age in the Land of Islam.

Another factor reinforcing OPEC solidarity is the predominantly
Arab nationality of OPEC oil. Close to seventy percent of this
oil belongs to OAPEC countries who also exercise effective control
over OPEC policies, primarily because of Saudi Arabia and the other high reserves per capita producers. The Arab connection is
one further reinforcing bond within OPEC, particularly since the
recent political events of Iran which brought back Iranian oil
and foreign policy in line with Arab policy, notably regarding
the thorny issue of the Arab-Israeli conflict. In this last respect OAPEC countries view OPEC as a precious and perhaps as the
main leverage instrument to bring about concessions and hopefully
a political settlement to the Palestine-Israel issue. Accordingly,
OAPEC countries have a vital interest in the maintenance and
health of OPEC.

In the light of the foregoing it appears that an imminent breakup
of OPEC due to internal dissentions or due to futile oil imports
quota and oil taxation schemes is even more unlikely today than
it was in 1974, because there are several powerful forces: economic, ideological, and political that tend to reinforce the cohesion of OPEC. To illustrate quantitatively the importance of
the cohesive bonds uniting the OPEC group, it may be in order to
mention that OPEC countries spend four times more than OECD countries in terms of a percentage of their GNP on foreign aid. Moreover, OPEC foreign aid is primarily concentrated on oil-poor Arab
countries. Egypt, for example, receives 2 billion dollars annually
from Saudi Arabia alone.

This convincing demonstration of Islamic-Arab solidarity by the
dominant elements, as well as the majority of the OPEC coalition,
leads to the conclusion that it would be foolhardy to forecast or
model the future price and supply of OPEC oil on the basis of
purely economic considerations. How to take into account the
ideological-cum-political factors as well as the economic factors
affecting the future price and production of OPEC oil is a great
challenge to energy system analysts and model builders. It seems
to me that a first step ought be to assign relative weights to
the economic, ideological and political factors affecting the
price and the production of OPEC oil. Here again the relative
weights vary with the various categories of OPEC member countries.

All OPEC members are vitally interested in ensuring their own economic development and are anxious to extend oil production income long enough to ensure the development of a viable economy after the oil resources are depleted. OPEC countries, notably Algeria, have spearheaded the Group of Seventy (countries) which is demanding in increasingly forceful terms a more equitable economic order. This "New Economic Order" mentality is shared by all OPEC countries and is gaining momentum. The quest for development and for a more equitable share unquestionably exerts an upward pressure on oil prices, in the absence of political considerations. Its effect on production rates varies depending on the "reserves per capita" status of the member country. Low reserves per capita countries such as Indonesia or Algeria tend to operate at low production life indices, i.e. to deplete their reserves fast in order to generate the cash flow required for pressing development commitments. High reserves per capita countries prefer to produce a smaller fraction of their reserves per year, i.e. to operate at high production life indices. This is the case of Saudi Arabia, Kuwait, the Emirates and, to a lesser extent, Libya. This tendency is further accentuated by the fact that the major oil reserves of OPEC are located in thinly populated countries who have lower requirements for petroleum income.

Considering the very high production potential of these same countries, notably Saudi Arabia and the relatively modest needs for petroleum income, these countries enjoy an unusually wide margin of freedom to raise or lower production rates in order to exert economic leverage. This leverage has been successfully applied by Saudi Arabia to impose its leadership on the OPEC coalition, notably on oil price levels, by adjusting the Saudi production rate to achieve the equilibrium price favoured by the Saudis. Since the price of oil and oil supply levels are of crucial importance to the rest of the world -in particular the United States- OPEC countries led by Saudi Arabia and with the cooperation of Khomeini's Iran are increasingly in a strong leverage position in international politics. Fortunately for the OECD countries, Saudi Arabia, the leader of OPEC and OAPEC, is a dove and has tended to moderate the demands for higher prices from the low reserves per capita countries, and to blunt acute oil shortages by increasing oil production when needed. These favours to the OECD countries and the United States in particular are in my opinion in good part motivated by political-cum-ideological considerations. Being anticommunist and pro-capitalist, Saudi Arabia dreads the spectre of a severe depression in OECD countries which would further the cause of communism. On the other hand, Saudi Arabia and the other high reserves per capita countries are anxious to see a satisfactory resolution of the Arab-Israeli problem. As long as the Saudis perceive that progress towards a solution of this problem is tangible, they do not hesitate to make oil price and production level favours to OECD and Uncle Sam. In case of a setback, as they perceived the Camp David accords to be, the Saudis tend to sit back

and let the pressure of an oil hungry market push the oil price
up, as happened recently after Iran drastically reduced its oil
production by several million barrels per day. Admittedly, Saudi
Arabia could not have intervened overnight like a central banker
as it would take several months of lead time and a crash programme
to expand Saudi oil production capacity by several million barrels
per day. Note, however, that the Saudis did not oppose the last
oil price increase in early 1979 and did not embark on a crash
oil production capacity expansion programme with the purpose of
forcing a roll back of the price of oil to pre-1979 levels. This
was a departure from previous Saudi policy; and part of the explanation lies in Saudi Arabia's displeasure with the Camp David
accords which the Saudis perceived as a step backward in the quest
for a global peace settlement in the Middle East. The relative
weight to be assigned to this factor is larger than customarily
perceived by planners and politicians in North America, judging
by recent Saudi pronouncements (...Yamani, Prince Fahd) and particularly by Saudi Arabia's behaviour as leader of OPEC.

Where do we go from here? I expect that the price of oil over the
medium term will continue to seek 1974 levels in constant dollars,
i.e. in buying power of the barrel of oil. The buying power of
the OPEC barrel of oil was allowed to fall substantially below
its 1974 value over the past years, primarily because of Saudi
Arabia's will to "reward" the Carter administration for its peace
initiatives in the Middle East, and also because of concern for
the economic health of the Western World. The events of Iran and
the Camp David accords discussed earlier triggered a substantial
upward adjustment of the price of oil in 1979, to 1974 levels,
with the blessing of Saudi Arabia. The resulting adverse effects
of this large price adjustment and the indications that President
Carter is too weak domestically to exert effective pressure in
the Middle East towards a global settlement leads to the prediction that the Saudis will not seriously oppose frequent oil price
adjustments, up to 1974 levels in real terms, until a new US president with sufficient credibility takes on the task of moving the
Middle East towards a global peace solution. This new policy offers also the advantage that smaller, although more frequent oil
price adjustments will be less disruptive to the world economy.
One last advantage in Saudi Arabia's perspective would be that the
increase in world oil demand which was fueled by the decline of
the real price of oil will be moderated by higher, stable prices
and will encourage energy conservation and the development of
alternative sources of energy.

In short, the 1974 prediction in my Canadian Journal of Petroleum
Technology paper is still valid for the coming years regarding the
future price of oil, which is to remain close to its 1974 value
in real terms, and regarding the cohesion and viability of OPEC.

James Sweeney

For the last couple of days, I've been asking Ibrahim to excuse me from this panel. I keep telling him that I, myself, am confused about world oil prices and what will happen next. But Ibrahim won't listen to my pleas. As I warned him, I will be forced to share my confusion with you. Let me begin.

Initially, I do not feel confident that we have agreed on the correct mental model which accurately describes OPEC behaviour. The next step -going from that mental model to a mathematical model- may be even more difficult. There are many modelling alternatives we could pursue and it is not clear to me which are the best ones. How should OPEC decision rules be represented? How can we best represent the future evolution of the world energy system under different OPEC prices? A great deal of uncertainty surrounds both areas of inquiry.

Let's talk about OPEC decision models. There are a number of mental models, as well as formalized models, which people implicitly adopt in framing the issues. OPEC is treated in a variety of ways. A catalogue would go something like this.

First, OPEC is seen as an economic optimizer.

Second, OPEC is seen as a "satisfier", following certain rules of thumb. An example of this approach would be Gately's models which postulate a set of rules to describe OPEC dynamics. These rules, of course, may be inconsistent with those models which postulate optimizing behaviour of OPEC, particularly those which assume a world of perfect certainty (including perfect foresight).

Third, is the group of "absorptive capacity" models. These models are based on the notion that some OPEC countries set target revenues as part of their development plans and set prices to reach those targets. I personally do not support these mental models, but they are used extensively, particularly in qualitative arguments. One example often cited is that of Iran, where it is asserted that failure to recognize a political absorptive capacity helped lead to the Shah's downfall.

Fourth is the theory of a producibility constraint (physical or other)--OPEC countries simply set a quantity trajectory and the price trajectory naturally follows. Setting a quantity trajectory can be absolutely consistent with optimizing behaviour, if one chooses a trajectory which maximizes profit. Whether price is set or quantity is set is irrelevant in a model which assumes no uncertainty, no policy response, etc. The issue that differentiates this class of conceptual models from the optimizing models is that changing external conditions (demand, etc.) will motivate a change in the supply trajectory projected using optimizing models. No such change occurs under this class. For example, some analysts at the U.S. Department of Energy and at the CIA conjecture that OPEC faces a maximum oil production constraint, or that OPEC, for non-economic reasons, has decided to produce oil at a predetermined rate over time. The result is that price projections are simply determined by the crossing of the demand function with the fixed supply.

Within the class of optimizing models there are several subcategories. Earlier speakers implicitly treated OPEC as a short run revenue maximizer. Within such models maximization could occur for OPEC as a whole, for coalitions of countries, or for individual countries. The smaller the price-setting subgroup assumed, the lower will be the calculated optimal OPEC price. For example, studies we conducted in the Federal Energy Administration suggested that if the maximization were for OPEC as a whole, then prices could be expected to sharply increase. However, if the Saudis as residual suppliers were to maximize their own revenue, then reduced prices could be expected.

OPEC could also be treated as maximizing a discounted stream of revenues net of costs, as in the Pindyck models or in Marshalla's work which treats OPEC as solving a Cournot game, or Stackleberg game.

One could also construct alternative models of world oil supply and demand. We have models of the oil market only, and models that examine oil within the context of the world energy situation. Some models have static demand functions, others dynamic. Fringe suppliers might be represented as moving along a (dynamic or static) demand curve or as solving an intertemporal optimization problem. The world oil market may be represented as a bathtub with a flow in and a flow out, or as a logistic network of tankers, pipelines, and refineries.

This still says nothing about problems of parameter estimation once the model is conceptually specified.

Why is all this important? Primarily because these various models all give different answers to the questions of future price trajectories and to estimation of impacts of policies.

Let me give you an example. As a mental experiment, let's assume that through a magical conservation effort, the United States were to reduce its demand function for oil by 3 million barrels per day. Would this increase or decrease the world price of oil? If you used a model of OPEC as a maximizer of revenues minus cost or as a revenue maximizer, this action would decrease the world price of oil, assuming the short run elasticity of demand to be less than unity. If you used an absorptive capacity model and the elasticity of demand were less than unity, then reducing the demand for oil would first reduce revenues. The OPEC response would be to raise price in order to again obtain the target level of revenues.

So here we have two models, both proposed by serious thinkers, which will give different estimates of the direction of impact on world oil price of such a simple change as a downward translation of the energy demand function. If this example is too easy, try using several conceptual models to examine the oil price impacts occuring in response to oil importing countries imposing a quota on imports (remember Tokyo?). Such a policy not only reduces the demand for energy but also reduces the elasticity of demand facing OPEC.

The point is that even the direction of world oil price change in response to policy options depends upon which conceptual model best describes the world. Similarly, the base case price trajectory depends upon the specifics of the model, as was beautifully shown by Leif Ervik in his presentation.

Fundamentally, then, I am less confident about world oil price projections than are the earlier speakers. I do not believe that we have definitive answers yet. We need to examine the alternative fundamental pricing models of OPEC behaviour, not limiting ourselves to economic paradigms, to examine various possible representation of world oil supply and demand, and only finally to debate the specific parameters for these models.

By way of example, remind yourself of Bob Deam's compelling case of why the world price of oil can be expected not to increase, but to decline.

In contrast, the Salant model, being used by the U.S. Department of Energy, Office of Policy, Planning, and Evaluation contains some assumptions parallel to Deam's work. However, it assumes that gas and oil are perfectly substitutable by modelling only one fuel: "Goil", in contrast to Deam's model of imperfect substitution. It assumes that OPEC is an intertemporal optimizing monopolist and that the fringe suppliers are perfectly competitive optimizers. There are no transportation costs or constraints and no lags. Those assumptions tend toward as low an oil price projection as possible. However, in that model the price of oil still increases above the current actual world prices. The difference seems to be in the representation of supply and demand for energy, including the assumption that there is a limited supply of gas and increasing demand for that fuel.

Even though I am quite confused about the right model, I still would be dishonest not to admit to my beliefs. I find it surprising

and counter-intuitive to imagine a world with a decreasing real
price of oil over the next several decades, except possibly in
a short run transitory situation which could be associated with
the cycles Charlie Blitzer discussed. I fully expect real oil
prices to go up because it appears that even if no monopoly power
is exercised by OPEC in the medium-run, prices must rise to keep
demand from exceeding productive capacities.

My final conclusion is that I am confused and I hope that I have
managed to share my confusion with all of you.

Robert Deam

I'll start off by saying that the price of marker crude or the
marginal form of energy is in fact determinate; it used to be
determined by the oil industry, but it is now determined by Saudi
Arabia. It is not determined by OPEC, but by Saudi Arabia. The
price is indeterminate in the short run. But then, we are talking
of the FOB price of the marker crude, and therefore, the FOB prices of all other crudes. There are several crude prices: there
is the FOB, there is the CIF, there is the "refinery fence" price,
or there is the derived prices of refinery products to the consumer. If you'll remember James Sweeney and myself show how we can
bring the FOB prices of crudes down by in fact increasing the cost
to the consumer. And in the short term that's what would be possible. In other words, there is no reason since it is a near-
determinate system why the consumer can set the FOB price of crude
rather than Saudi Arabia. I think we are all in perfect agreement
on that; and that's what we had in those two presentations. In
the long run, the price of crude is indeterminate. Since the cost
of crude is so small we can forget about it; and its highest price
in the long run must be the price of the substitute. The price
of the substitute will be the cost of the substitute, if we are
in any sort of equilibrium. This morning I suggested a low cost
substitute which would control the price of crude downwards, which
I suggested would emerge probably in a period of up to twenty
years. In twenty years, then, if my theory is right the marker
crude won't be light Arabian crude, anyway. It would be natural
gas in Australia or Russia. There is the "marker price of energy"
and this in turn would be controlled by some low cost, high volume
source of energy. What this cost will be depends upon the inven-
tion function. You can be an optimist or a pessimist in this world.
It is true that throughout history man has had more and more,
cheaper and cheaper energy for himself. Traditional wisdom suggests that we have had a change of state, and we are going to have
less and less, dearer and dearer energy. However, I doubt it, as
I believe man's ingenuity and his inventive function will produce
a low cost form of energy. There is one thing that we must remember; that there is no shortage (in the long term) of energy. There

may be a shortage of energy <u>at the price</u>, but there is no long
term shortage of energy and really we are coming back to the new
sets of prices of crude, a new set of prices of light Arabian crude.
Is it Saudi Arabia or is it the International Energy Agency, I
can't say, and just because that crude can be set low that does
not necessarily mean that it is not a higher price to the consumer. But if we have a supply substitution, it <u>does</u> mean a lower
price to the consumer, and a lower price to the producer of the
crude.

Leif Ervik

Before lunch I presented some of the results of our model and now
I would like to go through some of the insights from the model.
One of the insights is that 16 sectors is enough to grasp the long
term development pattern in transportation requirements. Secondly, the tanker crisis can be best solved from the supply side
(i.e. the number of tankers available for charter), and one should
not hope for the demand side to solve it. Thirdly, enhanced oil
recovery is the major uncertainty factor in the long term development of crude oil production. Fourthly, any constant demand elasticity causes the price to blow up. Therefore, it becomes increasingly important, as Blitzer pointed out, to have a feedback
to the supply of other energy. It also becomes very important to
know what taxes are doing, whether they move together with oil
price, whether they stay constant, or whether they are indeed
reduced when oil price increases. The impact of taxes is significant. Finally, there is substantial uncertainty as to the price
development, especially after 1995.

The price in this period will be determined more by the availability of alternatives than by the cost of the alternatives. One
thing we should bear in mind here is the size of the numbers that
we are talking about. The present discounted value of OPEC oil,
when using 7% discount rate, is approximately the same number as
the present capital stock in the whole world today. The present
discounted value of OPEC does not change very much with their own
strategy. That is especially true if one uses low discount factors, but makes a significant difference if discount factors are
larger than 10%. When high discount rates are employed it is
important to attain high price and quantity <u>now</u> compared to high
revenue later. If low discount factors are used, however, there
are so many compensating mechanisms that present discounted value
will not change so much. I will give you some of the intuitive
explanations for that. One is that, if the price is <u>low</u> alternative energy supply will be lower and therefore the price higher
later. A low price now is correlated with high OPEC quantities
now and therefore OPEC's quantity will be lower later. So you
have a sort of a price-quantity ratio at any point in time, a

price-quantity relation over time, and the response of all the other suppliers and all consumers. So all in all, the present discounted value of OPEC does not change all that much: remember though we are talking about numbers in the order of magnitude of the total world capital stock, so although the percentage change in such large numbers is not great, the absolute change may be many billions of dollars.

Now when we look at the graph (see Fig. 12 in Ervik's paper) which I feel is fairly representative for the present state of knowledge, then there is a very low chance that the price will decrease – though I admit that our study has not considered Deam's methanol scenario. There is also a very low likelihood that the price will be much higher than $ 50 at the end of the simulation. A middle-of-the-road development (we are now talking about 1975 prices) is the development in which I showed a gradual increase to the order of $ 30 (in 1979 $) around the turn of the century and that is my present belief. I will go home and look at Deam's story and see if it makes a significant impact and next time your hear the speech, it will be incorporated.

RESPONSES

Charles Blitzer

I'll try and go in the reverse order starting with the graph (Fig. 12 in Ervik's paper) we just saw. I think it conforms more or less to my own expectations in terms of the quantities. What I'm relying here is the empirical fact that we observe estimates of the price of the alternative somehow to be 50% above the price of the crude oil, no matter what the price of the crude oil is and so I have no basic problems with Ervik's results. They conform to my own intuition of what things will be. James Sweeney managed to give us a very coherent view of a confused set of issues. I like to comment on just one or two of them. On the choice of his example of one model giving a higher price or a lower price if U.S. reduced the demand, I think it's that sort of an experiment which can differentiate the useful model from the not so useful model. It's these experiments that will throw out some and leave others in the potential set.

I have some comments on Joseph Debanne's points. First, the fact that the cost does not affect the price. Well, it's true the cost of producing a barrel of oil in the Persian Gulf may be 50 cents a barrel, which is perhaps an over-estimate. Nonetheless, the cost of alternatives certainly <u>does</u> affect the price of oil, and that is something which should be taken into account in looking at long run price projections. The proposition that the price of oil would go to $ 1 a barrel if OPEC broke up; I personally do not

believe in that. My own research in this area actually is inconclusive on the part of whether or not OPEC is acting optimally really as a real cartel. And here, I like to bring in the notion
which Sweeney mentioned that physical capacity constraints exist.
We all know that there are limits to how rapidly oil fields could
be expanded. None of the optimizing models which I know of take
this into account, and if they did, it is not at all clear that
the optimal policy will not be for an individual country to act
like a price taker. If that were to happen, using realistic assumptions of what the technical capacities of OPEC are, in other words
the capacity without blowing its fields, my own numerical experiments indicate that the following policy will emerge: Prices shoot
up for a while, take advantage of a very low short run elasticity,
give a lot of profits to OPEC and then prices slowly decline to
discourage investment alternatives; this continues for a relatively
long period for which the cycle emerges again. This price cycle
is consistent with relatively constant production at near technical levels. Now the numbers I used some years ago might not be
exactly consistent any more with today's, but still it has yet to
be proved that the optimal behaviour of the cartel is cartel behaviour. Although this is something still to be tested empirically,
there is in any case no evidence that every member of the cartel
producing at maximum capacity will be the equivalent of the breakup of the cartel, and that prices will go down to $ 1 a barrel.
My hunch is that they would not decrease at all. Finally, on the
point of freezing the prices of oil, that the low price of oil is
in the interest of the consumers, is exactly the flip side of the
question from the view point of the cartel. Low prices may be in
the U.S. interest today, but looked at in the long run, low prices
today discourage the U.S. and OECD and other consuming nations from
making the technical change in the supply and the demand side which
ensure its own long run economic viabilities. It is not exactly
clear to me at all that low prices are in the consumers' interest.

James Sweeney

There are a couple of points one can make in response to Charles
Blitzer concerning absorptive capacity models and the empirical
tests used to determine whether or not a model makes sense. The
current situation in Iran helps to illustrate that short run absorptive capacity models are simply not viable: they do not correspond with observable realities. However, one might argue that
a more subtle medium run oriented version of an absorptive capacity model could be consistent with empirical realities. I will
not push that point, though.

Although I am backtracking, I would like to comment on something very
significant that Leif Ervik pointed out. To paraphrase him, in
most scenarios the discounted revenues received by OPEC nations are,

for the most part, relatively insensitive to their oil pricing strategies, if I have understood him correctly.

That implies to me that while we could develop a model of OPEC as strictly optimizing, there is ample room for political forces and other objectives beyond traditional economic considerations. These other objectives could be pursued without imposing major economic losses (from optimality) on the OPEC nations. The price response to external changes thus could not be expected to coincide with projections from traditional economic models. If this Ervik conclusion is in fact correct, then the real conceptual advances in modelling OPEC behaviour will likely be based upon paradigms other than the traditional economic paradigms. I think that it will be very useful to explore this point of view more fully and to decide to what extent Ervik's conclusion can be supported.

Joseph Debanne

I probably should respond to the last remarks and at the same time answer some other remarks of my colleagues in the panel. I'll start by clarifying, if it needs to be clarified, what they said about the relation between the cost and the price of oil. There is no relation between what a barrel of oil has cost you and the price at which you will sell it. The barrel of oil could have cost you a million dollars. You would sell it at the going price in oil to minimize your loss or to maximize your gain. There is no relationship between the two.

On the subject of the actual cost of producing it (which is usually very small), you'll abandon that well the day when the price of oil goes below the production cost. When it comes to the alternative cost of fuels, of course this sets up the ceiling and what I said was that this ceiling was misestimated at about $ 12 a barrel equivalent in 1974. In effect, that seems to be a rather good guess because inflation has really allowed this opportunity cost to go up since that time.

Now let me come to the issue that was brought up by I think Sweeney and Blitzer, the issue of productive capacity. Productive capacity is usually measured in terms of what is known as the "production life index". Usually when you discover a field, in the initial period the production life index is high. Saudi Arabia now has a production life index of almost 100 years, 75 certainly. That is, at the present rate of production they can produce for 75 to 100 years. With time as you deplete your resources, this production index goes down. In North America, this production life index is between 10 and 11, it has been there for the last 10 years. Despite all the money we have spent, we just could not build it up. In all due respect to Ervik, I remember very well Adelman's statement

that when it was only $ 4, if you allow us $ 8 in U.S., secondary
recovery and new reserves will be such that we will have a glut of
oil that we'll have to protect the producer against a drop in the
price of oil. So you went to the Federal Energy Administration
and negotiated an international floor price of oil so as to protect
the producer. And yet this was done. I really think that this
issue of production life index rushing capacity is working in the
interest of the cartel. The associates of Saudi Arabia and in this
respect I agree with Robert Deam, are less and less powerful. Even
Iran was producing at fully capacity. It has a production life
index of 18 years, and the production life index was going down.
Algeria was at the limit. There are others close to the limit.
Maybe Libya has cut down a little bit. But by and large, that is
why these people will have less and less manouverability as far
as being able to say "I am going to increase my production life
index". They can not increase their production life index. Saudi
Arabia is really a far and out world price leader; Saudi Arabia
and that particular region. And this actually reinforces the power
of the cartel.

Robert Deam

I am a minority of one. I can see that. Most people cited that
the cost of producing alternative energy is higher than the cost
of producing an oil product. I do not want to use the word "crude",
because crude is useless stuff. It is no use until it's an end
product. It is cheaper to convert natural gas to a substitute for
gasoline or number 2 furnace fuel oil component. That is, to convert fuel oil to those components. The price of refining is set
not by distillation of crude oil at all; it is set by the conversion of fuel oil to the right product. And it's interesting in
other peoples' models, if they used a lower cost of producing the
marginal fuel oil, motor gasoline from the crude, then one wonders
whether the price trajectory would go up. The next thing is of
course volume. Thinking of the proven reserves of exportable natural gas, of the assumption of OPEC keeping a constant volume
does not matter very much when there's the equivalent of OPEC just
waiting to be developed with much cheaper capital cost than it even
takes to build refineries and produce more crude. And one wonders
whether that taken into account alters one's expectations of prices
Of course, in due course, gas will be used up just as oil will be
used up, in which case we have to think of yet another alternative
If I knew what that other alternative was, I'd be down at the patent office filing papers. It's true that the thermodynamic bon-
ding of hydrogen, or the hydrocarbons in coal which are bound in
a carbon lattice are very very weak. If that could be extracted
some way, that is with minimum energy, the coal would yield roughly 50% carbon, 50% liquid hydrocarbons. I do not know how that
works. As I said, if I did know how it worked, I would go out and

474

patent it. But I cannot believe in due course man won't find a
method which makes the cost of converting at least half the coals
into liquids abominably cheap, compared with the sledge hammer
that we're using today (in the thermodynamic sense) of hydrogenation or gasification in the Fischer-Tropps practice.

So, I rather think that there have been some erroneous basic assumptions; i.e. that the cost of alternative energy is higher than
the cost of derived products. We must not think of crude; nobody
consumes crude, they consume derived products. I told you this
morning that the cost of making the right products from natural
gas with capital and operating cost and everything else is cheaper
than going from fuel oil. There's a relative abundance of exportable gas, at least equal to the proven reserves of the Middle East
crude; and one wonders whether we would be having this debate today if the production capacity of the Middle East was not thirty
million barrels a day.

That is effectively what we have got facing us, to secure energy
with minimum capital. I cannot believe that under those circumstances crude price or energy prices are going to rise. If I can
make one point, it was suggested that lower prices may not be in
the long term consumer interest. That, as a matter of fact, is
obviously right for crude oil tax; because it puts up the price
of crude, reduces the CIF price of OPEC crude which reduces the
transfer of wealth from the consumers to the producers.

Leif Ervik

First, an observation on who is the majority and who is the minority in the question of future price development. I think that
this question is mainly dependent on the most recent price changes.
That brings me back to the fundamental question about alternative
energy. It takes a long time before people believe that whatever
they see in the market place will be a long term price. Therefore,
four or five years or even more elapse before people believe enough
in the present price in order to make investments on that basis.
I believe increasingly as observing the phenomenon in the real
world that the delay in bringing alternative energies on stream
will be more important than the cost of alternative energy in the
long run. If Deam gives many lectures around, he will say "Now,
do not believe these figures on the spot market" and finally a
lot of people would be saying that. I am not saying he is right
or wrong, I am just saying it will take a long time before people
believe whatever they see, and that will lead to a longer delay
before alternatives are commercially available.

I said the present discounted value will not change too much. I
have not tested that conclusion against a strategy of playing on

the short term elasticity, though I expect the conclusion would still hold. I showed you earlier today what the consequences would be if OPEC reduced its reserve production ratio to 25 years instead of some 40 today for OPEC as a whole. A sudden reduction in the ratio would lead to an immediate price drop. However, if we were to reduce the RPR gradually - more in keeping with the technical reality - then the price would remain constant until 1986-1987, but after that increase to the level it would otherwise have had at around the turn of the century. So, again, I do not believe that OPEC strategy makes all that much difference in the long run. What makes the difference is the supply and demand of all the important countries.

DISCUSSION

The question of how the world economy was affected by higher oil prices was raised. One view was that the short term effect for oil importing nations was the charge in price times the quantity imported. This could of course be a big number and Sweeney felt that developed countries (DC's) would be better able to respond than lesser developed ones (LDC's). Blitzer disagreed on this last point, however. Latest figures from the World Bank showed that the growth rate of LDC's was hardly changed after the jump in the price of oil in 1973; it was reduced from 6.1% to 6%. Growth rates in DC's had been reduced much more significantly from about 4% to 2%. Blitzer felt that the reasons for this were to be found in the flexibility of the LDC's economy - they had a high turnover of capital stock, their ability to finance the increased costs via aid - it was also easier for them to obtain oil dollars, and the fact that they were developing new export markets in oil exporting countries.

Several speakers disagreed with Blitzer. They felt that the flexibility in LDC's economies was imaginary and that OPEC aid was at best thinly spread. It was pointed out that LDC's would now have to develop energy technologies which were more capital intensive than oil-based technologies and as capital was scarce in LDC's, this would have an adverse effect. Doubt was expressed at relating changes in the post-1973 growth rates to oil price alone; many other factors could be important determinants. Another speaker was sceptical at the validity of conclusions based upon percentage changes in a statistically uncertain base; experience in Turkey had shown that statistics could be very misleading.

The idea of a two-tiered pricing system for oil was put forward - one price for DC's and a lower one for LDC's. Deam doubted the viability of such a scheme. He maintained that oil companies would arrange transactions such as to maximize profits.

The panel was asked its opinion on the future role of the Soviet block and other socialist countries. Debanne felt that most Soviet oil production would be used to satisfy growing domestic expectations. This feeling was backed up by Ervik who referred to the results of his simulation model. This showed both Eastern Europe and China exporting small amounts of oil until the 1990's and then becoming net oil importers. The figures were, however, modest in relation to domestic consumption.

It was put to the panel that an optimal pricing strategy for oil was simply the discounted value of the future back-stop technology. Such a strategy would maximize profits, and was once proposed by W. Nordhaus. This approach assumed that one knew the point in time when back-stop technology would be introduced and that one knew the price of this technology. One should also take present extraction costs into consideration so that if one were to apply this piece of asset management theory, price would consist of extraction costs plus royalty where royalty were the discounted value of the difference between the cost of back-stop technology and today's technology.

Ervik had concluded from his study of the world oil market that the present value of OPEC oil remained relatively unchanged by different depletion policies and production profiles. The point was made that it would be interesting to extend the analysis to examine the oil value of the individual OPEC members. This might give an indication of how cohesive OPEC would be in the future. Deam reiterated that the Saudi Arabians were the real price setters In OPEC. They manipulated price so as to maintain their liftings and maximize revenues. Saudi Arabia had had a balance of trade deficit last year for the first time. Their motivation to maximize revenues was just as great as that of other OPEC members with high absorptive capacity in their economies.